Actuator Technologies and Control: Materials, Devices and Applications

Actuator Technologies and Control: Materials, Devices and Applications

Guest Editor

Paolo Mercorelli

MDPI

Basel • Beijing • Wuhan • Barcelona • Belgrade • Novi Sad • Cluj • Manchester

Guest Editor
Paolo Mercorelli
Institute for Production
Technology and Systems (IPTS)
Leuphana Universität Lüneburg
Lüneburg
Germany

Editorial Office
MDPI AG
Grosspeteranlage 5
4052 Basel, Switzerland

This is a reprint of the Special Issue, published open access by the journal *Actuators* (ISSN 2076-0825), freely accessible at: https://www.mdpi.com/journal/actuators/special_issues/D6D8S58IWR.

For citation purposes, cite each article independently as indicated on the article page online and as indicated below:

Lastname, A.A.; Lastname, B.B. Article Title. *Journal Name* **Year**, *Volume Number*, Page Range.

ISBN 978-3-7258-7821-5 (Hbk)
ISBN 978-3-7258-7822-2 (PDF)
https://doi.org/10.3390/books978-3-7258-7822-2

Contents

About the Editor

Paolo Mercorelli

Paolo Mercorelli (Member, IEEE) received his Ph.D. in Systems Engineering from the University of Bologna, Italy, in 1998. In 1997, he was a Visiting Researcher for one year with the Department of Mechanical and Environmental Engineering at the University of California, Santa Barbara, USA. From 1998 to 2001, he was a Postdoctoral Researcher at Asea Brown Boveri Corporate Research, Germany, where his research resulted in three patents. During the same period, he was awarded a three-year fellowship under the Marie Curie Actions Research Fellowship Program, one of the most competitive and prestigious European funding schemes sponsored by the European Commission. From 2002 to 2005, he was a Senior Researcher at the Institute of Automation and Informatics, Germany, where he led the Control Group. From 2005 to 2011, he served as Associate Professor of Process Informatics at Ostfalia University of Applied Sciences, Germany. Since 2012, he has been Full Professor and Chair of Control and Drive Systems at the Institute for Production Technology and Systems, Leuphana University of Lueneburg, Germany. His current research interests include applications of Kalman filters, robotics, wavelets, geometric control, and sliding mode control. He has received numerous Best Paper Awards at international conferences. From 2019 to 2025, he has been listed among the world's top 2% of scientists, included in both the career-long and single-year rankings, according to the Elsevier database and Stanford University (Ioannidis, J.; Boyack, K.; Baas, J., 2020, updated science-wide author databases of standardized citation indicators, PLOS, October 16). Since 2019, he has been Editor-in-Chief of *Transactions on Computer Research*, an Open Access journal published by the World Scientific and Engineering Academy and Society. Since 2022, he has been Editor-in-Chief of the Section "Control Theory and Mechanics" in *Mathematics*, an Open Access journal published by MDPI.

Preface

Actuation technologies form the critical bridge between control decisions and physical action, and their importance continues to grow as engineering systems evolve toward higher levels of autonomy, intelligence, and integration. As emphasized in the introductory article of this Reprint, actuators are no longer viewed as isolated hardware elements but as essential components of complex, interconnected systems required to operate reliably in uncertain and dynamic environments.

This Reprint is devoted to the control and design of next-generation actuators, with a focus on emerging concepts that combine advanced materials, innovative architectures, and intelligent control strategies. Its scope spans fundamental developments and applied research, including new actuator materials, miniaturized and micro-scale devices, precision measurement and performance assessment, as well as applications in robotics, manufacturing, transportation, and biomedical systems. Particular attention is given to the integration of actuators within modern control frameworks, where adaptability, robustness, and real-time responsiveness are key requirements.

The motivation behind this Reprint arises from the rapid and interdisciplinary progress in actuation technologies. Advances in materials science, sensing, and computation, together with breakthroughs in control theory and system integration, are reshaping how actuators are designed and utilized. These developments call for a unified platform that captures current achievements while also addressing the challenges associated with reliability, scalability, and intelligent operation in complex environments.

The aim of this Reprint is to provide a comprehensive and coherent collection of contributions that reflect the state of the art and emerging directions in actuator research. It seeks to foster a deeper understanding of the interplay between physical design and control, and to promote innovative solutions that meet the demands of modern engineering applications.

This Reprint is addressed to researchers, engineers, and practitioners working in control systems, mechanical and electrical engineering, materials science, and related disciplines. It is particularly intended for those engaged in the design and control of advanced actuation systems, as well as for readers interested in interdisciplinary approaches to next-generation engineering technologies.

We hope that this Reprint will serve as both a valuable reference and a source of inspiration, supporting continued advances in the control and design of intelligent, high-performance actuators.

Paolo Mercorelli
Guest Editor

actuators

MDPI

Editorial

Actuator Technologies and Control: Materials, Devices and Applications

Paolo Mercorelli

Institute for Production Technology and Systems (IPTS), Leuphana Universität Lüneburg, 21335 Lüneburg, Germany; paolo.mercorelli@leuphana.de

1. Motivation and Technical Context

Actuation technologies lie at the heart of modern engineering systems, providing the means by which computational intelligence and control decisions are translated into purposeful physical action. As technological systems increasingly move toward higher levels of autonomy, connectivity, and functional integration, the role of actuators is undergoing a profound transformation. No longer viewed as isolated hardware components, actuators are now integral elements of complex, intelligent systems, required to operate reliably and efficiently while interacting with uncertain environments, humans, and other machines.

This evolution is driven by growing expectations for adaptability, robustness, and performance across diverse application domains, including robotics, transportation, manufacturing, healthcare, and cyber–physical infrastructures. Advances in sensing, control theory, materials, and system integration are converging to enable actuators that can adjust their behavior in real time, compensate for disturbances and uncertainties, and seamlessly cooperate with higher-level decision-making frameworks. As a result, actuation research has become inherently interdisciplinary, bridging traditional boundaries between mechanical design, electronics, control, computation, and application-specific requirements.

A central pillar of this transformation is the development of advanced modeling and control methodologies for smart material and precision actuators. Smart material-based systems, such as piezoelectric, magnetostrictive, and shape memory actuators, offer high bandwidth and compact form factors yet present significant nonlinearities and hysteresis effects that complicate control design. Smith, in [1], highlights the fundamental modeling and control challenges associated with smart material actuation, emphasizing the need for accurate constitutive models and robust compensation strategies. In the domain of microscale actuation, Kanchan et al. [2] provide a comprehensive review of piezoelectric microactuators, detailing their design principles, dynamic characteristics, and control approaches, which are essential for applications in micro-manipulation and precision engineering.

Precision actuation systems, particularly those employed in high-performance positioning and manufacturing processes, must operate under stringent accuracy requirements despite disturbances and parameter variations. Disturbance observer-based control techniques have emerged as powerful tools for enhancing robustness and disturbance rejection, as discussed by Kim et al. [3]. Such approaches are increasingly complemented by data-driven and learning-based strategies. In [4], Li et al. explore data-driven control methods for intelligent actuator systems, demonstrating how machine learning and real-time data analytics can enhance adaptability and performance in complex and uncertain environments. In particular, this study proposes a data-driven modeling method for nonlinear factors.

As actuators become embedded within networked and cyber–physical systems, resilience and security considerations gain prominence. Network-induced delays, packet

 https://doi.org/10.3390/act15040218

losses, and potential cyber threats pose significant challenges to stable and reliable operation. Jiao et al, in [5] address foam-based soft actuators are lightweight and highly compressible, which make them an attractive option for soft robotics In this context, Gorial, in [6], review and discuss smart actuation solutions. In fact, recent advances in series elastic actuator (SEA) research have focused on improving both mechanical design and control strategies to enhance compliance, robustness, and interaction performance in robotic systems. In [7], Sariyildiz et al. proposed a novel variable stiffness SEA that integrates an adjustable stiffness mechanism with a dedicated control framework to achieve improved force and position tracking. The design enables adaptability to varying task requirements and enhances safety in human–robot interaction by allowing modulation of the actuator stiffness.

Beyond mechanical design improvements, several studies have focused on advanced control methods to address the dynamic challenges introduced by actuator elasticity. For instance, ref. [8] introduced a robust elastic structure-preserving control approach aimed at achieving high impedance rendering while maintaining the intrinsic elastic properties of SEAs. By explicitly preserving the actuator's elastic dynamics, the proposed method improves robustness against disturbances and modeling uncertainties.

Adaptive control techniques have also been investigated to handle parameter variations and resonance phenomena in compliant actuators. In [9], an $\mathcal{L}1$ adaptive resonance ratio control strategy was developed to guarantee fast transient performance while mitigating resonance effects. The controller demonstrates improved tracking accuracy and robustness under uncertain system parameters.

Finally, disturbance rejection remains a critical challenge in SEA systems operating in dynamic environments. To address this, ref. [10] proposed a robust motion control architecture that separates resonance suppression from robustness compensation, improving disturbance rejection while maintaining stable actuator behavior during interaction tasks.

Another major trend concerns safe and compliant interaction between actuators and humans. In collaborative robotics and assistive technologies, actuators must balance performance with safety and adaptability. Variable stiffness actuators (VSAs) have emerged as a promising solution, enabling modulation of mechanical impedance to achieve both precision and safe interaction. Reference [11], by Agostini et al., investigates energy efficiency and dynamic behavior through simulation and experiments of a compact electro-hydrostatic actuator system consisting of an electric motor, external gear pump/motors, hydraulic accumulator, and differential cylinder.

Within this broader and rapidly evolving context, the ten papers, assembled in this Special Issue, present representative and timely contributions that capture the current momentum of the field. The works span advanced control and optimization methodologies, resilient and secure actuation in networked systems, innovative and compliant actuator concepts, precision actuation technologies, and application-oriented sensing and actuation solutions. Together, these contributions not only reflect the state of the art in actuation research, but also point toward future directions in which actuators will play an increasingly central role in enabling intelligent, reliable, and adaptable engineering systems.

2. Contributions of the Special Issue

A major theme of this Special Issue is the development of advanced control and optimization strategies for electric drive systems. In review paper [12], Santos et al. provide a systematic review of smart insole prototypes incorporating artificial intelligence and wireless communication, underscoring the growing role of actuator and sensor technologies in biomechanics and healthcare applications.

https://doi.org/10.3390/act15040218

In [13], Jiang et al. propose a torque oscillation attenuation method for permanent magnet synchronous motors based on sliding-mode control combined with an equivalent input disturbance approach, enhanced through continuous-domain ant colony optimization. Complementing this contribution, Kethiri et al. address efficiency optimization in BLDC motor drives through adaptive flux control strategies based on incremental conductance and fuzzy logic, demonstrating significant reductions in power losses under varying operating conditions [14].

Robustness and reliability are further explored in the context of cyber–physical systems and industrial robotics. Hassine et al. [15] introduce an event-triggered observer-based control framework that enhances resilience against Markovian cyberattacks, including denial-of-service and false data injection attacks, while reducing communication load. In parallel, Boldsaikhan and Birney present an optimal real-time toolpath planning method for industrial robots equipped with sparse sensing, enabling accurate adaptation to unknown surface variations with limited sensing and computational resources [16].

Accurate modeling and experimental characterization of actuators remain essential for control design and system integration. Abouseda et al. present in [17] an experimental study on a BLDC motor using system identification techniques to capture its dynamic behavior under fixed load torque and variable speed conditions, while also evaluating efficiency and battery state-of-charge effects.

Beyond traditional electric drives, several papers focus on novel actuator concepts and compliant mechanisms. Els et al. investigate the design of fluid elastomeric actuators for soft robotic end-effectors, emphasizing FEM-based optimization and application-specific design guidelines [18]. Xu et al. introduce in [19] a lightweight twisted string actuator with continuously variable transmission that automatically adapts its transmission ratio in response to external loads, demonstrating strong potential for robotic and wearable applications. Razek and Bernard provide a comprehensive review of piezoelectric actuators and sensors, highlighting their high precision and reliability in medical therapeutics, robotic interventions, and structural sensing [20].

The importance of application-driven actuation and sensing systems is illustrated in [21] by Craig and Ahmadian, who present the design and prototyping of a deployable actuation system for railcar-mounted sensors intended for automated railway inspection. In summary, the contributions in this Special Issue highlight the interdisciplinary nature of contemporary actuation research, spanning advanced control, optimization, modeling, emerging actuator technologies, and real-world applications. Together, these works provide valuable insights and directions for the development of efficient, resilient, and intelligent actuation systems.

Funding: This research received no external funding.

Institutional Review Board Statement: Not applicable.

Informed Consent Statement: Not applicable.

Data Availability Statement: No new data were created or analyzed in this study. Data sharing is not applicable to this article.

Conflicts of Interest: The authors declare no conflicts of interest.

References

1. Smith, R.C. *Smart Material Systems: Model Development*; Society for Industrial and Applied Mathematics: Philadelphia, PA, USA, 2025. [CrossRef]
2. Kanchan, M.; Santhya, M.; Bhat, R.; Naik, N. Application of Modeling and Control Approaches of Piezoelectric Actuators: A Review. *Technologies* **2023**, *11*, 155. [CrossRef]

https://doi.org/10.3390/act15040218

3. Kim, N.; Oh, D.; Oh, J.-Y.; Lee, W. Disturbance-Observer-Based Dual-Position Feedback Controller for Precision Control of an Industrial Robot Arm. *Actuators* **2022**, *11*, 375. [CrossRef]
4. Li, B.; Xie, X.; Ren, G.; Fan, D.; He, J. A data-driven dynamic modeling method for servo actuators. *Mech. Sci.* **2026**, *17*, 33–44. [CrossRef]
5. Jiao, Z.; Zhuang, Z.; Cheng, Y.; Deng, X.; Sun, C.; Yu, Y.; Li, F. Lightweight Dual-Mode Soft Actuator Fabricated from Bellows and Foam Material. *Actuators* **2022**, *11*, 245. [CrossRef]
6. Gorial, I.I. A review of smart actuation technologies in modern mechatronic systems. *Eng. Technol. J.* **2025**, *43*, 670–682. [CrossRef]
7. Sariyildiz, E.; Mutlu, R.; Roberts, J.; Kuo, C.H.; Ugurlu, B. Design and Control of a Novel Variable Stiffness Series Elastic Actuator. *IEEE-ASME Trans. Mechatron.* **2023**, *28*, 1534–1545. [CrossRef]
8. Lee, H.; Lee, J.; Keppler, M.; Oh, S. Robust Elastic Structure Preserving Control for High Impedance Rendering of Series Elastic Actuator. *IEEE Robot. Autom. Lett.* **2024**, *9*, 3601–3608. [CrossRef]
9. Min, F.; Wang, G.; Chen, X. $\mathcal{L}1$ Adaptive Resonance Ratio Control for Series Elastic Actuator with Guaranteed Transient Performance. *ISA Trans.* **2024**, *154*, 285–298. [CrossRef] [PubMed]
10. Ito, J.; Murakami, T. Disturbance Rejection Improvement in Robust Motion Control of Series Elastic Actuator. In Proceedings of the IEEE RAS/EMBS International Conference on Biomedical Robotics and Biomechatronics (BioRob), Heidelberg, Germany, 1–4 September 2024; pp. 1599–1604. [CrossRef]
11. Agostini, T.; De Negri, V.; Minav, T.; Pietola, M. Effect of Energy Recovery on Efficiency in Electro-Hydrostatic Closed System for Differential Actuator. *Actuators* **2020**, *9*, 12. [CrossRef]
12. Santos, V.M.; Gomes, B.B.; Neto, M.A.; Freitas Rodrigues, P.; Amaro, A.M. A Systematic Review on Smart Insole Prototypes: Development and Optimization Pathways. *Actuators* **2025**, *14*, 408. [CrossRef]
13. Jiang, R.; Yin, X.; She, J.; Wang, F.; Kawata, S. Torque Oscillation Attenuation in PMSM Using Equivalent-Input-Disturbance-Based Sliding-Mode Control. *Actuators* **2026**, *15*, 85. [CrossRef]
14. Kethiri, M.F.; Charrouf, O.; Betka, A.; Salman, M.; Boccaletti, C. Minimizing Power Losses in BLDC Motor Drives Through Adaptive Flux Control: A Real-Time Experimental Study. *Actuators* **2025**, *14*, 395. [CrossRef]
15. Hassine, E.; Thabet, A.; Gasmi, N.; Bel Haj Frej, G. An Event-Triggered Observer-Based Control Approach for Enhancing Resilience of Cyber–Physical Systems Under Markovian Cyberattacks. *Actuators* **2025**, *14*, 412. [CrossRef]
16. Boldsaikhan, E.; Birney, C. Optimal Realtime Toolpath Planning for Industrial Robots with Sparse Sensing. *Actuators* **2025**, *14*, 279. [CrossRef]
17. Abouseda, A.I.; Doruk, R.; Emin, A.; Akdeniz, O. Modeling, Dynamic Characterization, and Performance Analysis of a 2.2 kW BLDC Motor Under Fixed Load Torque Levels and Variable Speed Inputs: An Experimental Study. *Actuators* **2025**, *14*, 400. [CrossRef]
18. Els, D.; van Niekerk, T.; Mercorelli, P.; Welgemoed, J. A Fluid Elastomeric Actuator Design for Soft Robots. *Actuators* **2025**, *14*, 417. [CrossRef]
19. Xu, C.; Liu, T.; Dong, S.; Wang, Y.; Wang, X. A Lightweight, Simple-Structure, Low-Cost and Compliant Twisted String Actuator Featuring Continuously Variable Transmission. *Actuators* **2024**, *13*, 477. [CrossRef]
20. Razek, A.; Bernard, Y. Potential of Piezoelectric Actuation and Sensing in High Reliability Precision Mechanisms and Their Applications in Medical Therapeutics. *Actuators* **2025**, *14*, 528. [CrossRef]
21. Craig, C.; Ahmadian, M. Actuation and Control of Railcar-Mounted Sensor Systems. *Actuators* **2025**, *14*, 289. [CrossRef]

https://doi.org/10.3390/act15040218

actuators

Review

Potential of Piezoelectric Actuation and Sensing in High Reliability Precision Mechanisms and Their Applications in Medical Therapeutics

Adel Razek * and Yves Bernard

Group of Electrical Engineering—Paris (GeePs), CNRS, University of Paris-Saclay and Sorbonne University, F91190 Gif sur Yvette, France; yves.bernard@centralesupelec.fr
* Correspondence: adel.razek@centralesupelec.fr

Abstract: The present contribution aims to analyze and highlight the potential of piezoelectric materials in actuation and sensing duties, obtaining reliable high-precision outcomes in cutting-edge applications including medical interventions. This involves high-precision actuations of robotized procedures, as well as monitoring and controlling various physical phenomena via structural sensing. The characteristics of these applications offer enhanced precision machinery and robotic tools, medical robotic precise interventions, and high-accuracy structural sensing. The paper exposed, analyzed, reviewed and discussed different subjects related to piezoelectric actuators, involving their displacement and positioning strategies, piezoelectric sensors, medical applications of piezoelectric actuators and sensors, including robotic actuation for medical interventions, and structural sensing in the monitoring of wearable healthcare tools. Discussions among others on the advantages and limitations of piezoelectric sensors and actuators in general, as well as future research perspectives in medical involvements, are also presented at the end of the article. The specific features in the illustrated applications reflect crucial behaviors in robotic actuation for medical interventions, structural sensing in the monitoring of healthcare wearable tools, and the control of various structural physical occurrences.

Keywords: piezoelectric; actuators and sensors; motion and positioning accuracy; procedure actuation; image-assisted interventions; structural sensing; wearable devices; digital monitoring

1. Introduction

Motion and positioning are frequently exercised in advanced energy mechanisms, e.g., robotic junctions, spatial arrangements, precision machinery, etc. Such duties can be accomplished by means of actuators exhibiting high degrees of displacement resolution and positioning accuracy along with swift responses, high stiffness and actuation potency, simple configuration, flexible stroke, endurance to electromagnetic (EM) interference, and scalable arrangements with unimportant sizes. On the other hand, the accurate sensing of different quantities can play a significant role in such advanced tools, particularly for monitoring and controlling various structural and physical occurrences. Actuators and sensors utilizing piezoelectric materials are on-track contenders for such features [1–8]. Actuators can be incorporated for movement and/or positioning in energy implements [2,4], or for setting up self-ruling robots [1,5]. Such actuators operate by obeying the piezoelectric inverse effect. Actually, a piezoelectric material creates electric potential under an applied compression, which is the direct effect, while an applied electric potential on the material generates mechanical deformation, which is the inverse effect. The direct effect is operated

https://doi.org/10.3390/act14110528

in structural sensors and energy transducers as well as damping and energy harvesting systems, while the inverse one is achieved using actuating tools. Such actuating tools can be exploited by themselves or via an amplified action depending on the actuated strength application [9–11]. Such amplification displays double-conversion electric–mechanic (inverse piezoelectric effect—small displacement) and mechanic–mechanic (displacement magnifying) structures, typifying the movement fashioned by the amplified actuator.

The recent precision revolution impacted sophisticated innovative industrial processes and medical therapeutics in both actuation and sensing tasks. Both occurrences are related to displacement resolution and positioning accuracy, but are mostly involved in robotized procedures as well as pressure detection. In both industrial and medical domains, piezoelectric materials can contribute to effectively accomplishing their duties. The industrial applications involved are mainly related to precision machinery and robotic tools [12–14] as well as pressure sensing [15]. In the case of healthcare, the concerned therapeutics are related to wearable pressure detection tools [16] and robotic interventional surgery and drug delivery [17]. In this context, chronologically, the strategies involved were robotized laparoscopic, computerized robotic, and image-guided robotic interventions [18]. Furthermore, in addition to the important role of piezoelectric materials in sensing and actuation, their voltage-dependent actuating action allows the performance of energy harvesting and structural health monitoring. This occurrence fashions piezoelectric sensors and actuators valuable for monitoring and controlling various physical observable facts, such as vibration, tremor, position, shift, pressure, deformation, etc. [19,20]. In addition, the more specific involvement of piezoelectric materials could be found in different recent applications; for instance, an ultra-high load-to-weight ratio [21] and long-range positioning with a nanometer resolution [22].

Various research has been published on specific applications of the studied topic. This contribution aims to evaluate and merge some related approaches, focusing on analyzing and demonstrating the potential of piezoelectric reliable actuation and sensing tasks in revolutionary industrial and medical fields.

The aim of this paper is the assessment of piezoelectric material use in actuation and sensing duties, performing reliable towering precision solutions in recent industrial and medical involvements. This work particularly concerns the actuation of robotic procedures in medical interventions, as well as structural detection and assistance in wearable medical tools. The high-reliability precision mechanisms used in these specific cases in the medical field are directly linked to patient safety and staff ease. The different themes addressed in this paper, although autonomous, are supported by examples from the literature, allowing a deeper comprehension.

In the present contribution, after a general introduction on actuation and sensing devices involving piezoelectric materials, different related characteristics will be exposed, analyzed, reviewed, and discussed. Section 2 concerns piezoelectric actuators, including different categories of actuators and traveling wave piezoelectric robots as well as displacement and positioning strategies focusing on the analysis of displacement resolution and positioning accuracy. Section 3 is related to piezoelectric sensors, including their use in monitoring and controlling various structural physical occurrences. Section 4 is devoted to medical applications of piezoelectric actuators and sensors. These concern robotic actuation for medical interventions and structural sensing in monitoring of healthcare wearable tools. Section 5 deals with discussions related to details on points from the previous sections, including comments on the advantages and limitations of piezoelectric sensors and actuators as well as research perspectives. Section 6 is devoted to conclusions and a summary of possible future work.

2. Piezoelectric Actuators

Topical progresses in technology have resulted in a rapid increase in revolutionary high-accuracy positioning and directing skills. Piezoelectric actuators exhibit resolution dominance in the nanometer-range, swift response, and invulnerability to EM interference, outclassing their pairs, such as magnetostrictive and shape–memory alloy actuators [23–25], in the uses of precision engineering, movement yield, medical handling, and microfluidics management. Furthermore, the mechanical scalability and outstanding load aptitude, make piezoelectric actuators an promising candidate in the fields of automation, robotics, industrial assessment, healthcare, defense, and aerospace.

2.1. Different Categories of Piezoelectric Actuators

Piezoelectric actuators are often classified in different manners related to, for instance, material, performance, structure, vibration condition, cutting-edge performance, etc. In general, from such classifications, an outcome indicates the different adaptabilities of an actuator to a specific application. However, while selecting a viable kind of actuator for diverse use situations, assorted issues ought to be deliberated, accounting for accuracy, velocity, power intake, mechanical intricacy, control complication, and price [6,26]. The classifications of piezoelectric actuators have been investigated in many published works, e.g., [6,8,27], and they are not in the scope of the present work.

The present contribution focuses on actuated procedures exhibiting high degrees of displacement resolution, positioning accuracy, and swift response. We will consider, in this context, different robotic procedures, depending on the intended application, utilizing piezoelectric materials.

The first concerns robotic structures actuated directly by piezoelectric incorporated materials [28] in the form of bonded pieces. These are traveling wave (TW) beams and plates which are generally employed in miniaturized form in applications involving controlled precise displacements of small masses on a smooth surface or other mediums, in general [5,28,29], or of liquids in conduits as mini or micro pumps [4,30,31]. Such actuators are generally used in precision processes that need accuracy, repeatability, and reliability, like in robotics and automation as they require great load facility, a higher move span, and soft displacement. Figure 1 shows an example of a TW piezoelectric integrated minirobot moving on a smooth surface [5]. Figure 2 shows the same robot loaded by a small masse moving on a rough surface [5].

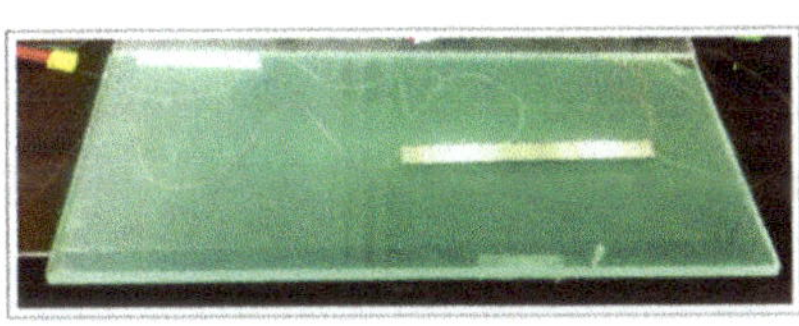

Figure 1. Example of a TW piezoelectric integrated minirobot moving on a smooth surface [5].

Figure 2. The same robot as in Figure 1, loaded by a small masse moving on a rough surface [5].

The second concerns actuated robotic joints using actuators functioning on the conversion of the piezoelectric material deformation into indirect displacement through complex configurations using techniques such as repeating and/or stepping, permitting larger strokes and higher degrees of freedom [32]. Such a category includes stepping actuators [8,33] and ultrasonic actuators [27]. These are characterized by rapid responsiveness, high efficiency, and invulnerability to EM interference, making them particularly suitable for medical and aerospace applications. Stepper actuators have slower movement but offer better mechanical flexibility and lower cost. Figure 3 illustrates, in the context of actuated robotic joints, an example of a prototype of an actuated robotic arm [3].

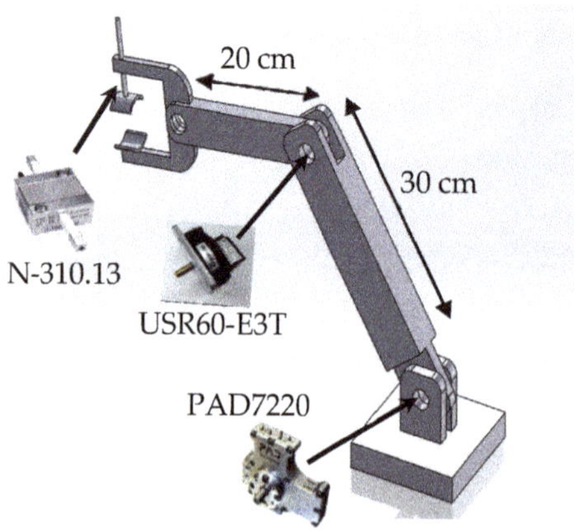

Figure 3. Illustration of actuated robotic joints in a prototype of robotic arm actuated by PZM technologies [3].

Note that piezoelectric materials can be used, in addition to the abovementioned actuation procedures, in different applications, e.g., piezo surgery, which is an osteotomy technique requiring the use of micro vibrations of blades at ultrasonic frequency. Piezo surgery is therefore an ultrasonic transduction obtained by the contraction and expansion of piezoelectric ceramic [34]. Another example is the use of ultrasound transducers for hyperthermia tumor treatment. This technique uses intracavitary ultrasound applicators acting as a multi-element ultrasound transducer that distributes adapted power deposited in the tissue volume by controlling each element, thus allowing selective tissue eradication, which is an alternate to conventional surgery [35].

2.2. *TW Piezoelectric Robotic Structures*

Generally, a structural wave can be generated in a material by a vibration source interacting with that substance. This wave propagates in the material, transporting energy from one position to another. The ability to generate and propagate a wave motion in a restricted material can be reached by specific actuation means. TWs can be obtained in one-dimensional mode in beams actuated at their ends, or in two-dimensional mode in plates with actuators positioned at different positions depending on the intended propagation. These actuators generate controlled structural vibrations consistent with their specific excitations, which define the features of the fashioned progressive waves.

2.2.1. TW Piezoelectric Beam Robots

Regarding piezoelectric ultrasonic actuators, consisting of a fixed excited stator and a moving slider, the linear version of these actuators [35] initiates the concept of traveling wave (TW) piezoelectric beam robots. In this case, the entire robotic beam moves forward by itself rather than moving the slider as in the case of an ultrasonic actuator. The involved motion can be produced by a single or dual excitation mode. In reality, pure TWs can only exist in very long structures. Furthermore, in finite configurations, e.g., beams, the

TW vibration is partly restored when it hits the edges. The mentioned excitation modes allow for the circumvention of the wave reflection. In the case of one-mode excitation, two piezoelectric transducers are placed at each end of the beam, with one acting as an actuator generating at resonance frequency as the TW, and the other acting as a sensor permitting the management of vibrations via their active control regulation down the beam or via electric power dissipation through a passive RL circuit. In the two-mode case, again, two transducers are placed at each end of the beam but both act as an actuator producing beam vibration, developing a TW via active control by simultaneously applying in the two transducers two neighboring beam natural mode shapes at the same frequency but phased at 90°. TW and motion direction can be reversed by the exchange of the two transducers in the one-mode excitation and switching the phase of the two transducers from 90° to −90° in the two-mode case [5,36]. Figure 4 illustrates the schematics of a piezoelectric TW beam robot and a prototype of such a beam minirobot [5].

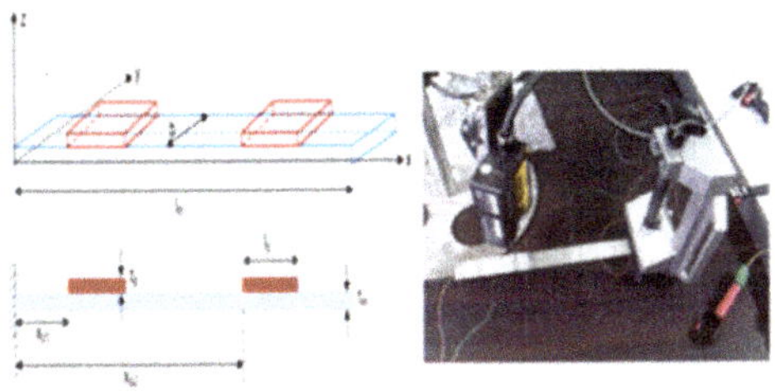

Figure 4. Schematic representation of the shape of a piezoelectric TW beam robot (**left**) and a prototype of such a beam minirobot (**right**) [5].

2.2.2. TW of Piezoelectric Patches Bonded on Thin Structures

Thin structures enfolding piezoelectric materials are largely utilized for vibration control [37], for structural damage and fatigue reveal [38], for designing motors with vibrations at the micrometer level [39], etc.

Small-scale robots utilizing piezoelectric patches or sheets attached on thin structures are operated for various uses concerning beam and plate structures and generally feature miniature robots. A significant class of these relates to beam robots that allow linear motion involving actions of two piezoelectric patches joined on the two beam ends. As discussed in the previous section, such behaviors can be of actuator–sensor nature or an actuator–actuator one [36]. In the case of miniature plate robots, they permit movements in different directions contingent on the actions and locations of piezoelectric patches bonded on the plate [29]. Figure 5 illustrates a representation of a beam robot with two piezoelectric patches on its ends performing in mode 1 (actuator–sensor: vibrating–absorbing) and mode 2 (actuator–actuator: vibrating–vibrating) [5]. In this figure, an aluminum strip is equipped with two bonded patches. In the first case, (a), one of the patches creates the wave, which is absorbed by the second. Different damping techniques can be used. In the second case, (b), the two patches are powered by a voltage 90° out of phase between the left and right patches.

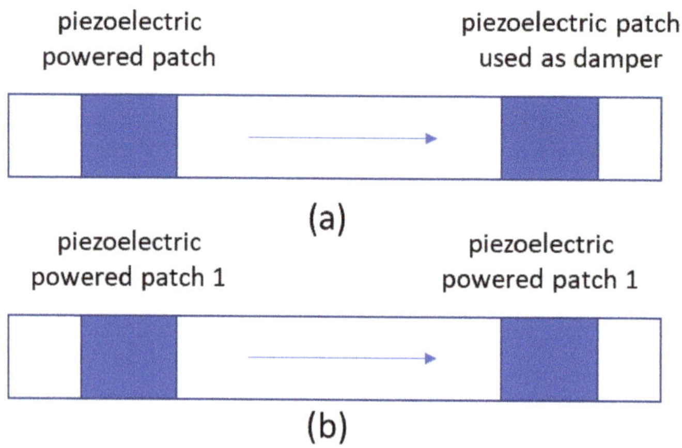

Figure 5. Schematic illustration for piezoelectric beam robot excitation modes: (**a**) actuator–sensor, (**b**) actuator–actuator [5].

2.3. Displacement and Positioning

The inverse piezoelectric effect, as mentioned before, is the ruling law for piezoelectric actuators. Compared to EM ones that reflect good execution in far-reaching force/movement with relatively important displacements, piezoelectric actuators can permit motions of mm or nm. Due to their distinctive benefits of swift response, great motion resolution, high stiffness, actuation potency, and scalable arrangement, as well as endurance to EM interference, piezoelectric materials are often employed in precision displacement, including integration into other systems where space is limited [40].

Piezoelectric actuators are used in precision motion and vibration control applications. Hence, they provide position and speed control, high accuracy and repeatability, and reduced unwanted vibration for smooth, safe, and efficient operation. It should be noted that the displacement and positioning realized by piezoelectric actuation can be detected by piezoelectric sensors, as will be shown in the next section. The combination of highly accurate displacement and positioning actuation and sensing permits reliable robotic management for procedures involving high degrees of sensitivity in zone-restricted actions, like in neurosurgery, restricted drug delivery, etc. [17,18]. In reality, the actuation accuracy in a delimited area can be controlled by sensor detection. Piezoelectric devices allow the detection, actuation, or integration of two tasks; this operation corresponds to a multifunctional tool capable of integrating different tasks on its own component (see Section 5.3).

3. Piezoelectric Sensors

Piezoelectric sensors create an electric charge when submitted to mechanical stress and put forward elevated sensitivity to dynamic variations and valuable process within a wide-ranging frequency. They are generally utilized in structural integrity monitoring, ultrasonic assessing, vibration testing, and are largely used in controlling various industrial structural physical occurrences [19].

3.1. Force Detecting

Piezoelectric force sensors are broadly known for their great sensitivity and well linear response, making them perfect for several force-sensing requests. They function following the principle of direct piezoelectric effect and present a wide-ranging sensing range, from m. newton to k. newton, satisfying varied uses like in healthcare and medical interventions [41–44].

3.2. Structural Integrity Supervising

Structural integrity monitoring is a vital use of piezoelectric sensing in mechanical and civil domains. It is the practice of constantly supervising the structural health of an edifice and identifying any variations or deficiency that might arise [45,46]. A piezoelectric sensor fits well for such supervision since it is particularly sensitive to slight variations in strain. Such supervision is exploited for the integrity checking of bridges, constructions, and other edifices for detecting strain changes and further parameters indicating structural weakening or degradation [47–50]. Moreover, it is frequently exploited to sense and examine structural vibrations triggered by blustery weather, earthquakes, etc. [51,52].

3.3. Movement and Position Detection

As mentioned above, displacement and positioning achieved by piezoelectric actuation can also be detected by piezoelectric sensors in real time due to their great sensitivity and swift response time in many applications involving high precision. Thus, the sensor can identify minor displacement or position changes, which are vital for accurate movement and distance control in applications such as precision machinery, automotive sensing, robotics, and domestic smart appliances [53,54]. Note that an actuator can use self-sensing to detect changes in displacement or position.

4. Examples of Medical Applications of Piezoelectric Actuators and Sensors

Various recurring applications promote a novel well-being owing to health-connected tactics that call attention to safety, comfort, and salutary outcomes. Topical medical advancements have made it possible to determine the sources of many illnesses and institute methodologies to handle them from diagnosis by providing therapeutic assistance and surgical interventions. The efficacy of these treatments is directly connected to the abovementioned patient's health-linked tactics, which suggest that procedures should be minimally invasive (MI) and accurately supervised.

Two main therapeutic categories in this context call for piezoelectric actuators and sensors, namely robotic actuation for medical interventions and structural sensing in the monitoring of healthcare wearable tools. Moreover, as mentioned before, some other therapeutics directly use piezoelectric materials, e.g., piezo surgery [34] and ultrasound transducers for hyperthermia [35].

4.1. Robotic Actuation for Medical Involvements

In medical robotic procedures, as discussed in Section 2.1, robotic structures actuated directly by incorporated piezoelectric materials can be used in miniature form or through robotic joints using actuators by converting the deformation of piezoelectric materials into indirect displacement, which can be used in medical interventions. Directly actuated self-propelled miniature robotic structures will be discussed in Sections 5.2 and 5.7.3. Actuated robotic joints for medical interventions are detailed in the following paragraphs.

4.1.1. Robotic Medical Interventions

Recently, medical interventional procedures have resulted in fast-developing specialties wherever groundbreaking expertise are rapidly introduced and disseminated across diverse surgical fields and skillsets. In addition, interventional approaches have advanced significantly and are constantly improving through open, laparoscopic, and robotic interventions. Progress in practiced medical procedures have generated numerous benefits to patients. Alternatively, procedural intricacy has augmented, and the responsibilities of surgeons have altered considerably; see, for example, [55].

From the time the first surgery was performed, the open practice has usually been exercised to "see fully" and is still utilized today in circumstances associated with intricate frameworks and challenging procedures. On the other hand, the invasive nature of open interventions presents several risks in different specific circumstances and MI procedures would be preferred.

Laparoscopic intervention has meaningfully renovated the long-recognized open procedure due to the various advantages it offers to patients through the MI technique [56,57]. The MI technique has transformed the approach to specific zones of the body by delivering an enlarged view through using a small camera and a light on the end of a miniscule instrument [58,59]. Furthermore, the laparoscopic technique is associated with abridged postoperative pain and quicker recovery, allowing for shorter clinical periods [58–60]. Moreover, it grants significant profits. For instance, it allows for better aesthetics of lesions and a reduced threat of impediment [58] and can play a specific role in diagnostic laparoscopy [61]. Nevertheless, laparoscopic processes exercising lengthened instruments in addition to 2-D visualization revelations might result in some ergonomic problems during operation [56,62] and possible increased postoperative risk [63]. A robotic MI laparoscopic procedure can avoid such problems. In fact, particular processes that are complex or widespread might require open or more sophisticated robot-supported laparoscopic (for sew up, and tissue deal in) procedures [64]. Consequently, only a computerized robotic MI intervention can avoid all of the abovementioned limitations.

In the case of computerized robotics, interventional ingenuity and ergonomics are enhanced due to 3-D vision, robotically amplified degree of freedom, contributing to broader growth in MI intervention practices, e.g., [65–67]. The practice of robotic intervention has progressed from minimal inert actions for instrument withdrawal, or occupying machines such as rail-fixed implements or camera directing, to dynamic robotic arrangements whose movement extent of mechanisms permits improved execution and enhanced precision for ending and suturing while utilizing a MI methodology in autonomous procedures, e.g., [68–70], and with staff in the loop, e.g., [71–73]. Furthermore, it eliminates tactile shakes and laparoscopic intervention pivot effects. Robotic wrist implements propose ample degrees of freedom to surmount the restrictions of laparoscopic tools, which typically do not allow its pointer to reach the tissue anterior and authorize suturing in problematic ergonomic postures.

The above analysis shows that a patient's health and well-being depends on diverse issues linked to the extent of tissue invasiveness, pursuing precision, intervention duration, healing rapidity, etc. These aspects are connected, in addition to the patient, to the therapeutic team in terms of dexterity, ergonomics, and execution ease. The abilities of the staff involved in the different interventional procedures are fairly state-specific, regarding the patient, staff, interventional intricacy, expense, etc.

The expectation of secure, self-ruling MI intervention [74] with skill-substituted tactile and visual abilities, altogether with "staff in the loop", was a dream that has been realized today through digital intelligence. Thus, in addition to patient well-being, staff ease, and MI benefits, the expected accurate positioning and visual ability could be achieved by reliable interventional robotic procedures assisted by imaging scanners. In fact, image-assisted robotic intervention looks to be an obvious advancement of laparoscopic and robotic interventions, alongside skill augmentation and surgeons embracing improved posture during the whole intervention [17,18,75,76]. Moreover, such image-guided robotic interventional procedures are well-adapted for intricate surgeries [77–80] or restricted drug distributions [81–83], and both call for actions in a circumscribed area to safeguard healthy tissues from touching the troubled zone.

4.1.2. Interventional Robotic Actuation

The various robotic procedures discussed in the previous section, namely robotic laparoscopy, computerized robotics, and image-assisted robotics, all require the actuation means necessary for robotic movements. Different actuator technologies are available, with the most common being pneumatic, hydraulic, EM, and smart actuators such as piezoelectric, shape–memory alloy, electroactive polymer, magnetostrictive, and photomechanical actuators. Their difference lies in the type of energy conversion into motion. They present different specific characteristics and applications related to the robotic force, speed, precision, environment, etc.

The different abovementioned interventional robotics can use the most well-adapted actuation technology based on the nature of the intended intervention related to the displacement resolution, positioning accuracy, response speed, stiffness, actuation potency, configuration complexity, stroke flexibility, endurance to EM interference, size scalability, etc.

We have seen in the last section that secure, self-ruling MI intervention with accurate positioning and visual ability could be accomplished by consistent interventional robotic procedures assisted by imaging scanners. Such security, in this context, is related to the intervention nature and duration that is closely allied to the scanner technology. For relatively long imaging intervals such as medical interventions, magnetic resonance imaging (MRI) and ultrasound scanners are generally employed [77–80]. Note that the robotic tools in this context perform close to or inside the scanner.

MRI scanners are progressively and increasingly used in surgery and drug administration, primarily due to their superior ability to distinguish tumors, or affected areas in general, from healthy tissue during procedures related to tumor removal [84–87] or drug administration [17,18]. Moreover, they can be used in all tissue categories, unlike ultrasound scanners, which are limited to body parts devoid of air and bone.

As mentioned earlier, piezoelectric actuation, in addition to the specific requirements of robotics, have a resolution above the nanometer scale, rapid responsiveness, and invulnerability to EM interference, thus surpassing their counterparts in the smart actuation category, as well as common pneumatic, hydraulic, and EM actuators. These characteristics are perfectly compatible with MRI scanners as well as ultrasound scanners.

It is worth noting that the different robotic procedures using piezoelectric materials, depending on the intended medical application, can be divided into two categories: robotic structures actuated directly by incorporated piezoelectric materials and robotic joints actuated by converting piezoelectric deformation into indirect displacement. In fact, the first category corresponds to traveling wave beams and plates, which are generally used in miniaturized form in applications involving precise and controlled displacements of small masses on a surface or other mediums in general [5,28,29,36] or liquids in pipes such as mini or micro pumps [4,30,31]. These robots are generally used in precision processes that require repeatability, reliability, high load capacity, greater range of motion and smooth displacement; see Sections 2 and 5.2. Figure 6 shows the functioning principal of a TW piezoelectric mini-pump accurately delivering controlled liquid. Figure 7 illustrates the prototype constituents of the mini-pump of Figure 6 [4].

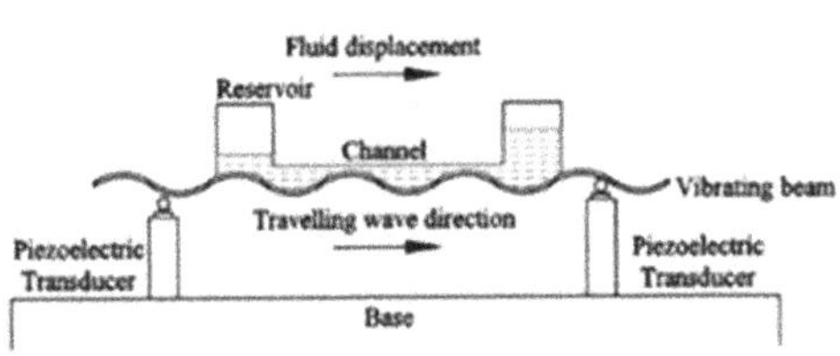

Figure 6. Schematics of a TW piezoelectric mini-pump moving accurately controlled fluid [4].

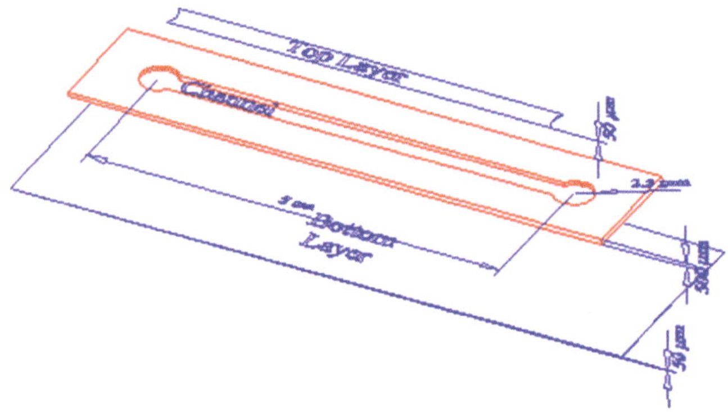

Figure 7. Structural prototype elements (schematic representation of the shape) of a fluid-controlled delivery using a TW mini-pump [4].

Figure 7 shows one of the possibilities for using a traveling wave. In this figure, the bottom layer is the one in which the traveling wave propagates. We deposit a PDMS channel which is then covered by the top layer. When liquid is placed in the 2.5 mm wide and 5 cm long channel, it is driven by the vibrations of the bottom layer. We have created a pump that can drive small quantities of liquid.

Additionally, the second category uses techniques such as repetition and/or stepping, allowing larger strokes and higher degrees of freedom (DOF) [32]; these include stepper actuators [8,33] and ultrasonic actuators [27], which are particularly suitable for medical interventions as will be shown in the next section.

4.1.3. MRI-Assisted Robotic Actuation

As mentioned in the previous section, MRI scanners require an environment that is immune to EM interference to operate. This includes protection from exposure to external EM fields (EMFs) and the exclusion of EMF-sensitive materials in the scanner scaffold. Without these mandatory precautions, serious disturbances would alter the image [17,18]. Thus, robotic machines, housed in the scaffold near the involved body tissues, should generally be MRI compatible, i.e., free of materials sensitive to EMFs, such as magnetic or massive conductive materials. Most robotic mechanisms, including interventional tools and structures, can be constructed with MRI-compatible materials. Regarding robotic actuation tasks, few actuators of acceptable performance offer such compatibility, such as pneumatic and piezoelectric actuators. As mentioned previously, the latter outperform in several respects, particularly in terms of responsiveness [17,18,75,76].

Actuation piezoelectric machineries reflecting MRI compatibility come in different styles of configuration, composition, construction, and use [88,89]. Different examples could be found in the literature for stepping actuation [8,33], stick-slip multi DOF [32] with flexure hinge and bionic imitating body movement actuations [90–92], and for ultrasonic [27] standing wave miniature linear and multidimensional motion actuators [93–95]. Furthermore, cases of piezo robotic hands for micro-to-macro motion manipulation and adaptive miniature piezoelectric robots are given in [1,96–98] and cases of micro-motion robotic surgery tool and image-based biopsy using OCT control are given in [99,100]. Cases concerning the role of robotics in image-guided interventions and in healthcare robotics are given in [101,102].

Figure 8 schematically summarizes the above conclusions relative to scanner and robotic actuating means in an image-assisted robotic intervention.

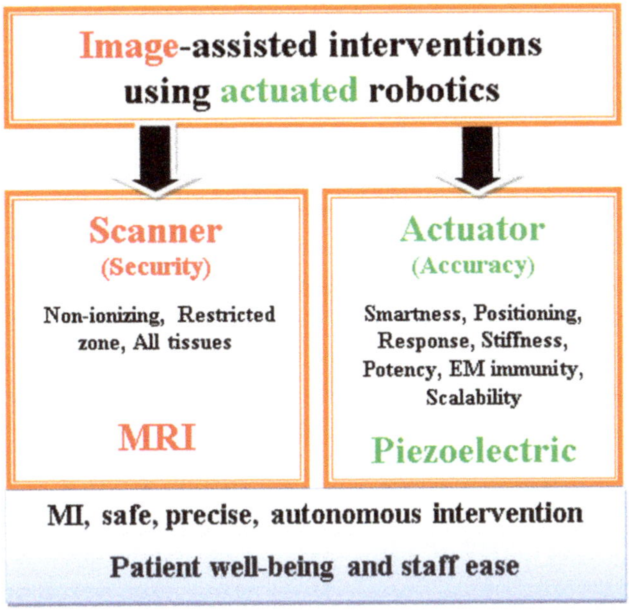

Figure 8. Summarized selection strategies relative to scanner and robotic actuating means in an image-assisted robotic intervention.

4.2. Structural Sensing and Monitoring of Healthcare Wearable Tools

Wearable healthcare tools that are close to or integrated in the body generally correspond to portable, removable, or incorporated tools. They can perform passive or focused duties engaged in body tissues. Such tasks are related to sensing or assisting activities.

Sensing activity includes detection [103–105], diagnostic [106,107], monitoring [108–110], and control [111,112], and can be involved in diagnosis support, health monitoring and prognosis, heart rate monitoring, respiratory monitoring, measurement of blood pressure, personalized pain medicine, etc.

Assisting activity comprises stimulating, such as pacemakers and defibrillators [113], drug release, such as implanted delivery devices [114], monitoring, such as implants in the spinal cord and head [115,116], and assisting, such as cardiac devices assisting continuous blood-flow [117], and MRI-guided robot-assisted surgery and interventions [17,78].

These two activities could be distinctive or networked and autonomous or remotely steered and perform continuously in real time. In addition to tasks of surveillance, forecast, maintenance, stimulation, etc., of wearable tools, these allow the administration of post-treatment situations after preceding treatments. This includes by-passing relocations, transpositions, transfers, etc., replacing face-to-face treatment with an incorporated joined watch over the line of attack.

Sensing duties require portable pressure sensors that reflect non-invasive, real-time, and uninterrupted supervision of health vitals. They permit healthcare staff to collect consistent and precise information. By monitoring pressure, several biological issues, such as wrist pulse, heart rate, and joint and muscle behavior, can be evaluated. Portable pressure sensors present the benefit of permitting patients to carry out their everyday tasks while staying monitored. As mentioned before, piezoelectric pressure sensors possess high sensitivity and large linear response, making them suitable for several wide-ranging pressure-sensing requests, like in the management of dynamic signal detection. For instance, they are used in accurate wrist pulse signal acquisition [41], assessment of Parkinson's tremor [42], and robot-assisted MI surgeries [43,44].

Assisting duties are MI and use, as mentioned before, in addition to sensor control, and actuating means. As stated above, they could be integrated in tissues for health upkeep and stimulation of diverse body parts, or enclosed in tissues for image-assisted medical interventions. Both integrated and enclosing devices are robotically based and can, respectively, directly use piezoelectric patches [4,5,28–31] or be indirectly actuated by piezoelectric actuators [8,27,33]. Such activity has been largely developed in Section 4.1 "Robotic Actuation for Medical Involvements".

5. Discussion

In the analyses concerned in the preceding sections, a number of points merit further discussion:

5.1. Advantages and Limitations of Piezoelectric Sensors and Actuators

The recent progression of piezoelectric sensors and actuators (in general and for medical applications) is related to groundbreaking materials with enhanced assets such as augmented stability, endurance, and electric behavior, and to cutting-edge manufacturing techniques such as nanotechnology that allowed for the development of nanogenerators, resulting in the conversion of mechanical vibrations at low-frequency to electrical output.

In the instance of the above-discussed wearable sensors, soft pliable materials such as PVDF (polyvinylidene fluoride) and flexible piezoelectric composites (FPCs) allow for efficient heath supervision in real time, movement sensing, and biomechanical motion harvesting, all possibly integrated in intelligent fabrics [16,118]. Moreover, they have encouraging projections in addition to wearable devices, such as in the fields of biology, aerospace, electromechanical, etc. [119].

In the case of an efficient precision actuator, thanks to an innovative laser processing manufacturing technique, high-frequency vibration control has been made possible [120].

Regarding the limitations of piezoelectric sensors and actuators, these are mainly related to functional and material behaviors.

For instance, in the case of wearable sensing tools, piezo materials tackle a balance of structural elasticity and electric yield. For example PVDF present good compliance but weak piezoelectric modulus compared to inflexible ceramics, which are easily broken and hence unreliable for dynamic wearables. Furthermore, endurance under repetitive mechanical pressure poses further issues [19]. Additionally, ceramics such as PZT (Lead Zirconate Titanate) contain Lead (Pb), which poses additional issues related to Section 5.7.2 on biocompatibility, biodegradability, and non-toxicity.

5.2. Piezoelectric Microrobots and Bio-Inspired Concepts

Piezoelectric actuation is ideally suited for control techniques for executing movements using miniaturized configurations. Microactuators thus play a key role in micromachining technology and in exploring or modifying the microscopic world with objects ranging from cells to molecules. These microscopic entities can be maneuvered by interacting with a microrobot, or a high-precision macrorobot combined with a suitable end effector. Microrobots are particularly suitable, due to their lightness and flexibility, for biological and medical fields, like drug distribution and disease diagnosis.

Piezoelectric microrobots are potentially adapted to perform in difficult atmospheres, such as inside the human body. Another peculiarity lies in their execution, which is often based on bio-inspired concepts. For example, a robot with a thin and flexible plate structure actuated by specifically located fixed piezoelectric patches [5,29] can illustrate movements of crawling on the surface, flying in the air, or swimming in water. Different instances of bio-inspired robotic actuation could be found in the literature, such as a multilegged piezoelectric robot [121], earthworm [122], rhinoceros beetle [123], grasshopper [124], rotatory galloping gait [125], squirrel's galloping gait [126], octopus-crawling [127], and untethered levitation [128].

5.3. Piezoelectric Multifunctional Deeds and Flexible Wearable Tools

Piezoelectric materials exhibit direct (sensing) and inverse (actuating) conversions as stated previously. Such electromechanical two-way conversion is not limited to piezoelectricity, but piezoelectric materials matched to their competitors displaying such conversion present towering electromechanical effectiveness and notable scalability, making them applicable for miniaturization. Thus, they are broadly employed in sensing, actuating, or integrated purposes [129].

Along with sensing and actuating, in wearable devices, for example, energy harvesting generated by human movement activities can be realized by piezoelectric flexible material (PFM) devices that are well-adapted to great deformations [130].

A multifunctional tool is normally able to integrate different tasks through its own components. Such amalgamation can lead to further compacted outcomes. The natural multifunctional integration combines the sensing and harvesting tasks, resulting in a self-driven sensor. Figure 9 shows schematics of a multifunctional piezoelectric device involving sensing and harvesting functions, following mechanical-to-electrical conversion and inversely for actuating functions.

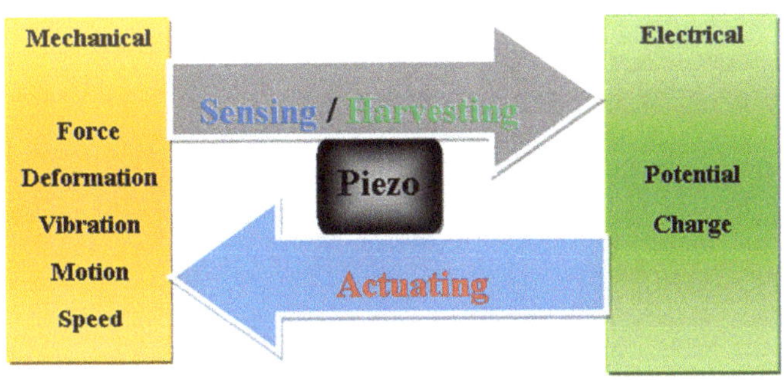

Figure 9. Schematics of multifunctional piezoelectric device involving sensing and harvesting functions as well as actuating functions.

5.3.1. Wearable Tools and Flexibility

As mentioned above, structural flexibility plays a significant role in wearable tools, which are generally miniature and weightless to ensure they are portable. This flexibility concerns not only PFM but also neighboring support sheets and electrodes, thus providing a reliable multifunctional flexible piezoelectric tool. Such wearable flexibility reflects the adaptation to dynamic movements of humans [130].

Energy harvesting in wearable devices can involve heart rate [131,132], respiration [133,134], pulse [135], and deformation within the gastric cavity [136].

5.3.2. PFM Main Categories

PFM can be obtained through polymers [137,138], composites [139,140], and inorganic thin films [141,142]. These three categories of PFM possess mutual characteristics of flexibility and piezoelectric but their performing features display substantial distinctions. Organic polymers are naturally fully flexible, but possess the lowest piezoelectric features; inorganic thin films hold the best features of crystalline piezoelectric thin layers, but are physically more brittle; and composites that are a crystal–polymer blend have behavior that falls somewhere between the previous two, where flexibility and piezoelectric properties can be adjusted based on structural design or blend ratios. These distinctive performance characteristics suggest the operational adaptability of each PFM in specific situations. For example, organic PFM polymers may be more suitable for motion management on complex surfaces or substrates, while inorganic PFMs meet higher sensitivity requirements. Also, thin-film or composite PFMs are better at meeting actuation and energy harvesting requirements.

5.4. Endurance of Piezoelectric Medical Devices to EM Interference

The operation of various medical instruments can be disrupted by EM interference. This can be caused by external EMF radiation on the instrument or by the inclusion, attachment, or insertion of objects sensitive to EMFs, due to their effects on the instrument's own fields, as is the case of MRI scanners [17]. Moreover, such interference could be produced by a combined EMF radiation and presence of EMF-sensitive matters in the tool structure, such as the case of wearable tools [143]. In both cases, the use of actuation or sensing devices incorporating materials with reliable resistance to EM interference is recommended. Piezoelectric materials used in MRI-assisted medical procedures (see Section 4.1.3), whether in microrobotics or for the actuation of robotic interventions, as well as for the detection of portable tools, offer a dependable solution in this context. These actuation or detection devices are mainly made, as mentioned above, of dielectric piezoelectric materials insensitive to EMFs, but furnished with thin conductors, allowing for electrical conversion. These electrodes have a theoretical sensitivity to EMFs, a skinny nature, and can be structurally adjusted [17], making it possible to largely alleviate their effects. In fact, the importance of the eddy currents, induced in these conductors and responsible for the disturbances, depends on the surface of the conductor perpendicular to the field concerned, so if the orientation of the device is such that such a surface corresponds to the trivial thickness of the electrode, there would be no problem.

Verification of the immunity of a piezoelectric device to EM interference, when necessary, can be carried out by an EM compatibility (EMC) analysis; see, for example, [144]. In general, such immunity exists when the EMF distribution in the host application (MRI scaffold field or wearable sensing tool radiated field) involving the piezoelectric device would be the same with and without the device (see Figures 8 and 9 in [17]). In fact, for example, in an MRI scanner, the radiofrequency field in the scanner scaffold, which is directly associated with living tissue images, displays a 3-D field distribution depending on the targeted problem details. The perturbation of such distribution alters the image through artifacts, leading to a wrong interpretation. The introduction of external objects in the scaffold alongside living tissues could perturb the field distribution. Thus, if the field distribution remains the same with and without a piezoelectric device, this indicates the device is insensitive to EMFs and is hence immune to EM interference.

5.5. Digital Monitoring of MRI-Assisted Robotic Interventions

As mentioned earlier, a safe and autonomous MI intervention, with precise positioning and good visual capability, could be efficiently performed by a consistent interventional robotic procedure, assisted by imaging scanners, and actuated by an adequate positioning device. Moreover, for relatively long intervention intervals and a better ability to distinguish affected areas from healthy tissues, MRI assistance would be suitable for procedures related to tumor ablation or drug delivery in all tissue categories. Additionally, robotic components, including actuation devices in such an MRI-assisted procedure, would be immune to EM interference, favoring the choice of piezoelectric devices known for their precise positioning.

5.5.1. Closed-Loop Controlled MRI-Assisted Autonomous Scenery

In fact, patient safety is linked to the therapeutic limitation attributed to the targeted area, which mainly depends on actuation accuracy of the interventional device and its spatial positioning. Thus, a cooperative arrangement includes the MRI scanner, interventional tool, location and duties assigned, robotic actuation, and control alongside the imaged, restricted troubled zone, all performed in a closed-loop controlled autonomous procedure as exemplified in Figure 10.

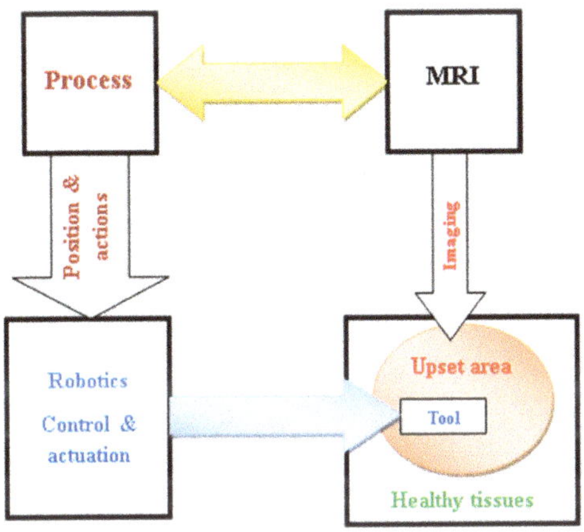

Figure 10. Schematics of a closed-loop controlled autonomous cooperative procedure including the MRI scanner, interventional tool, location and duties assigned, robotic actuation, and control alongside the imaged, restricted troubled zone.

The involved accuracy in this control procedure would be dependent on the different issues comprising the complexity level of its interacting components, their linked uncertainty features, and unforeseen exterior threat events, together with MRI compatibility issues related to robotic immunity to EM interference, including the piezoelectric actuating device. It is only by managing such potentially tormenting problems that reliable performance can be achieved.

5.5.2. Digital Twin Administration of MRI-Assisted Interventions

The depreciation of the menacing distresses, together with respect to particular personalized data, are necessary for reliable operational MRI-assisted controlled procedures. These might be achieved by supervising the implicated parameters in a corresponding real–virtual couple by way of a digital twin (DT) implement [145].

A DT consists of a physical element, a virtual picture of that element, and a quasi-real-time, two-way data flow between those two elements. In other words, a DT is considered as an integration of information into a real-life experiment and its digital copy, thus forming a pair of bidirectional routines. This methodology is performed in the administration of intricacy in controlled processes [146] and organized as a real–virtual couple permitting self-adapting tasks. Hence, the real part distributes treated detected information to its virtual part; however, the latter conveys control instructions to the real part. Such self-adaptation corresponding assists, in addition to complexity monitoring, in decreasing uncertainties within the pair and unforeseen hazards in the perturbing dynamics of MRI-assisted robotic control. In fact, the matching of physical and model sides allows each side to correct and adjust the other. This self-adaptation aids in reducing uncertainties (physical and computational) as well as unforeseen hazards. The assistance in managing complexity comes from representation of interconnected physical phenomena involved in MRI-assisted controlled robotic interventions (the scanner fields, robot instruments, actuation system, and living tissues) by a coupled mathematical model. The matching of these physical interconnected phenomena with their coupled model permits, again, their self-adaptation and hence, adjustment.

It is worth noting that the DT concept has been progressively proposed recently in healthcare, nursing, and extended administration; see, e.g., reviews illuminating therapies, supervising, and administrations [147–152].

A comprehensive DT administration of an MRI-assisted robotic control through exchanges between its real–virtual wings involves the delivery of sensed treated data by the real wing, which is paralleled and adapted by exterior Internet of Things "IoT" information as well as the historical learnt data. The result, after a data analysis form training, is transferred, together with a suitable proposal of model reduction, to the DT virtual wing. In fact, a rapid matching among the DT wings dictates a realistic model with a short computation interval. Consequently, the complete model, which accurately characterizes the real process, would be abridged, permitting restrained execution time while conserving the representation of the physical process. Figure 11 shows the topographies of a monitoring DT of a MRI-guided robotic intervention. This DT monitoring can be exploited for training of medical personnel, forecasts by means of real phantoms along with their digital replicas, or a physical patient–virtual model, containing self-decision matching plus keeping staff "in the loop". It should be noted that the DT concept is also applicable to the monitoring of industrial robotic procedures, including piezoelectric actuations [145].

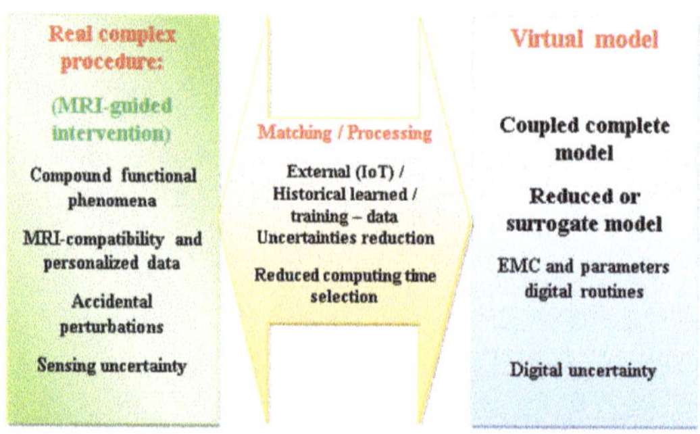

Figure 11. Summarized topographies of a monitoring DT of a MRI-guided robotic intervention.

5.5.3. Digital Augmented DT in MRI-Assisted Interventions

The abovementioned involvement of staff associated with robotics permits a cutting-edge MRI-guided monitoring of medical interventions, therefore decreasing the patient risk and guaranteeing a dependable result for staff [153–155]. Furthermore, artificial intelligence (AI) practices in these treatments support staff by reducing the data-acquisition and post-processing complexities and accomplish repeated scheduled training duties [156,157]. Additionally, the intervention can be meaningfully enhanced via extended staff–robot links, progressing the whole organization across augmented reality (AR)-supported robotic activities. Hence, AR joined to a MRI can diminish intricate interventional risks such as tissue injury, bleeding, and distress post-intervention. Moreover, DTs can execute a significant task in AR-aided interventional robotics. Therefore, the likely disorder origin and its intervention method can be precisely identified through personal patient examination via deep learning databanks. Likewise, a number of further benefits of fused AR-DT are linked to enhanced suturing precision, fastening, and repairing, which are comparable to manual tasks [158–162].

DT practice is a type of digital treatment and is commonly utilized in personalized medical therapies that can be used in nursing, recognition, tutoring, or interventions. DTs are frequently connected, in addition to AI and AR tools, with virtual reality (VR); for instance, VR training improves the skill needed to reproduce daily training circumstances while providing the facility to correctly measure performance. Also, the preparation of DTs permits staff to perceive the development of the disorders and adjust cure strategies to choose the best appropriate therapy. Such preparation in personalized scheduling supports the progression of early diagnosis and search for novel cures or interventions [163,164]. It should be noted that such digital treatment contributes to the high-reliability precision of medical therapies allowing for patient safety and staff ease.

5.6. Matching of Performance Indicators with Application Scenarios

In the present paper the potential of piezoelectric actuation and sensing in high reliability precision in medical therapeutics was investigated. We have seen that these devices surpass, in this context, their competitors', particularly for high reliability precision involved in intricate tasks. This section illustrates some comparisons of device performances and medical scenarios. Table 1 shows a comparison of piezoelectric actuators with traditional electromagnetic, hydraulic, or shape–memory alloy actuation methods in terms of resolution, response time, driving force, and MRI compatibility.

Table 1. Comparison of performances of different actuating devices involved in medical interventions.

Actuator	Resolution	Response Time	Driving Force	MRI Compatibility
piezoelectric	nanometer	microsecond	Newtons	High
electromagnetic	tens of micrometers	millisecond	Newtons	Low
hydraulic	micrometer	millisecond	Newtons	High
shape–memory alloy	nanometer	second	Newtons	Medium

Table 2 illustrates a comparison of different actuators in the two scenarios of computerized and MRI-assisted robotic interventions.

Table 2. Comparison of actuators in two interventional scenarios.

Actuator	Computerized Robotic Common Intervention	MRI-Assisted Robotic Intricate Intervention
Piezoelectric	V. Good	V. Good
Electromagnetic	Good	No
Hydraulic	Moderate	Moderate
Shape–memory alloy	Good	No

Table 3 illustrates a comparison between piezoelectric sensors and other types related to EM interference.

Table 3. Illustration of piezoelectric sensors regarding EM interference.

Sensors	Sensing Without EMF Exposures	Sensing in EMF Exposure Environment
Piezoelectric	Good	Good
Others	Good	Need shields

5.7. *Future Research Perspectives on Piezoelectric Implications in the Medical Field*

This section is devoted to possible future perspectives of the involvement of piezoelectric devices in medical applications, and particularly in robotic interventions and wearable tools investigated in this paper. When evaluating such involvement, we refer to piezoelectric materials, their device structures, their intended requests, and their specific performance. The main features include piezoelectric performance, structure flexibility, multifunctional behavior, adaptation upon request, embedded biocompatibility, and biodegradability [165–167].

5.7.1. Wearable and Implantable Medical Tools

An important goal in wearable and implantable tools is to focus, via technological tasks, on improving both structure flexibility and piezoelectric performance as well as biocompatibility in case of embedded concerns. These allow multifunctional and miniaturized tools that require less energy intake and are more safe. Moreover, in implanted tools, actuating and harvesting duties would be possible by the incorporation of energy storage means.

5.7.2. Biocompatibility, Biodegradability, Non-Toxicity, and Piezoelectric Biomaterials

In many medical applications, supervision, monitoring, or intervention tasks are performed within or near living tissues. Most of these tasks impose biocompatibility and non-toxicity, and in some cases, biodegradability. Furthermore, the presence of natural piezoelectric effects in several parts of the body, such as bones, tendons, skin, and other tissues, highlights the potential of biomaterials in biomedicine to improve or substitute biological tasks [168]. Thus, biosecurity, biocompatibility, non-toxicity, and the absence of immune rejection are guaranteed. Augmented medicine and sustainable evolution related to human requests [169,170] put forward the exploration of piezoelectric biomaterials in medical applications.

5.7.3. Dependable Self-Moving Miniature Robots

Continuous developments in fabricating and assembling strategies have led to improvements in autonomous (self-moving) robots, including different forms (beam, plate, rod, conduit, etc.) with a piezoelectric actuation systems (bonded, embedded, loaded, etc.). The bonded form, favored by flexible and widely available mechanical designs, has generated a high demand for precision tools, especially in medical robotics. Potential explorations of miniature robotics in this context are possible, including increased degrees of freedom, transport and positioning capabilities [171], extended torque and speed range [172], large working stroke and thrust force [173], millisecond-scale response time, millinewton output force, high-speed operation, sub-micrometer-level resolution [174], and centimeter-scale reconfigurable robots [175].

5.7.4. MRI Compatibility in Image-Guided Robotic Interventions

As mentioned previously, in addition to the intrinsic benefits of image-assisted robotic interventions (closed-loop control procedure) related to patient and staff comfort, MRI and robots each have their own advantages. MRI allows for better distinction between affected areas and healthy tissue during tumor ablation or drug administration in all tissue categories, and exhibits non-ionizing behavior ensuring patient safety over relatively long intervention intervals. Robotic action reflects precise movement and positioning directly related to its actuation device. The combination of MRI and robots requires robotic components immune to EM interference (MRI compatibility). Piezoelectric actuation devices perfectly provide this precision and EM immunity [17,18,78].

The issue of MRI compatibility generally concerns body-embedded devices as implants and near-body instruments as all robotic components, and its deficiencies, can lead to image artifacts that compromise the outcome of the procedure. Further research on MRI-compatible materials and structural features, including piezoelectric devices, remains necessary [176,177]. Furthermore, since image artifacts are generally unavoidable [178], their assessment is important for image analysis and the adoption of a fitting correction strategy, based on body pre-scans, scan parameters, and shimming. Further research on artifact correction methods, image processing techniques, and scanning routines assisted by digital monitoring strategies are needed (see Section 5.5.3).

5.7.5. Summarized Illustration of Future Research Perspectives

Figure 12 illustrates a summary of the abovementioned future research perspectives.

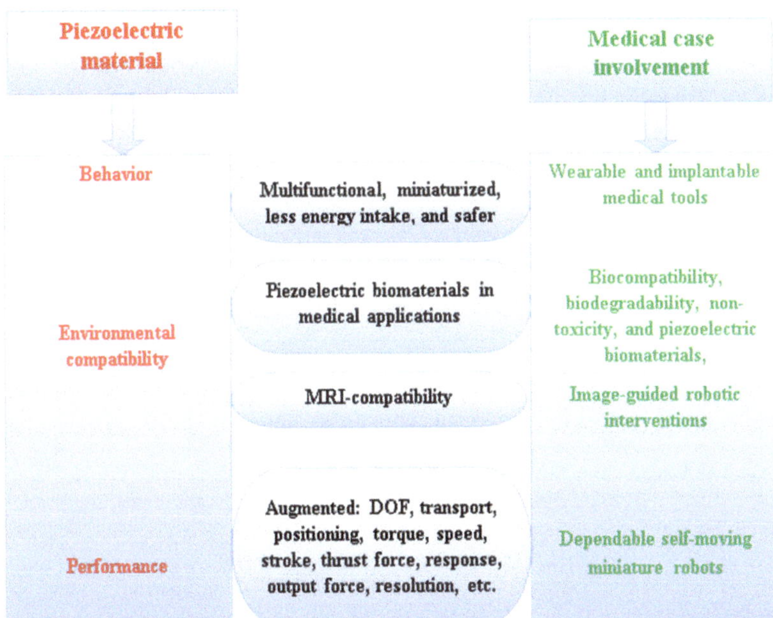

Figure 12. Summarized illustration of future research perspectives (medical case involvements) related to piezoelectric material features (behavior, environmental compatibility, and performance).

6. Conclusions

The present contribution analyzed and emphasized the possibilities of piezoelectric strategies in precision duties and their involvement in the health field. The paper underlined the potential of such strategies in medical interventional procedures and monitoring through wearable healthcare tools. The high-reliability precision medical procedures in question have been shown to be associated with improved patient safety and staff ease.

In interventional tasks, the article illustrated the role of piezoelectric robotic actuation involving precise control of displacement and positioning. Moreover, the interventional commission performed by miniaturized piezoelectric robots, based on bio-inspired concepts, was shown to be potentially suitable for operation inside the human body. Both tasks exhibit accurate MI interventions that can be used in surgery or implanted drug delivery.

Regarding wearable health tools, the article highlighted the potential of flexible piezoelectric materials in miniaturized monitoring tools involving sensing, actuation, and energy harvesting. These tools provide non-invasive, real-time, and uninterrupted monitoring of vital indications, as well as assistive functions of MI using actuation means in addition to sensor control.

Different perspectives for future research on piezoelectric implications in the medical field suggest further investigations on the following (for details, see Section 5.7):

- Wearable and implantable medical tools;
- Biocompatibility, biodegradability, non-toxicity, and piezoelectric biomaterials;
- Dependable self-moving miniature robots;
- MRI compatibility in image-guided robotic interventions.

Author Contributions: A.R. and Y.B. have contributed equally to all items. All authors have read and agreed to the published version of the manuscript.

Funding: This research received no external funding.

Informed Consent Statement: Not applicable.

Data Availability Statement: The original contributions presented in this study are included in the article. Further inquiries can be directed to the corresponding author.

Conflicts of Interest: The authors declare no conflicts of interest.

References

1. Li, J.; Deng, J.; Zhang, S.; Chen, W.; Zhao, J.; Liu, Y. Developments and challenges of miniature piezoelectric robots: A review. *Adv. Sci.* **2023**, *10*, 2305128. [CrossRef] [PubMed]
2. Yuan, Z.; Zhou, S.; Hong, C.; Xiao, Z.; Zhang, Z.; Chen, X.; Zeng, L.; Wu, J.; Wang, Y.; Li, X. Piezo-actuated smart mechatronic systems for extreme scenarios. *Int. J. Extrem. Manuf.* **2025**, *7*, 022003. [CrossRef]
3. Brahim, M. Modeling and Position Control of Piezoelectric Motors. Ph.D. Thesis, University of Paris XI, Gif-sur-Yvette, France, 2017. Available online: https://theses.hal.science/tel-01689921v1/file/72356_BRAHIM_2017_diffusion.pdf (accessed on 1 October 2025). (In English)
4. Hernandez, C. Realization of Piezoelectric Micro Pumps. Ph.D. Thesis, University of Paris XI, Gif-sur-Yvette, France, 2010. (In French)
5. Hariri, H. Design and Realization of a Piezoelectric Mobile for Cooperative Use. Ph.D. Thesis, University of Paris XI, Gif-sur-Yvette, France, 2012. Available online: https://theses.hal.science/tel-01124059v1/file/2012PA112321.pdf (accessed on 1 October 2025). (In English)
6. Zhou, X.; Wu, S.; Wang, X.; Wang, Z.; Zhu, Q.; Sun, J.; Huang, P.; Wang, X.; Huang, W.; Lu, Q. Review on piezoelectric actuators: Materials, classifications, applications, and recent trends. *Front. Mech. Eng.* **2024**, *19*, 6. [CrossRef]
7. Maqbool, S.A.; Touqeer, M.; Esmaeilzadeh, B.; Yang, S.; Meng, W.; Wang, J.; Feng, Q.; Hou, Y.; Lu, Q. A compact multi-degree-of-freedom piezoelectric motor with large travel capability. *Rev. Sci. Instrum.* **2025**, *96*, 043702. [CrossRef] [PubMed]
8. Wang, S.; Rong, W.; Wang, L.; Xie, H.; Sun, L.; Mills, J.K. A survey of piezoelectric actuators with long working stroke in recent years: Classifications, principles, connections and distinctions. *Mech. Syst. Signal Process.* **2019**, *123*, 591–605. [CrossRef]
9. Wei, W.; Ding, Z.; Wu, J.; Wang, L.; Yang, C.; Rong, X.; Song, R.; Li, Y. A miniature piezoelectric actuator with fast movement and nanometer resolution. *Int. J. Mech. Sci.* **2024**, *273*, 109249. [CrossRef]
10. Ling, M.; Zhang, C.; Chen, L. Optimized design of a compact multi-stage displacement amplification mechanism with enhanced efficiency. *Precis. Eng.* **2022**, *77*, 77–89. [CrossRef]
11. Ma, S.K.; Yang, Y.L.; Cui, Y.G.; Wu, G.H.; Wei, Y.D. A purely centered and non-redundant piezoelectric stick-slip rotary stage with force amplification. *Mech. Syst. Signal Process.* **2024**, *220*, 111689. [CrossRef]
12. Zhong, B.W.; Liao, Z.; Hu, H.Z.; Liu, S.L.; He, C.G.; Sun, L.N. A review of recent studies on piezoelectric stick-slip actuators. *Precis. Eng.* **2025**, *94*, 175–190. [CrossRef]
13. He, L.; Liu, X.; Huang, Z.; Liu, F.; Tian, H.; Dong, Y.; Ge, X. Inertial impact rotary piezoelectric motor with non-reversing properties and a clamping mechanism. *Smart Mater. Struct.* **2025**, *34*, 055028. [CrossRef]
14. Wang, X.; An, D.; Qin, Z.; Wang, C.; Liu, Y.; Yang, Y. Nonlinearity Characterization of Flexible Hinge Piezoelectric Stages Under Dynamic Preload via a Force-Dependent Prandtl–Ishlinskii Model with a Force-Analyzed Finite Element Method. *Actuators* **2025**, *14*, 411. [CrossRef]
15. Zhi, C.; Shi, S.; Wu, H.; Si, Y.; Zhang, S.; Lei, L.; Hu, J. Emerging Trends of Nanofibrous Piezoelectric and Triboelectric Applications: Mechanisms, Electroactive Materials, and Designed Architectures. *Adv. Mater.* **2024**, *36*, e2401264. [CrossRef] [PubMed]

16. Zhi, C.; Shi, S.; Si, Y.; Fei, B.; Huang, H.; Hu, J. Recent Progress of Wearable Piezoelectric Pressure Sensors Based on Nanofibers, Yarns, and Their Fabrics via Electrospinning. *Adv. Mater. Technol.* **2023**, *8*, 2201161. [CrossRef]
17. Razek, A. Image-Guided Surgical and Pharmacotherapeutic Routines as Part of Diligent Medical Treatment. *Appl. Sci.* **2023**, *13*, 13039. [CrossRef]
18. Razek, A. From Open, Laparoscopic, or Computerized Surgical Interventions to the Prospects of Image-Guided Involvement. *Appl. Sci.* **2025**, *15*, 4826. [CrossRef]
19. Mangi, M.A.; Elahi, H.; Ali, A.; Jabbar, H.; Bin Aqeel, A.; Farrukh, A.; Bibi, S.; Altabey, W.A.; Kouritem, S.A.; Noori, M. Applications of piezoelectric-based sensors, actuators, and energy harvesters. *Sens. Actuators Rep.* **2025**, *9*, 100302. [CrossRef]
20. White, A.; Little, I.; Artyuk, A.; McKibben, N.; Kouchi, F.R.; Chen, C.; Estrada, D.; Deng, Z. On-demand fabrication of piezoelectric sensors for in-space structural health monitoring. *Smart Mater Struct.* **2024**, *33*, 055053. [CrossRef]
21. Liang, S.; Wang, S.; Haoyu, G.; Zhang, Y.; Shao, S.; Xu, M. A micro-piezoelectric inertial robot with an ultra-high load-to-weight ratio: Design and experimental evaluation. *Smart Mater. Struct.* **2025**, *34*, 075017. [CrossRef]
22. Guo, Z.-R.; Tan, H.-S.; Chang, C.-S.; Hwang, I.-S.; Hwu, E.-T.; Liao, H.-S. The Design of a Closed-Loop Piezoelectric Friction–Inertia XY Positioning Platform with a Centimeter Travel Range. *Actuators* **2025**, *14*, 265. [CrossRef]
23. Gao, C.D.; Zeng, Z.H.; Peng, S.P.; Shuai, C.J. Magnetostrictive alloys: Promising materials for biomedical applications. *Bioact. Mater.* **2021**, *8*, 177–195. [CrossRef]
24. Kumar, D.; Daudpoto, J.; Chowdhry, B.S. Challenges for practical applications of shape memory alloy actuators. *Mater. Res. Express* **2020**, *7*, 073001. [CrossRef]
25. Berhil, A.; Barati, M.; Bernard, Y.; Daniel, L. Accurate sensorless displacement control based on the electrical resistance of the shape memory actuator. *J. Intell. Mater. Syst. Struct.* **2023**, *34*, 1097–1103. [CrossRef]
26. Kanchan, M.; Santhya, M.; Bhat, R.; Naik, N. Application of Modeling and Control Approaches of Piezoelectric Actuators: A Review. *Technologies* **2023**, *11*, 155. [CrossRef]
27. Tian, X.; Liu, Y.; Deng, J.; Wang, L.; Chen, W. A review on piezoelectric ultrasonic motors for the past decade: Classification, operating principle, performance, and future work perspectives. *Sens. Actuators A Phys.* **2020**, *306*, 111971. [CrossRef]
28. Ding, Z.; Cui, M.; Wu, J.; Wei, W.; Rong, X.; Li, Y. Development of an Untethered Self-Moving Piezoelectric Actuator with Load-Carriable, Fast, and Precise Movement Driven by Piezoelectric Stack Plates. *IEEE Trans. Ind. Electron.* **2025**, *99*, 11635–11646. [CrossRef]
29. Hariri, H.; Bernard, Y.; Razek, A. 2-D Traveling Wave Driven Piezoelectric Plate Robot for Planar Motion. *IEEE/ASME Trans. Mechatron.* **2018**, *23*, 242–251. [CrossRef]
30. Yang, Z.; Dong, L.; Wang, M.; Liu, G.; Li, X.; Li, Y. A wearable insulin delivery system based on a piezoelectric micropump. *Sens. Actuators A Phys.* **2022**, *347*, 113909. [CrossRef]
31. Hernandez, C.; Bernard, Y.; Razek, A. Design and manufacturing of a piezoelectric traveling-wave pumping device. *IEEE Trans. Ultrason. Ferroelectr. Freq. Control* **2013**, *60*, 1949–1956. [CrossRef]
32. Spanner, K.; Koc, B. Piezoelectric Motors, an Overview. *Actuators* **2016**, *5*, 6. [CrossRef]
33. Ghenna, S.; Bernard, Y.; Daniel, L. Design and experimental analysis of a high force piezoelectric linear motor. *Mechatronics* **2023**, *89*, 102928. [CrossRef]
34. Suárez-Pérez, L.; Peralta-Mamani, M.; Velázquez-Cayón, R.T. A comparison of piezoelectric surgery and conventional techniques in the enucleation of cysts and tumors in the jaws: A systematic review and meta-analysis. *Med. Oral Patol. Oral Cir. Bucal* **2025**, *30*, e333–e344. [CrossRef]
35. Zubair, M.; Uddin, I.; Dickinson, R.; Diederich, C.J. Delivering Volumetric Hyperthermia to Head and Neck Cancer Patient-Specific Models Using an Ultrasound Spherical Random Phased Array Transducer. *Bioengineering* **2025**, *12*, 14. [CrossRef] [PubMed]
36. Hariri, H.; Bernard, Y.; Razek, A. A traveling wave piezoelectric beam robot. *Smart Mater. Struct.* **2014**, *23*, 025013. [CrossRef]
37. Cui, M.; Liu, H.; Jiang, H.; Zheng, Y.; Wang, X.; Liu, W. Active vibration optimal control of piezoelectric cantilever beam with uncertainties. *Meas. Control* **2022**, *55*, 359–369. [CrossRef]
38. Roy, G.; Panigrahi, B.; Pohit, G. Crack identification in beam-type structural elements using a piezoelectric sensor. *Nondestruct. Test. Eval.* **2020**, *36*, 597–615. [CrossRef]
39. Li, C.; Lu, C. Investigation of a flat-type piezoelectric motor using in-plane vibrations. *Rev. Sci. Instrum.* **2025**, *96*, 015003. [CrossRef]
40. Čeponis, A.; Mažeika, D.; Vasiljev, P. Flat Cross-Shaped Piezoelectric Rotary Motor. *Appl. Sci.* **2020**, *10*, 5022. [CrossRef]
41. Gao, T.; Qiu, X.; Xu, P.; Hu, Z.; Yan, J.; Xiang, Y.; Xuan, F.Z. Piezoelectret-based dual-mode flexible pressure sensor for accurate wrist pulse signal acquisition in health monitoring. *Measurement* **2025**, *242*, 116283. [CrossRef]
42. Xie, Q.; Han, L.; Liu, J.; Zhang, W.; Zhao, L.; Liu, Y.; Chen, Y.; Li, Y.; Zhou, Q.; Dong, Y.; et al. Kirigami-Inspired Stretchable Piezoelectret Sensor for Analysis and Assessment of Parkinson's Tremor. *Adv. Healthc. Mater.* **2025**, *14*, 2402010. [CrossRef]

43. Hao, Y.; Zhang, H.; Zhang, Z.; Hu, C.; Shi, C. Development of Force Sensing Techniques for Robot-Assisted Laparoscopic Surgery: A Review. *IEEE Trans. Med. Robot. Bionics* **2024**, *6*, 868–887. [CrossRef]
44. Wang, W.; Wang, J.; Luo, Y.; Wang, X.; Song, H. A Survey on Force Sensing Techniques in Robot-Assisted Minimally Invasive Surgery. *IEEE Trans. Haptics* **2023**, *16*, 702–718. [CrossRef]
45. Jiao, P.; Egbe, K.-J.I.; Xie, Y.; Matin Nazar, A.; Alavi, A.H. Piezoelectric Sensing Techniques in Structural Health Monitoring: A State-of-the-Art Review. *Sensors* **2020**, *20*, 3730. [CrossRef]
46. Çakir, F.H.; Er, Ü.; Tekkalmaz, M. Monitoring the wear of turning tools with the electromechanical impedance technique. *J. Intell. Mater. Syst. Struct.* **2022**, *34*, 1341–1352. [CrossRef]
47. Hoshyarmanesh, H.; Abbasi, A. Structural health monitoring of rotary aerospace structures based on electromechanical impedance of integrated piezoelectric transducers. *J. Intell. Mater. Syst. Struct.* **2018**, *29*, 1799–1817. [CrossRef]
48. Elahi, H. The investigation on structural health monitoring of aerospace structures via piezoelectric aeroelastic energy harvesting. *Microsyst. Technol.* **2021**, *27*, 2605–2613. [CrossRef]
49. Wang, Y.; Qiu, L.; Luo, Y.; Ding, R.; Jiang, F. A piezoelectric sensor network with shared signal transmission wires for structural health monitoring of aircraft smart skin. *Mech. Syst. Signal Process.* **2020**, *141*, 106730. [CrossRef]
50. Ahmadi, J.; Feirahi, M.H.; Farahmand-Tabar, S.; Fard, A.H.K. A novel approach for non-destructive EMI-based corrosion monitoring of concrete-embedded reinforcements using multi-orientation piezoelectric sensors. *Constr. Build. Mater.* **2021**, *273*, 121689. [CrossRef]
51. Bani-Hani, M.A.; Almomani, A.M.; Aljanaideh, K.F.; Kouritem, S.A. Mechanical modeling and numerical investigation of earthquake-induced structural vibration self-powered sensing device. *IEEE Sens. J.* **2022**, *22*, 19237–19248. [CrossRef]
52. Yu, Y.; Qin, X.; Hussain, S.; Hou, W.; Weis, T. Pedestrian Counting Based on Piezoelectric Vibration Sensor. *Appl. Sci.* **2022**, *12*, 1920. [CrossRef]
53. Zhang, Y.; Miao, B.; Wang, G.; Zhou, H.; Zhang, S.; Hu, Y.; Wu, J.; Yu, X.; Li, J. ScAlN Film-Based Piezoelectric Micromechanical Ultrasonic Transducers with Dual-Ring Structure for Distance Sensing. *Micromachines* **2023**, *14*, 516. [CrossRef]
54. Zhang, Q.; Wang, Y.; Li, D.; Xie, J.; Tao, K.; Hu, P.; Zhou, J.; Chang, H.; Fu, Y. Multifunctional and Wearable Patches Based on Flexible Piezoelectric Acoustics for Integrated Sensing, Localization, and Underwater Communication. *Adv. Funct. Mater.* **2023**, *33*, 2209667. [CrossRef]
55. Ahmed, O.; Wang, M.; Zhang, B.; Irving, R.; Begg, P.; Du, X. Robotic Systems for Cochlear Implant Surgeries: A Review of Robotic Design and Clinical Outcomes. *Electronics* **2025**, *14*, 2685. [CrossRef]
56. Tetteh, E.; Wang, T.; Kim, J.Y.; Smith, T.; Norasi, H.; Van Straaten, M.G.; Lal, G.; Chrouser, K.L.; Shao, J.M.; Hallbeck, M.S. Optimizing ergonomics during open, laparoscopic, and robotic-assisted surgery: A review of surgical ergonomics literature and development of educational illustrations. *Am. J. Surg.* **2024**, *235*, 115551. [CrossRef] [PubMed]
57. Alkatout, I.; Mechler, U.; Mettler, L.; Pape, J.; Maass, N.; Biebl, M.; Gitas, G.; Laganà, A.S.; Freytag, D. The Development of Laparoscopy-A Historical Overview. *Front. Surg.* **2021**, *8*, 799442. [CrossRef] [PubMed]
58. Barrios, E.L.; Polcz, V.E.; Hensley, S.E.; Sarosi, G.A., Jr.; Mohr, A.M.; Loftus, T.J.; Upchurch, G.R., Jr.; Sumfest, J.M.; Efron, P.A.; Dunleavy, K.; et al. A narrative review of ergonomic problems, principles, and potential solutions in surgical operations. *Surgery* **2023**, *174*, 214–221. [CrossRef] [PubMed]
59. Bittner, R. Laparoscopic surgery—15 years after clinical introduction. *World J. Surg.* **2006**, *30*, 1190–1203. [CrossRef] [PubMed]
60. Bracale, U.; Corcione, F.; Pignata, G.; Andreuccetti, J.; Dolce, P.; Boni, L.; Cassinotti, E.; Olmi, S.; Uccelli, M.; Gualtierotti, M.; et al. Impact of neoadjuvant therapy followed by laparoscopic radical gastrectomy with D2 lymph node dissection in Western population: A multi-institutional propensity score-matched study. *J. Surg. Oncol.* **2021**, *124*, 1338–1346. [CrossRef]
61. Bizzarri, N.; Pedone Anchora, L.; Teodorico, E.; Certelli, C.; Galati, G.; Carbone, V.; Gallotta, V.; Naldini, A.; Costantini, B.; Querleu, D.; et al. The role of diagnostic laparoscopy in locally advanced cervical cancer staging. *Eur. J. Surg. Oncol.* **2024**, *50*, 108645. [CrossRef]
62. Pérez-Salazar, M.J.; Caballero, D.; Sánchez-Margallo, J.A.; Sánchez-Margallo, F.M. Comparative Study of Ergonomics in Conventional and Robotic-Assisted Laparoscopic Surgery. *Sensors* **2024**, *24*, 3840. [CrossRef]
63. Li, S.Y.; Wang, Y.; Xin, C.; Ji, L.Q.; Li, S.H.; Jiang, W.D.; Zhang, C.M.; Zhang, W.; Lou, Z. Laparoscopic surgery is associated with increased risk of postoperative peritoneal metastases in T4 colon cancer: A propensity score analysis. *Int. J. Color. Dis.* **2025**, *40*, 2. [CrossRef]
64. Taghavi, K.; Glenisson, M.; Loiselet, K.; Fiorenza, V.; Cornet, M.; Capito, C.; Vinit, N.; Pire, A.; Sarnacki, S.; Blanc, T. Robot-assisted laparoscopic adrenalectomy: Extended application in children. *Eur. J. Surg. Oncol.* **2024**, *50*, 108627. [CrossRef] [PubMed]
65. Williamson, T.; Song, S.E. Robotic Surgery Techniques to Improve Traditional Laparoscopy. *J. Soc. Laparosc. Robot. Surg.* **2022**, *26*, e2022.00002. [CrossRef] [PubMed]
66. Rivero-Moreno, Y.; Echevarria, S.; Vidal-Valderrama, C.; Pianetti, L.; Cordova-Guilarte, J.; Navarro-Gonzalez, J.; Acevedo-Rodríguez, J.; Dorado-Avila, G.; Osorio-Romero, L.; Chavez-Campos, C.; et al. Robotic Surgery: A Comprehensive Review of the Literature and Current Trends. *Cureus* **2023**, *15*, e42370. [CrossRef] [PubMed]

67. Lima, V.L.; de Almeida, R.C.; Neto, T.R.; Rosa, A.A.M. Chapter 72—Robotic ophthalmologic surgery. In *Handbook of Robotic Surgery*; Zequi, S.C., Ren, H., Eds.; Academic Press: Cambridge, MA, USA, 2025; pp. 701–704. [CrossRef]
68. Rivero-Moreno, Y.; Rodriguez, M.; Losada-Muñoz, P.; Redden, S.; Lopez-Lezama, S.; Vidal-Gallardo, A.; Machado-Paled, D.; Cordova Guilarte, J.; Teran-Quintero, S. Autonomous Robotic Surgery: Has the Future Arrived? *Cureus* **2024**, *16*, e52243. [CrossRef] [PubMed]
69. Han, J.; Davids, J.; Ashrafian, H.; Darzi, A.; Elson, D.S.; Sodergren, M. A systematic review of robotic surgery: From supervised paradigms to fully autonomous robotic approaches. *Int. J. Med. Robot.* **2022**, *18*, e2358. [CrossRef] [PubMed]
70. Lee, A.; Baker, T.S.; Bederson, J.B.; Rapoport, B.I. Levels of autonomy in FDA-cleared surgical robots: A systematic review. *NPJ Digit. Med.* **2024**, *7*, 103. [CrossRef]
71. Wan, Q.; Shi, Y.; Xiao, X.; Li, X.; Mo, H. Review of Human–Robot Collaboration in Robotic Surgery. *Adv. Intell. Syst.* **2024**, *7*, 2400319. [CrossRef]
72. Liu, T.; Wang, J.; Wong, S.; Razjigaev, A.; Beier, S.; Peng, S.; Do, T.N.; Song, S.; Chu, D.; Wang, C.H.; et al. A Review on the Form and Complexity of Human–Robot Interaction in the Evolution of Autonomous Surgery. *Adv. Intell. Syst.* **2024**, *6*, 2400197. [CrossRef]
73. Schreiter, J.; Schott, D.; Schwenderling, L.; Hansen, C.; Heinrich, F.; Joeres, F. AR-Supported Supervision of Conditional Autonomous Robots: Considerations for Pedicle Screw Placement in the Future. *J. Imaging* **2022**, *8*, 255. [CrossRef]
74. Dagnino, G.; Kundrat, D. Robot-assistive minimally invasive surgery: Trends and future directions. *Int. J. Intell. Robot. Appl.* **2024**, *8*, 812–826. [CrossRef]
75. Chinzei, K.; Hata, N.; Jolesz, F.A.; Kikinis, R. Surgical Assist Robot for the Active Navigation in the Intraoperative MRI: Hardware Design Issues. In Proceedings of the 2000 IEEE/RSJ International Conference on Intelligent Robots and Systems (IROS 2000) (Cat. No.00CH37113), Takamatsu, Japan, 31 October–5 November 2000; pp. 727–732. [CrossRef]
76. Tsekos, N.V.; Khanicheh, A.; Christoforou, E.; Mavroidis, C. Magnetic resonance-compatible robotic and mechatronics systems for image-guided interventions and rehabilitation: A review study. *Annu. Rev. Biomed. Eng.* **2007**, *9*, 351–387. [CrossRef]
77. Faoro, G.; Maglio, S.; Pane, S.; Iacovacci, V.; Menciassi, A. An artificial intelligence-aided robotic platform for ultrasound-guided transcarotid revascularization. *IEEE Robot. Autom. Lett.* **2023**, *8*, 2349–2356. [CrossRef]
78. Su, H.; Kwok, K.W.; Cleary, K.; Iordachita, I.I.; Çavuşoğlu, M.C.; Desai, J.P.; Fischer, G.S. State of the art and future opportunities in MRI-guided robot-assisted surgery and interventions. *Proc. IEEE Inst. Electr. Electron. Eng.* **2022**, *110*, 968–992. [CrossRef] [PubMed]
79. Padhan, J.; Tsekos, N.; Al-Ansari, A.; Abinahed, J.; Deng, Z.; Navkar, N.V. Dynamic Guidance Virtual Fixtures for Guiding Robotic Interventions: Intraoperative MRI-guided Transapical Cardiac Intervention Paradigm. In Proceedings of the 2022 IEEE 22nd International Conference on Bioinformatics and Bioengineering (BIBE), Taichung, Taiwan, 7–9 November 2022; pp. 265–270. [CrossRef]
80. Singh, S.; Torrealdea, F.; Bandula, S. MR imaging-guided intervention: Evaluation of MR conditional biopsy and ablation needle tip artifacts at 3T using a balanced fast field echo sequence. *J. Vasc. Interv. Radiol.* **2021**, *32*, 1068–1074. [CrossRef] [PubMed]
81. Xu, L.; Pacia, C.P.; Gong, Y.; Hu, Z.; Chien, C.Y.; Yang, L.; Gach, H.M.; Hao, Y.; Comron, H.; Huang, J.; et al. Characterization of the targeting accuracy of a neuronavigation-guided transcranial FUS system in vitro, in vivo, and in silico. *IEEE Trans. Biomed. Eng.* **2023**, *70*, 1528–1538. [CrossRef]
82. Navarro-Becerra, J.A.; Borden, M.A. Targeted Microbubbles for Drug, Gene, and Cell Delivery in Therapy and Immunotherapy. *Pharmaceutics* **2023**, *15*, 1625. [CrossRef]
83. Delaney, L.J.; Isguven, S.; Eisenbrey, J.R.; Hickok, N.J.; Forsberg, F. Making waves: How ultrasound-targeted drug delivery is changing pharmaceutical approaches. *Mater. Adv.* **2022**, *3*, 3023–3040. [CrossRef]
84. Jia, X.; Zhang, Y.; Du, H.; Yu, Y. Experimental study of double cable-conduit driving device for MRI compatible biopsy robots. *J. Mech. Med. Biol.* **2021**, *21*, 2140014. [CrossRef]
85. Li, X.; Young, A.S.; Raman, S.S.; Lu, D.S.; Lee, Y.H.; Tsao, T.C.; Wu, H.H. Automatic needle tracking using Mask R-CNN for MRI-guided percutaneous interventions. *Int. J. Comput. Assist. Radiol. Surg.* **2020**, *15*, 1673–1684. [CrossRef]
86. Bernardes, M.C.; Moreira, P.; Lezcano, D.; Foley, L.; Tuncali, K.; Tempany, C.; Kim, J.S.; Hata, N.; Iordachita, I.; Tokuda, J. In Vivo Feasibility Study: Evaluating Autonomous Data-Driven Robotic Needle Trajectory Correction in MRI-Guided Transperineal Procedures. *IEEE Robot. Autom. Lett.* **2024**, *9*, 8975–8982. [CrossRef]
87. Wu, D.; Li, G.; Patel, N.; Yan, J.; Monfaredi, R.; Cleary, K.; Iordachita, I. Remotely Actuated Needle Driving Device for MRI-Guided Percutaneous Interventions: Force and Accuracy Evaluation. In Proceedings of the 2019 41st Annual International Conference of the IEEE Engineering in Medicine & Biology Society (EMBC), Berlin, Germany, 23–27 July 2019; pp. 1985–1989. [CrossRef]
88. Mohith, S.; Upadhya, A.R.; Navin, K.P.; Kulkarni, S.M.; Rao, M. Recent trends in piezoelectric actuators for precision motion and their applications: A review. *Smart Mater. Struct.* **2020**, *30*, 013002. [CrossRef]
89. Gao, X.; Yang, J.; Wu, J.; Xin, X.; Li, Z.; Yuan, X.; Shen, X.; Dong, S. Piezoelectric actuators and motors: Materials, designs, and applications. *Adv. Mater. Technol.* **2020**, *5*, 1900716. [CrossRef]

90. Qiao, G.; Li, H.; Lu, X.; Wen, J.; Cheng, T. Piezoelectric stick-slip actuators with flexure hinge mechanisms: A review. *J. Intell. Mater. Syst. Struct.* **2022**, *33*, 1879–1901. [CrossRef]
91. Wang, S.; Zhou, S.; Zhang, X.; Xu, P.; Zhang, Z.; Ren, L. Bionic stepping motors driven by piezoelectric materials. *J. Bionic. Eng.* **2023**, *20*, 858–872. [CrossRef]
92. Yang, Z.; Li, X.; Tang, J.; Huang, H.; Zhao, H.; Cheng, Y.; Liu, S.; Li, C.; Xiong, M. A bionic stick-slip piezo-driven positioning platform designed by imitating the structure and movement of the crab. *J. Bionic. Eng.* **2023**, *20*, 2590–2600. [CrossRef]
93. Liu, J.; Gao, X.; Jin, H.; Ren, K.; Guo, J.; Qiao, L.; Qiu, C.; Chen, W.; He, Y.; Dong, S.; et al. Miniaturized electromechanical devices with multi-vibration modes achieved by orderly stacked structure with piezoelectric strain units. *Nat. Commun.* **2022**, *13*, 6567. [CrossRef] [PubMed]
94. Fu, D.K.; Fan, P.Q.; Yuan, T.; Wang, Y.S. A novel hybrid mode linear ultrasonic motor with double driving feet. *Rev. Sci. Instrum.* **2022**, *93*, 025003. [CrossRef] [PubMed]
95. Li, Z.; Guo, Z.; Han, H.; Su, Z.; Sun, H. Design and characteristic analysis of multi-degree-of-freedom ultrasonic motor based on spherical stator. *Rev. Sci. Instrum.* **2022**, *93*, 025004. [CrossRef]
96. Zhang, S.; Liu, Y.; Deng, J.; Gao, X.; Li, J.; Wang, W.; Xun, M.; Ma, X.; Chang, Q.; Liu, J.; et al. Piezo robotic hand for motion manipulation from micro to macro. *Nat. Commun.* **2023**, *14*, 500. [CrossRef]
97. Wu, G.; Wang, Z.; Wu, Y.; Zhao, J.; Cui, F.; Zhang, Y.; Chen, W. Development and Improvement of a Piezoelectrically Driven Miniature Robot. *Biomimetics* **2024**, *9*, 226. [CrossRef]
98. Shen, J.; Fang, X.; Liu, J.; Liu, L.; Lu, H.; Lou, J. Design, manufacture, and two-step calibration of a piezoelectric parallel microrobot at the millimeter scale for micromanipulation. *Mech. Mach. Theory* **2025**, *206*, 105928. [CrossRef]
99. Chen, X.; Kiziroglou, M.E.; Yeatman, E.M. Onboard visual micro-servoing on robotic surgery tools. *Microsyst. Nanoeng.* **2025**, *11*, 112. [CrossRef]
100. Del Giudice, G.; Orekhov, A.L.; Shen, J.H.; Joos, K.; Simaan, N. Investigation of Micro-motion Kinematics of Continuum Robots for Volumetric OCT and OCT-guided Visual Servoing. *IEEE ASME Trans. Mechatron.* **2021**, *26*, 2604–2615. [CrossRef] [PubMed]
101. Altun, I.; Nezami, N. Role of Robotics in Image-Guided Trans-Arterial Interventions. *Tech. Vasc. Interv. Radiol.* **2024**, *27*, 101005. [CrossRef] [PubMed]
102. Babaheidarian, P.; Soltanattar, A.; Sajadi, S.K.; Rostamian, L.; Foroutani, L.; Soleymanpourshamsi, T.; Zarrazvand, E.; Farshid, A.; Dadpour, M.; Janbozorgi, S.; et al. Robotics in Healthcare. *Kindle* **2025**, *5*, 1–178. Available online: https://preferpub.org/index.php/kindle/article/view/Book50 (accessed on 24 August 2025).
103. Liu, E.; Cai, Z.; Ye, Y.; Zhou, M.; Liao, H.; Yi, Y. An Overview of Fslexible Sensors: Development, Application, and Challenges. *Sensors* **2023**, *23*, 817. [CrossRef]
104. Devi, D.H.; Duraisamy, K.; Armghan, A.; Alsharari, M.; Aliqab, K.; Sorathiya, V.; Das, S.; Rashid, N. 5G Technology in Healthcare and Wearable Devices: A Review. *Sensors* **2023**, *23*, 2519. [CrossRef]
105. Mukhopadhyay, S.C.; Suryadevara, N.K.; Nag, A. Wearable Sensors for Healthcare: Fabrication to Application. *Sensors* **2022**, *22*, 5137. [CrossRef]
106. Chakrabarti, S.; Biswas, N.; Jones, L.D.; Kesari, S.; Ashili, S. Smart Consumer Wearables as Digital Diagnostic Tools: A Review. *Diagnostics* **2022**, *12*, 2110. [CrossRef]
107. Escobar-Linero, E.; Muñoz-Saavedra, L.; Luna-Perejón, F.; Sevillano, J.L.; Domínguez-Morales, M. Wearable Health Devices for Diagnosis Support: Evolution and Future Tendencies. *Sensors* **2023**, *23*, 1678. [CrossRef]
108. Pantelopoulos, A.; Bourbakis, N.G. A survey on wearable sensor-based systems for health monitoring and prognosis. *IEEE Trans. Syst. Man. Cybern. Part C* **2010**, *40*, 1–12. [CrossRef]
109. Ahmed, M.R.; Newby, S.; Potluri, P.; Mirihanage, W.; Fernando, A. Emerging Paradigms in Fetal Heart Rate Monitoring: Evaluating the Efficacy and Application of Innovative Textile-Based Wearables. *Sensors* **2024**, *24*, 6066. [CrossRef] [PubMed]
110. Moon, K.S.; Lee, S.Q. A Wearable Multimodal Wireless Sensing System for Respiratory Monitoring and Analysis. *Sensors* **2023**, *23*, 6790. [CrossRef] [PubMed]
111. Khan Mamun, M.M.R.; Sherif, A. Advancement in the Cuffless and Noninvasive Measurement of Blood Pressure: A Review of the Literature and Open Challenges. *Bioengineering* **2023**, *10*, 27. [CrossRef]
112. Xing, Y.; Yang, K.; Lu, A.; Mackie, K.; Guo, F. Sensors and Devices Guided by Artificial Intelligence for Personalized Pain Medicine. *Cyborg Bionic Syst.* **2024**, *13*, 0160. [CrossRef]
113. Bhuva, A.N.; Moralee, R.; Brunker, T.; Lascelles, K.; Cash, L.; Patel, K.P.; Lowe, M.; Sekhri, N.; Alpendurada, F.; Pennell, D.J.; et al. Evidence to support magnetic resonance conditional labelling of all pacemaker and defibrillator leads in patients with cardiac implantable electronic devices. *Eur. Heart J.* **2022**, *43*, 2469–2478. [CrossRef]
114. Joo, H.; Lee, Y.; Kim, J.; Yoo, J.S.; Yoo, S.; Kim, S.; Arya, A.K.; Kim, S.; Choi, S.H.; Lu, N.; et al. Soft Implantable Drug Delivery Device Integrated Wirelessly with Wearable Devices to Treat Fatal Seizures. *Sci. Adv.* **2021**, *7*, eabd4639. [CrossRef]

115. Cheng, Y.; Xie, D.; Han, Y.; Guo, S.; Sun, Z.; Jing, L.; Man, W.; Liu, D.; Yang, K.; Lei, D.; et al. Precise management system for chronic intractable pain patients implanted with spinal cord stimulation based on a remote programming platform: Study protocol for a randomized controlled trial (PreMaSy study). *Trials* **2023**, *24*, 580. [CrossRef]
116. Thotahewa, K.M.S.; Redouté, J.; Yuce, M.R. Electromagnetic and thermal effects of IR-UWB wireless implant systems on the human head. In Proceedings of the 2013 35th Annual International Conference of the IEEE Engineering in Medicine and Biology Society (EMBC), Osaka, Japan, 3–7 July 2013; pp. 5179–5182. [CrossRef]
117. Gordon, J.S.; Maynes, E.J.; O'Malley, T.J.; Pavri, B.B.; Tchantchaleishvili, V. Electromagnetic interference between implantable cardiac devices and continuous-flow left ventricular assist devices: A review. *J. Interv. Card. Electrophysiol.* **2021**, *61*, 1–10. [CrossRef]
118. Gui, J.; Liu, X.; Dou, H. Effects of Processing Parameters on the Structure and Mechanical Property of PVDF/BN Nanofiber Yarns. *Polymers* **2025**, *17*, 1931. [CrossRef]
119. Gao, T.; Liao, Q.; Si, W.; Chu, Y.; Dong, H.; Li, Y.; Liao, Y.; Qin, L. From fundamentals to future challenges for flexible piezoelectric actuators. *Cell Rep. Phys. Sci.* **2024**, *5*, 101789. [CrossRef]
120. Yang, T.; Zhou, S.; Litak, G.; Jing, X. Recent advances in correlation and integration between vibration control, energy harvesting and monitoring. *Nonlinear Dyn.* **2023**, *111*, 20525–20562. [CrossRef]
121. Yu, H.; Liu, Y.; Deng, J.; Li, J.; Zhang, S.; Chen, W.; Zhao, J. Bioinspired Multilegged Piezoelectric Robot: The Design Philosophy Aiming at High-Performance Micromanipulation. *Adv. Intell. Syst.* **2022**, *4*, 2100142. [CrossRef]
122. Deng, J.; Liu, Z.; Li, J.; Zhang, S.; Liu, Y. Development of a Highly Adaptive Miniature Piezoelectric Robot Inspired by Earthworms. *Adv. Sci.* **2024**, *11*, e2403426. [CrossRef]
123. Suzuki, M.; Iida, Y.; Tsukui, Y.; Kusama, H.; Kinoshita, R.; Kusui, E.; Sunohara, Y.; Minegishi, R.; Sugiyama, Y.; Nishimura, Y.; et al. Automatic Holonomic Mobile Micromanipulator for Submillimeter Objects Inspired by the Rhinoceros Beetle. *Adv. Intell. Syst.* **2024**, *6*, 2300517. [CrossRef]
124. Wang, Y.; Zhang, B.; Deng, J.; Zhang, S.; Chen, W.; Liu, Y. A Grasshopper-Inspired Miniature Piezoelectric Actuator Integrated with Three Vibrating Units. *Int. J. Mech. Sci.* **2025**, *300*, 110458. [CrossRef]
125. Wu, J.; Ding, Z.; Wang, L.; Zhang, R.; Zhang, Y.; Rong, X.; Song, R.; Dong, H.; Zhao, J.; Li, Y. Development of an Untethered Ultrasonic Robot with Fast and Load-Carriable Movement Imitating Rotatory Galloping Gait. *IEEE/ASME Trans. Mechatron.* **2025**, *30*, 1001–1013. [CrossRef]
126. Li, X.; Ding, Z.; Jiang Wu, J.; Wei, W.; Wang, L.; Zhang, Y.; Li, D.; Rong, X.; Song, R.; Li, Y. A Miniature Pole-climbing Piezoelectric Robot with Fast and Load-Towable Movement Inspired by Squirrel's Galloping Gait. *IEEE Trans. Ind. Electron.* **2025**, *72*, 5221–5233. [CrossRef]
127. Zhu, B.; Wang, Y.Q. Octopus-crawling-inspired highly agile miniature piezoelectric robot with strong load-bearing capacity. *Int. J. Mech. Sci.* **2025**, *291*, 110130. [CrossRef]
128. Sunohara, Y.; Ueno, S.; Minegishi, R.; Sekine, C.; Kitamura, Y.; Sugiyama, Y.; Ando, S.; Fuchiwaki, O. Ultrafast Untethered Levitation Device Utilized Squeeze Film for Omni-Directional Transport. *Adv. Intell. Syst.* **2025**, 2401098. [CrossRef]
129. Du, P.; Chen, W.; Deng, J.; Liu, Y. Sensing–Actuating Integrated Ultrasonic Device for High-Precision and High-Efficiency Burnishing. *IEEE Trans. Ind. Electron.* **2025**, *72*, 5199–5209. [CrossRef]
130. Chen, B.; Feng, Z.; Yao, F.Z.; Zhang, M.H.; Wang, K.; Wei, Y.; Gong, W.; Röde, J. Flexible piezoelectrics: Integration of sensing, actuating and energy harvesting. *npj Flex. Electron.* **2025**, *9*, 58. [CrossRef]
131. Dagdeviren, C.; Yang, B.D.; Su, Y.; Tran, P.L.; Joe, P.; Anderson, E.; Xia, J.; Doraiswamy, V.; Dehdashti, B.; Feng, X.; et al. Conformal piezoelectric energy harvesting and storage from motions of the heart, lung, and diaphragm. *Proc. Natl. Acad. Sci. USA* **2014**, *111*, 1927–1932. [CrossRef] [PubMed]
132. Hwang, G.T.; Byun, M.; Jeong, C.K.; Lee, K.J. Flexible piezoelectric thin-film energy harvesters and nanosensors for biomedical applications. *Adv. Healthc. Mater.* **2015**, *4*, 646–658. [CrossRef] [PubMed]
133. Raj, N.P.M.J.; Alluri, N.R.; Vivekananthan, V.; Chandrasekhar, A.; Khandelwal, G.; Kim, S.J. Sustainable yarn type-piezoelectric energy harvester as an eco-friendly, cost-effective battery-free breath sensor. *Appl. Energy* **2018**, *228*, 1767–1776. [CrossRef]
134. He, Q.; Briscoe, J. Piezoelectric Energy Harvester Technologies: Synthesis, Mechanisms, and Multifunctional Applications. *ACS Appl. Mater. Interfaces* **2024**, *16*, 29491–29520. [CrossRef]
135. Kang, S.; Kim, S.H.; Lee, H.B.; Mhin, S.; Ryu, J.H.; Kim, Y.W.; Jones, J.L.; Son, Y.; Lee, N.K.; Lee, K.; et al. High-power energy harvesting and imperceptible pulse sensing through peapod-inspired hierarchically designed piezoelectric nanofibers. *Nano Energy* **2022**, *99*, 107386. [CrossRef]
136. Dagdeviren, C.; Javid, F.; Joe, P.; von Erlach, T.; Bensel, T.; Wei, Z.; Saxton, S.; Cleveland, C.; Booth, L.; McDonnell, S.; et al. Flexible piezoelectric devices for gastrointestinal motility sensing. *Nat. Biomed. Eng.* **2017**, *1*, 807–817. [CrossRef]
137. Murayama, N.; Nakamura, K.; Obara, H.; Segawa, M. The strong piezoelectricity in polyvinylidene fluroide (PVDF). *Ultrasonics* **1976**, *14*, 15–24. [CrossRef]

138. Mohammadpourfazeli, S.; Arash, S.; Ansari, A.; Yang, S.; Mallick, K.; Bagherzadeh, R. Future prospects and recent developments of polyvinylidene fluoride (PVDF) piezoelectric polymer; fabrication methods, structure, and electro-mechanical properties. *RSC Adv.* **2023**, *13*, 370–387. [CrossRef]
139. Verma, K.; Kumar, A.; Sharma, R. Development of flexible piezoelectric nanogenerator based on PVDF/KNN/ZnO nanocomposite film for energy harvesting application. *J. Mater. Sci. Mater. Electron.* **2024**, *35*, 1732. [CrossRef]
140. Sasmal, A.; Patra, A.; Maiti, P.; Sahu, B.; Krahne, R.; Arockiarajan, A. Microstructuring of conductivity tuned piezoelectric polydimethylsiloxane/(Ba0.85Ca0.15)(Ti0.90Hf0.10)O3 composite for hybrid mechanical energy harvesting. *Polym. Compos.* **2025**, *46*, 1–16. [CrossRef]
141. Kum, H.S.; Lee, H.; Kim, S.; Lindemann, S.; Kong, W.; Qiao, K.; Chen, P.; Irwin, J.; Lee, J.H.; Xie, S.; et al. Heterogeneous integration of single-crystalline complex-oxide membranes. *Nature* **2020**, *578*, 75–81. [CrossRef] [PubMed]
142. Ren, Z.; Deng, S.; Shao, J.; Yangyang, S.; Zhou, C.; Luo, J.; Wang, T.; Li, J.; Li, J.; Liu, H.; et al. Ultrahigh-power-density flexible piezoelectric energy harvester based on freestanding ferroelectric oxide thin films. *Nat. Commun.* **2025**, *16*, 3192. [CrossRef] [PubMed]
143. Razek, A. Biological and Medical Disturbances Due to Exposure to Fields Emitted by Electromagnetic Energy Devices—A Review. *Energies* **2022**, *15*, 4455. [CrossRef]
144. Razek, A. Assessment of a Functional Electromagnetic Compatibility Analysis of Near-Body Medical Devices Subject to Electromagnetic Field Perturbation. *Electronics* **2023**, *12*, 4780. [CrossRef]
145. Tao, F.; Sui, F.; Liu, A.; Qi, Q.; Zhang, M.; Song, B.; Guo, Z.; Lu, S.C.Y.; Nee, A.Y.C. Digital twin-driven product design framework. *Int. J. Prod. Res.* **2019**, *57*, 3935–3953. [CrossRef]
146. Grieves, M.; Vickers, J. Digital twin: Mitigating unpredictable, undesirable emergent behavior in complex systems. In *Trans-Disciplinary Perspectives on Complex Systems*; Springer: Cham, Switzerland, 2017; pp. 85–113. [CrossRef]
147. Sun, T.; He, X.; Li, Z. Digital twin in healthcare: Recent updates and challenges. *Digit. Health* **2023**, *9*, 20552076221149651. [CrossRef]
148. De Benedictis, A.; Mazzocca, N.; Somma, A.; Strigaroet, C. Digital twins in healthcare: An architectural proposal and its application in a social distancing case study. *IEEE J. Biomed. Health Inform.* **2022**, *27*, 5143–5154. [CrossRef]
149. Haleem, A.; Javaid, M.; Singh, R.P.; Suman, R. Exploring the revolution in healthcare systems through the applications of digital twin technology. *Biomed. Technol.* **2023**, *4*, 28–38. [CrossRef]
150. Mohamed, N.; Al-Jaroodi, J.; Jawhar, I.; Kesserwan, N. Leveraging digital twins for healthcare systems engineering. *IEEE Access* **2023**, *11*, 69841–69853. [CrossRef]
151. Ricci, A.; Croatti, A.; Montagna, S. Pervasive and connected digital twins-a vision for digital health. *IEEE Internet Comput.* **2022**, *26*, 26–32. [CrossRef]
152. Wickramasinghe, N.; Ulapane, N.; Sloane, E.B.; Gehlot, V. Digital Twins for More Precise and Personalized Treatment. *Stud. Health Technol. Inform.* **2024**, *310*, 229–233. [CrossRef] [PubMed]
153. Song, Y. Human digital twin, the development and impact on design. *J. Comput. Inf. Sci. Eng.* **2023**, *23*, 060819. [CrossRef]
154. Burattini, S.; Montagna, S.; Croatti, A.; Gentili, N.; Ricci, A.; Leonardi, L.; Pandolfini, S.; Tosi, S. An Ecosystem of Digital Twins for Operating Room Management. In Proceedings of the 2023 IEEE 36th International Symposium on Computer-Based Medical Systems (CBMS), L'Aquila, Italy, 22–24 June 2023; pp. 770–775. [CrossRef]
155. Hagmann, K.; Hellings-Kuß, A.; Klodmann, J.; Richter, R.; Stulp, F.; Leidner, D. A digital twin approach for contextual assistance for surgeons during surgical robotics training. *Front. Robot. AI* **2021**, *8*, 735566. [CrossRef] [PubMed]
156. Katsoulakis, E.; Wang, Q.; Wu, H.; Shahriyari, L.; Fletcher, R.; Liu, J.; Achenie, L.; Liu, H.; Jackson, P.; Xiao, Y.; et al. Digital twins for health: A scoping review. *npj Digit. Med.* **2024**, *7*, 77. [CrossRef]
157. Cusumano, D.; Boldrini, L.; Dhont, J.; Fiorino, C.; Green, O.; Güngör, G.; Jornet, N.; Klüter, S.; Landry, G.; Mattiucci, G.C.; et al. Artificial intelligence in magnetic resonance guided radiotherapy: Medical and physical considerations on state of art and future perspectives. *Phys. Med.* **2021**, *85*, 175–191. [CrossRef]
158. Seetohul, J.; Shafiee, M.; Sirlantzis, K. Augmented reality (AR) for surgical robotic and autonomous systems: State of the art, challenges, and solutions. *Sensors* **2023**, *23*, 6202. [CrossRef]
159. Avrumova, F.; Lebl, D.R. Augmented reality for minimally invasive spinal surgery. *Front. Surg.* **2023**, *9*, 1086988. [CrossRef]
160. Long, Y.; Cao, J.; Deguet, A.; Taylor, R.H.; Dou, Q. Integrating Artificial Intelligence and Augmented Reality in Robotic Surgery: An Initial dVRK Study Using a Surgical Education Scenario. In Proceedings of the International Symposium on Medical Robotics (ISMR), Atlanta, GA, USA, 13–15 April 2022; pp. 1–8. [CrossRef]
161. Fu, J.; Rota, A.; Li, S.; Zhao, J.; Liu, Q.; Iovene, E.; Ferrigno, G.; De Momi, E. Recent Advancements in Augmented Reality for Robotic Applications: A Survey. *Actuators* **2023**, *12*, 323. [CrossRef]
162. Qian, L.; Wu, J.Y.; DiMaio, S.P.; Navab, N.; Kazanzides, P. A review of augmented reality in robotic-assisted surgery. *IEEE Trans. Med. Robot. Bionics* **2020**, *2*, 1–16. [CrossRef]

163. Kukushkin, K.; Ryabov, Y.; Borovkov, A. Digital twins: A systematic literature review based on data analysis and topic modeling. *Data* **2022**, *7*, 173. [CrossRef]
164. Giansanti, D.; Morelli, S. Exploring the Potential of Digital Twins in Cancer Treatment: A Narrative Review of Reviews. *J. Clin. Med.* **2025**, *14*, 3574. [CrossRef] [PubMed]
165. Cheng, Y.; Wang, T.; Zhu, H.; Hu, X.; Mi, J.; Li, L.; Zhang, Y.; Yang, J.; Dong, L.; Li, Y.; et al. Molecular Engineering of Amino Acid Crystals with Enhanced Piezoelectric Performance for Biodegradable Sensors. *Angew. Chem. Int. Ed. Engl.* **2025**, *64*, e202500334. [CrossRef] [PubMed]
166. Cui, J.; Du, L.; Meng, Z.; Gao, J.; Tan, A.; Jin, X.; Zhu, X. Ingenious Structure Engineering to Enhance Piezoelectricity in Poly (vinylidene fluoride) for Biomedical Applications. *Biomacromolecules* **2024**, *25*, 5541–5591. [CrossRef]
167. Pourmadadi, M.; Ahmari, A.; Mirshafiei, M.; Omrani, Z.; Yazdian, F.; Rahdar, A.; Fathi-karkan, S.; Aboudzadeh, M.A. Polyvinylidene fluoride in biomedical applications: Properties, challenges, and future prospects. *Eur. Polym. J.* **2025**, *231*, 113889. [CrossRef]
168. Nain, A.; Chakraborty, S.; Barman, S.R.; Gavit, P.; Indrakumar, S.; Agrawal, A.; Lin, Z.H.; Chatterjee, K. Progress in the development of piezoelectric biomaterials for tissue remodeling. *Biomaterials* **2024**, *307*, 122528. [CrossRef]
169. Vinchurkar, K.; Bukke, S.P.N.; Jain, P.; Bhadoria, J.; Likhariya, M.; Mane, S.; Suryawanshi, M.; Veerabhadrappa, K.V.; Eftekhari, Z.; Onohuean, H. Advances in sustainable biomaterials: Characterizations, and applications in medicine. *Discov. Polym.* **2025**, *2*, 2. [CrossRef]
170. Chen, C.; Zhang, Y.; Zheng, Y.; Zhang, Y.; Liu, H.; Wu, J.; Yang, L.; Yang, Z. Topology in Biological Piezoelectric Materials. *Adv. Mater.* **2025**, *37*, e2500466. [CrossRef]
171. Liu, J.; Ding, Z.; Zhu, Z.; Wu, J.; Rong, X.; Song, R.; Li, Y. A Multi-DOF Self-Moving Piezoelectric Actuator with High Carrying/Positioning Capability by Constructing a Multi-Vibration-Tailored Non-uniformly Distributed Electrode. *IEEE/ASME Trans. Mechatron.* **2025**, 1–12. [CrossRef]
172. Xu, R.; Yang, Y.; Jin, B.; Wang, L. Adaptive multi-mode piezoelectric actuator with extended torque-speed bandwidth. *Int. J. Mech. Sci.* **2025**, *303*, 110675. [CrossRef]
173. Sun, J.; Deng, J.; Li, J.; Zhang, S.; Xun, M.; Liu, Y. Development of a Stick-Slip Piezoelectric Screw Actuator with High Thrust Force Density. *IEEE Trans. Ind. Electron.* **2025**, 1–10. [CrossRef]
174. Wang, Y.; Deng, J.; Li, J.; Zhang, S.; Chen, W.; Liu, Y. Progress, Challenges, and Prospects of Miniature Ultrasonic Motors. *Adv. Mater. Technol.* **2025**, e00711. [CrossRef]
175. Gao, Y.; Li, J.; Zhang, S.; Deng, J.; Chen, W.; Liu, Y. Centimeter-Scale Reconfiguration Piezo Robots with Built-in-Ceramic Actuation Unit. *Engineering* 2025, *in press*. [CrossRef]
176. Zhao, Z.; Bales, C.; Fischer, G. Design and Characterization of MRI-compatible Plastic Ultrasonic Motor. *arXiv* **2024**, arXiv:2409.04006. [CrossRef]
177. Li, H.; Li, X.; Wang, W.; Fang, H.; Xu, W.; Du, Z. Dual-foot asynchronous MRI-compatible piezoelectric actuator with backward motion suppression for neurosurgical puncture. *Sens. Actuators A Phys.* **2025**, *393*, 116857. [CrossRef]
178. Rafiee, M.J.; Eyre, K.; Leo, M.; Benovoy, M.; Friedrich, M.G.; Chetrit, M. Comprehensive review of artifacts in cardiac MRI and their mitigation. *Int. J. Cardiovasc. Imaging* **2024**, *40*, 2021–2039. [CrossRef]

actuators

MDPI

Systematic Review

A Systematic Review on Smart Insole Prototypes: Development and Optimization Pathways

Vítor Miguel Santos [1,*], Beatriz B. Gomes [2,3], Maria Augusta Neto [1], Patrícia Freitas Rodrigues [1] and Ana Martins Amaro [1]

[1] University of Coimbra, Centre of Mechanical Engineering, Materials and Processes (CEMMPRE-ARISE), Department of Mechanical Engineering, 3040-248 Coimbra, Portugal; augusta.neto@dem.uc.pt (M.A.N.); pf.rodrigues@uc.pt (P.F.R.); ana.amaro@dem.uc.pt (A.M.A.)

[2] University of Coimbra, Centre for the Study of Human Performance (CIPER), Faculty of Sports Sciences and Physical Education (FCDEF), 3040-248 Coimbra, Portugal; beatrizgomes@fcdef.uc.pt

[3] University of Coimbra, Centre of Investigation in Sports and Physical Activity (CIDAF), Faculty of Sports Sciences and Physical Education (FCDEF), 3040-248 Coimbra, Portugal

* Correspondence: vmfsantos@student.dem.uc.pt

Abstract: This review synthesizes research on smart insole prototypes and their designs, focusing on those incorporating artificial intelligence (AI) and a wireless communication/transmission system. The main objective of this work is to summarize existing studies, identify key trends, evaluate the performance of these innovative biomechanical tools, and recognize the factors that could lead to optimization. This comprehensive analysis includes studies from PubMed, Scopus, and Web of Science databases and other investigations on the critical themes to consider. It follows strict inclusion and exclusion criteria, ensuring the quality and accuracy of the overview. The findings emphasize significant progress in smart insoles, particularly in AI-enhanced prototypes, while addressing existing challenges and problems. This review helps guide potential future research and define practical application directions. The growing importance of biomechanics, especially on smart insoles, underscores the considerable potential of these innovations to monitor and improve human movement in both clinical and non-clinical settings, promising a future of more effective and personalized health and performance interventions. This protocol was registered with the International Platform of Registered Systematic Review and Meta-Analysis Protocols (INPLASY) on 6 February 2025 and was last updated on 6 February 2025.

Keywords: smart insole; biomechanics; artificial intelligence; machine learning; design; prototype

1. Introduction

Wearable sensors and smart devices have become a reality and a helpful tool that can provide essential information across various fields of study [1]. Smart insoles fall within this category of equipment, not only representing a cutting-edge development in wearable technology but also being specifically designed to monitor activity and enhance the performance of several motor functions through real-time data collection and analysis. This product is gaining popularity and becoming a preferable wearable device for everyday use. Furthermore, smart insoles can serve as an alternative to traditional technological gadgets such as the global positioning system (GPS) or smartphones when these are unavailable or inaccessible [2]. Additionally, they offer significant advantages in terms of low cost and reliability [3].

The increasing popularity of smart insoles was driven by several factors, including advancements in technology, user-friendly signal processing algorithms, and machine learning (ML) techniques, alongside advances in new sensor technologies [2]. By collecting valuable data regarding motion, muscle activity, nerve responses, and pressure, these devices can analyze patterns and actions, enabling practical interventions, ergonomic improvements, and even more efficient prediction of disease development [4]. With the integration of AI, these devices have the potential to revolutionize patient care and athletic performance in a non-intrusive way and with daily confidence [3]. The junction of these technologies can even help in the disease rehabilitation process and provide rehabilitation in the comfort of patients' homes [4]. However, despite these advantages, gaps and opportunities for improvement persist across all applications of this technology. For example, nowadays, the known technology of wearable sensors for sports performance purposes has a few obstacles to overcome, such as the high cost, the high energy consumption, and the fitting [1].

This systematic review aims to provide a comprehensive overview of the methodologies employed in smart insole prototypes, focusing on AI's potential to enhance functionality and accuracy. We examine the objectives, processes, and outcomes of existing studies, identify critical gaps in the current literature and propose directions for future research.

2. Materials and Methods

In this review, a strategic approach was taken to select three databases for the search: PubMed, Scopus, and Web of Science. The chosen keywords, "smart insole", "artificial intelligence" or "machine learning", and "prototype" or "design", led to a certain number of hits, as shown in Table 1. The table provides the parameters and the essential information of the keywords and combinations used for the searches in the selected databases.

Table 1. Keyword combinations used for searches in the selected databases.

Database Name	Platform	Date Coverage	Date of Search	Search Within	Source Title	Command	# (No.) of Results
Scopus	Elsevier	–2024	17 September 2024	Article title, Abstract, Keywords	All	"smart insole" AND "artificial intelligence" OR "machine learning" AND "prototype" OR "design"	12
Web of Science	Clarivate	–2024	17 September 2024	All Fields	All	"smart insole" AND "artificial intelligence" OR "machine learning" AND "prototype" OR "design"	6
PubMed	NLB/NIH	–2024	17 September 2024	All Fields	All	"smart insole" AND "artificial intelligence" OR "machine learning" AND "prototype" OR "design"	1

This study followed the PRISMA [5] guidelines for systematic reviews and the protocol for this systematic review was registered on INPLASY (NO. 202520031). Thirty-nine works were included in the review; nineteen were retrieved from the three databases, and thirty were added manually. These last works resulted from the suggestions of the respective websites used for the research, and the suggestions were probably created by the algorithms due to each library's search history. Including these studies increases the review work and gives a better perception, justification, and knowledge of the referred authors' practical investigations and theoretical works. In this search strategy, it was agreed that restrictive conditions should not be applied. Hence, it was decided to include articles, review articles, systematic review articles, conference papers, proceeding papers, and book chapters only written in English.

On the other hand, short communications were not included since, from a group perspective, they do not offer benefits and value to the study. Next, there were no restrictions or imposed conditions regarding the publication year. Studies were selected based on the predefined inclusion and exclusion criteria mentioned above, focusing on those that detailed AI-enhanced smart insoles' methodologies, manufacture, design and performance metrics. Figure 1 presents the PRISMA flow diagram, illustrating the screening procedure followed. The choice of the selected databases ensured that low-quality and low-impact articles were not considered.

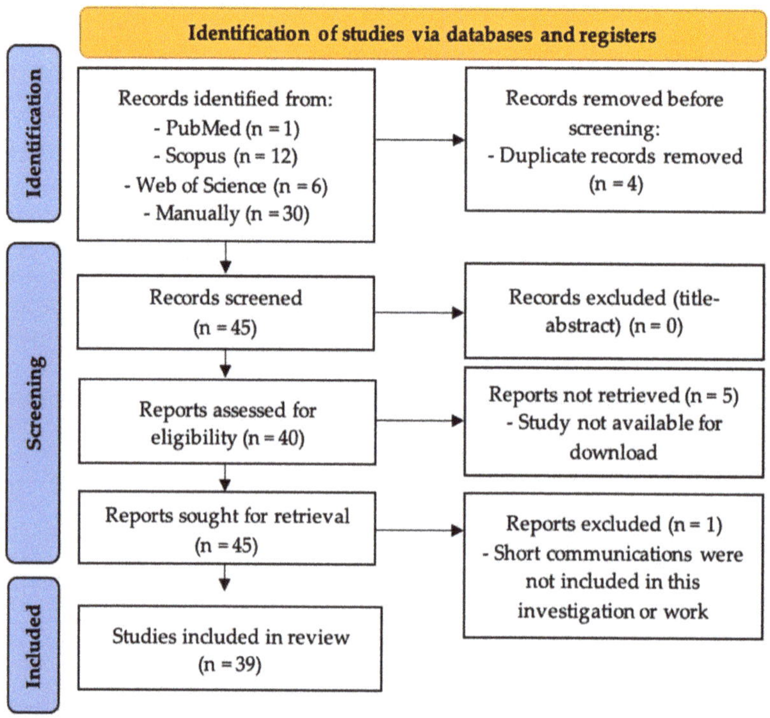

Figure 1. PRISMA flow diagram.

Four duplicate records were excluded from the total number of works searched, five reports were unavailable for download and consultation, and finally, one short communication was excluded in this review.

Figure 2 provides a detailed analysis of the type of documents used in this investigation. As stated previously, the specificity of the subject and the decision not to apply many conditions resulted in gathering various documents.

As explained previously, the specificity of the research and the decision not to apply many restrictions resulted in gathering several documents. The number of articles included in this work was twenty-four, followed by seven reviews, five conference papers, and three systematic reviews. Incorporating a more comprehensive range of source types is essential to encompass a more extensive array of academic views and perspectives, thus enhancing the strength and thoroughness of the study's conclusions. In the next figure it is possible to observe the year of publication of the works selected for this study.

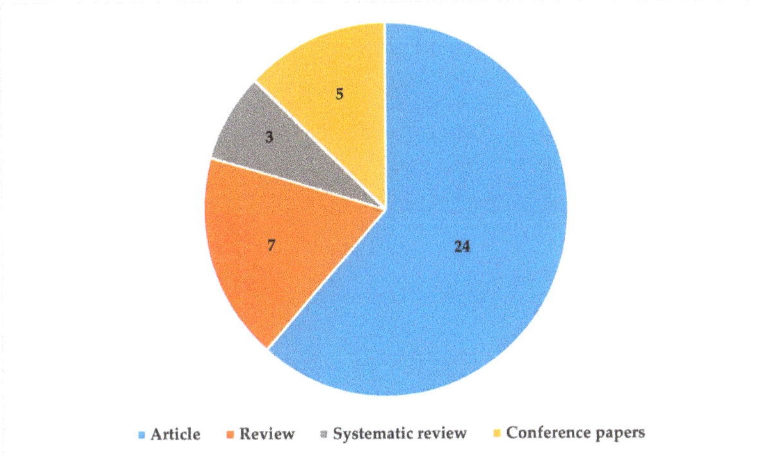

Figure 2. Information of the document types included in this systematic review.

The data shown in Figure 3 make it easy to conclude that interest in this subject increased exponentially in 2017 compared to 2012 and from 2022 to the present. Another fact is that the sum of the records in 2022, 2023, and 2024 represents about 67% of the total number of works. This last knowledge demonstrates that this study is in line with the research's interests.

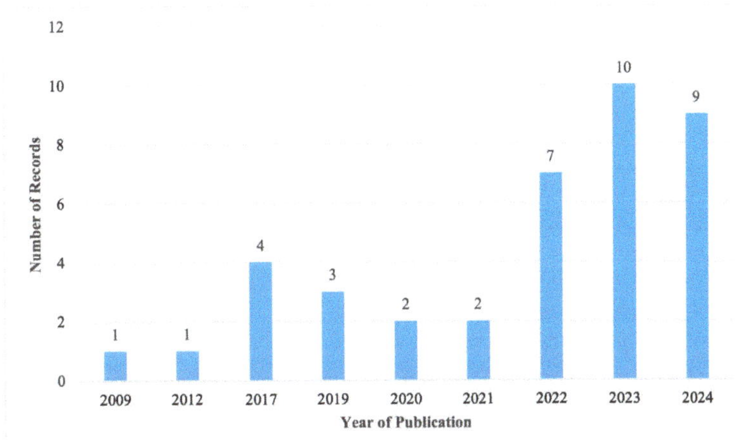

Figure 3. Information related to the publication year of the documents included in this systematic review.

In the first map generated with the VOSviewer software (Version 1.6.20 (0)), which is presented in Figure 4, the keywords included in the thirty-nine works included in the research are observed. The appearance of keywords like "wearable sensor", "plantar pressure", "ground reaction force", "model", "center of pressure (CoP)", "gait", "algorithm", "real time" and/or "development" serves as a strong validation of the chosen methodology of database search. However, it is important to note that other terms that appear with highlights, like "diabetic foot insole" and "research," could be present because most of the studies and prototypes have applications in the healthcare field. All the clusters presented have a specific color. It is evident that the red cluster focuses on technological development with applications in the healthcare field, including keywords such as "development", "wearable sensor", "algorithm", and "diabetic foot insole". The purple cluster features terms such as "plantar pressure", "shoe", "importance", and "real time", emphasizing the importance of real-time plantar pressure measurement. Keywords such as "part", "speed", "detail", and "PETG" belong to the yellow cluster and are related to the 3D printing and material properties. The green cluster includes "ground reaction force", "model", "research", and "sensorized insole", and the following keywords are linked to biomechanical analysis and modeling. Finally, the blue cluster links terms associated with gait analysis, such as "gait", "center", "CoP", "layout", and "flex sensor".

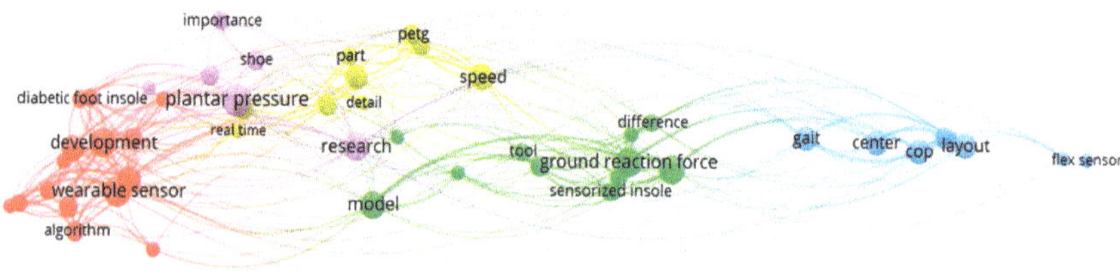

Figure 4. Complete counting keywords network visualization.

This graphic representation was achieved by creating a network visualization with three minimum occurrences, which provided a map with five clusters and a total of fifty-two terms. However, other words are inherent to the keyword search, and their presence confirms the depth of the subjects and the value of the gathered studies. The purpose of this study is related to sensor technology, applications, materials and designs, machine learning, and other types of artificial intelligence and prototypes. It is safe to conclude that most of the keywords presented in this map are incorporated in one form or another in the subjects previously referred to.

Figure 5 presents a keyword density visualization. This type of representation allows one to make quick observations and comparisons regarding the number of times (quantity) that a certain keyword appears in the selected works. So, to analyze correctly, it is important to be aware of the intensity and brightness of the color, the area length and the font size. In conclusion, it is safe to say that "ground reaction force", "plantar pressure", "development" and "wearable sensor" have a higher number of occurrences compared to the others. However, it is also visible through this type of image or representation that the keyword "algorithm" does not have many occurrences; on the other hand, it is possible to verify that the keyword "model" appears to have a higher frequency of appearance, and it can be considered that there is a direct link and connection between the two terms.

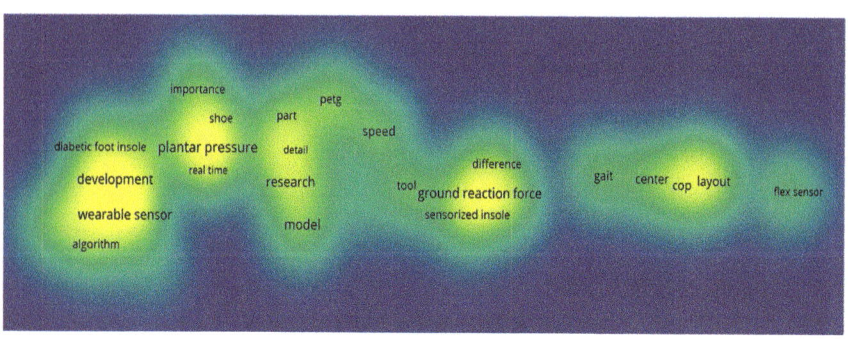

Figure 5. Complete counting keywords density visualization.

In Figure 6, the keyword network visualization images show remarkable similarities, underscoring the value of this data representation. The primary purpose of this visualization is to reveal the connections between keywords and the year of publication. A closer examination shows that the brightness of the color indicates the recency of the keyword's occurrence. For example, keywords like "diabetic foot insole" and "importance" are prominent in more recent works, while others like "algorithm", "model" and "real time" have been consistently present since 2022, providing a valuable historical context. This can occur for the same reasons stated in the Figure 4 analysis. However, it is noticeable that these terms are current and appear more frequently due to the diabetes pandemic and thus the number of patients suffering from such conditions.

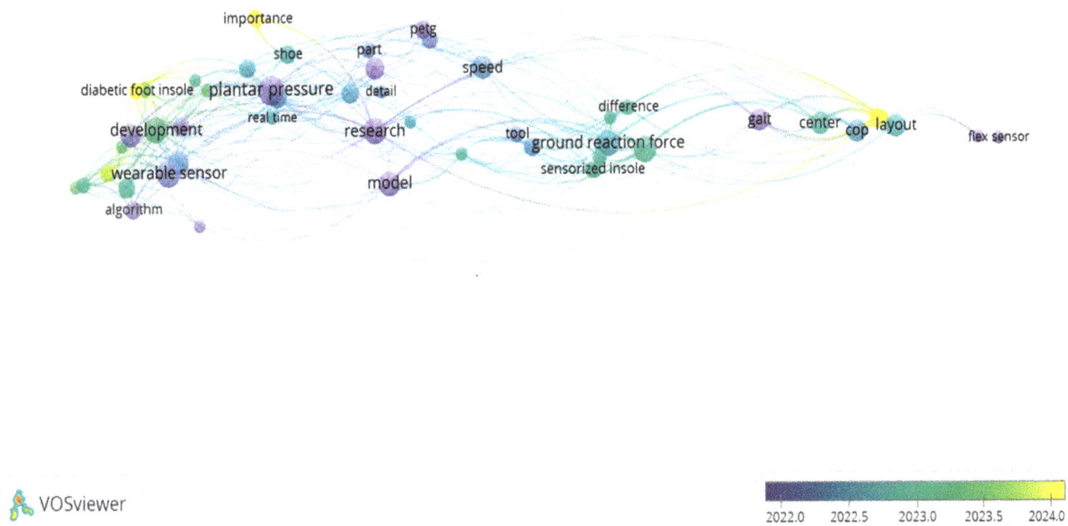

Figure 6. Complete counting overlay visualization.

3. Results

In all the investigations presented above in Table 2, some information is visible about the insole manufacturing materials, types and numbers of sensors, and hardware and software used, and if they integrated a wireless system or any type of AI. Five worked on devices already available in the market of the twenty-six studies, and the rest labored on their prototypes. Between all of them, ten integrated a wireless system, two implemented AI in the equipment and eight presented a smart insole system with wireless and AI.

According to the information provided in all investigations, studies and works, regarding the smart insole prototypes with wireless connection, the 3D-printed capacitive insole [6] and the PDMS-based capacitive sensor insole [7] demonstrated superior performance, making them the most effective for high-precision plantar pressure and gait monitoring applications. The 3D-printed capacitive insole demonstrated sensitivity of 1.19 MPa^{-1}, an operating pressure range of 41.5–872.4 kPa, excellent linearity ($R^2 = 0.993$), fast response and recovery times (142–160 ms), durability over 2280 cycles, and low hysteresis (<10%). It also offers practical solutions for stable operation across gait speeds of 30–70 steps/min. Complementarily, the PDMS-based capacitive insole showed high resolution, linear pressure response, and superior mechanical stability under both controlled loading and gait conditions, with a reported sensitivity of 0.020 pF/kPa for 3 mm thick sensors.

Table 2. Prototype specifications developed in retrieved studies.

Authors	Insole Manufacturing Materials	Types of Sensors	Number of Sensors	Microcontrollers (Hardware)	Power Supply (Hardware)	Communication/ Transmission (Hardware)	Software Programs	Wireless (Yes (Y)/No (N))	AI/ML Techniques (Yes (Y)/No (N))
Sorrentino et al. (2020) [8]	Flexible printed circuit boards/dielectric layer, 3D-printed support made of plastic with gaps and conductive Lycra	Capacitive (pressure and temperature)	336 (280 + 56) each insole	Microcontroller tactile board (MTB)	Connection to a host computer through a controller area network (CAN) bus interface	MTBs—host computer—CAN	MATLAB (process and visualize data) (The Mathworks Inc., Natick, MA, USA), YARP protocol (stream data), qpOASES (C++ software package)	N	N
Ho et al. (2022) [9]	Textile pressure sensor (W-290-PCN)	Capacitive	10 per insole	Capacitance to Digital Converter (CDC)	ND *	ND *	MATLAB R2021a (The Mathworks Inc., Natick, MA, USA)	Y	N
Eguchi, R. and Takahashi, M. (2023) [10]	FlexiForce A301 (Tekscan, South Boston, MA, USA), insole-shaped polyethylene terephthalate sheet using polytetrafluoroethylene tape	Force-sensing resistors	15 each	mbed NXP (LPC1768, Arm, Cambridge, UK)	ND *	Radio module (XBee, Digi International, Hopkins, MN, USA), op-amps (LMC6484AIM, Texas Instruments, Dallas, TX, USA), 12-bit analog-to-digital (A/D) converters (MCP3204T-BI/SL, Microchip Technology, Chandler, AZ, USA), a microSD card (Transcend, Taipei, Taiwan)	MATLAB (The Mathworks Inc., Natick, MA, USA)	N	Y

Table 2. *Cont.*

Authors	Insole Manufacturing Materials	Types of Sensors	Number of Sensors	Microcontrollers (Hardware)	Power Supply (Hardware)	Communication/ Transmission (Hardware)	Software Programs	Wireless (Yes (Y)/No (N))	AI/ML Techniques (Yes (Y)/No (N))
Haron et al. (2024) [11]	Flexible silicone insole, phenolic sheet stiffener	Strain gauge rosettes, FlexiForce normal pressure sensors	6 (3 + 3) each	Teensy 4.1 32-bit microcontroller (PJRC Electronic Projects, ARM Cortex-M7 processor, Portland, OR, USA)	3.7 V 3500 mAh Lithium Polymer battery (LP104567, EEMB, Moscow, Russian Federation), linear regulator (LDO, B08HQQ32M2, DollaTek, Hong Kong, China)	Analog-to-digital amplifier (HX711 ADC, HALJIA, Zhongai, China), ESP8266 UART WiFi adapter (Espressif Systems, Shanghai, China)	MATLAB (The Mathworks Inc., Natick, MA, USA)	Y	N
Peyer et al. (2017) [12]	F-Scan VersaTek wireless insole system (Tekscan, Boston, MA, USA)	Pressure sensors	960 each	mbed NXP LPC1768	ND *	F-Scan VersaTek wireless insole system, Motive's sync output	MATLAB R2014a (The Mathworks Inc., Natick, MA, USA)	Y	N
Lakho et al. (2022) [13]	Insole with 3 layers: A commercial shoe insole (not specified), a Polyvinyl Chloride (PVC) insole with the sensors and a fabric layer	Flexi force sensor and resistive flex sensor	5 (4 + 1) each	ND *	ND *	Bluetooth module and USB connection panel (not specified)	ND *	Y	N
Luna-Perejón et al. (2023) [7]	Insole with 3 layers: Ground layer (not specified), a flexible Polydimethylsiloxane (PDMS)-based dielectric layer, and last layer with the measurement electrodes	Capacitive sensors were made using PDMS as the dielectric material	12 each	STM32L432KC microcontroller with an Arm Cortex-M4 MCU	two 7.4 V LiPo batteries	Bluetooth module (HC-06), AD7147-1 capacitance-to-digital converter (CDC)	ND *	Y	N
Ciniglio et al. (2021) [14]	Pedar-X®, Novel, 100 Hz	Capacitive pressure sensors	99 each	ND *	ND *	ND *	Novel EmedLink, MATLAB 2018b (The Mathworks Inc., Natick, MA, USA)	ND *	N

Table 2. *Cont.*

Authors	Insole Manufacturing Materials	Types of Sensors	Number of Sensors	Microcontrollers (Hardware)	Power Supply (Hardware)	Communication/ Transmission (Hardware)	Software Programs	Wireless (Yes (Y)/No (N))	AI/ML Techniques (Yes (Y)/No (N))
Khandakar et al. (2022) [15]	(not specified)	Force-sensitive resistor (FSR) sensors and thermistor-based temperature sensors	24 (16 + 8) each	ESP32 TinyPico	Lithium Polymer (Li-Po) batteries	Bluetooth Low Energy (BLE)	Arduino software and a Python-based Real-Time Data Logger	Y	N
Lin et al. (2017) [16]	Advanced conductive eTextile fabric coated with a piezoelectric polymer. The insole also contained a flexible Printed Circuit Board (PCB) that housed the microcontroller	Pressure sensors and Inertial Measurement Unit (IMU)	48 + 3-axis accelerometer, 3-axis gyroscope, 3-axis magnetometer	CC2541 microcontroller (Texas Instruments)	Rechargeable battery (not specified)	BLE module	ND *	Y	Y | Action Manifold Learning (AML) framework
Matthies et al. (2017) [17]	Electrodes attached to a laser-cut polyacrylate sheet, a compressible foam plastic layer that acts as dielectric, a flexible plastic sheath as a protector of the electrodes and a silicone insole	Capacitive pressure sensors	6 each	Arduino Nano microcontroller	Lithium Polymer (Li-Po) battery	BLE module	Arduino IDE, CapSense library, Weka tools	Y	Y | ML techniques (classifiers such as, Random Forest (RF), Bayes Net (BN), Naive Bayes (NB), Instance Based Learner (IBK), and Sequential Minimal Optimization (SMO)
Ishtiaque et al. (2022) [18]	Plastic housing for the electronics	Pressure sensors	ND *	ND *	ND *	Narrowband Internet of Things (NB-IoT)	Custom mobile application	Y	Y | ML (Neural Networks (NNs))
Yang et al. (2017) [19]	Sennotech Insole X, Sennotech Inc.	Pressure sensors and IMU	16 + 3-axis accelerometer, 3-axis gyroscope, 3-axis magnetometer	Microcontroller Unit (MCU) (model not specified) and integrated with a 16 to 1 channel multiplexer (MUX)	Battery module (model not specified) rechargeable via USB	BLE module	Smartphone application with graphical user interface and cloud server hosts a TUG data analysis module	Y	Y | ML techniques (not specified)

Table 2. *Cont.*

Authors	Insole Manufacturing Materials	Types of Sensors	Number of Sensors	Microcontrollers (Hardware)	Power Supply (Hardware)	Communication/ Transmission (Hardware)	Software Programs	Wireless (Yes (Y)/No (N))	AI/ML Techniques (Yes (Y)/No (N))
Eldrige et al. (2019) [20]	NA	Pressure sensors	4 each	ND *	ND *	ND *	Smartphone application, ML-based analysis module, Edge computing software, Microsoft Kinect and OpenPose	Y	Y | ML techniques (RF, Decision Tree (DT), Support Vector Machine (SVM), k-Nearest Neighbors (k-NN), Gradient Boosting (GB))
Ivanov et al. (2019) [21]	Sensors were affixed on a flexible printed circuit board inserted between two flexible polyvinyl chloride sheets	Force sensors (FlexiForce A301) and inertial sensor (BMI160)	9 + 1 each	nRF52832 (by Nordic Semiconductor)	Rechargeable battery (2.5 V)	BLE, 12-bit ADC (ADS7042)	GaitPy	Y	Y | ML technique (RF classifier)
He et al. (2021) [22]	piezoresistive carbon nanotube conductive films, silver-plated conductive yarn (by Jinan Yumo Technology and Trade Co., Ltd. (Shandong, China)), Ethylene-vinyl acetate copolymer (PEVA) foam insoles (by Jinjiang Jinyi Shoe Material Co., Ltd. (Fujian, China)), Fusible interlining (by Dongguan Yufang Interlining Co., Ltd. (Guangdong, China))	Wearable textile-film sensors	5 each	ND *	ND *	ND *	LabView	N	N

Table 2. *Cont.*

Authors	Insole Manufacturing Materials	Types of Sensors	Number of Sensors	Microcontrollers (Hardware)	Power Supply (Hardware)	Communication/ Transmission (Hardware)	Software Programs	Wireless (Yes (Y)/No (N))	AI/ML Techniques (Yes (Y)/No (N))
Burch et al. (2023) [23]	Carbon nanotube (CNT)-based pressure sensors, textile substrate, conductive threads	Textile-based pressure sensors (piezoresistive sensors)	8 each	Arduino Uno Rev3, microSD card and adapter	ND *	ND *	Arduino IDE, MATLAB (The Mathworks Inc., Natick, MA, USA)	N	N
Jiang et al. (2024) [24]	NA	Pressure sensors, IMU, MoCap	230 + 2 + 2 each	ND *	ND *	BLE	ND *	Y	Y | ML technique (NN, Linear Discriminant Analysis (LDA))
Lueken et al. (2020) [25]	NA	Force-sensing resistors (FSRs)	4 FSR402 sensors each	CC1310 wireless microcontroller (by Texas Instruments)	230 mAh lithium-ion battery	868 MHz radio communication interface, 12-bit ADC MCP3204 (Microchip Technology Inc., Chandler, AZ, USA)	ND *	Y	N
Samarentsis et al. (2020) [6]	3-layer sensors: Protopasta CDP11705 composite conductive PLA (top and bottom), Filaflex 70A TPU; 2-layer insoles: Filaflex 82A (upper and lower layer)	3D-printed capacitive pressure sensors	16 each	2x Arduino Nano system boards (Arduino LLC, Boston, MA, USA)	ND *	Stepper motor (Nema 23 57BYGH115 from Wantai motors), motor driver (TB6600), microdisplay (1300 OLED Display Module from Waveshare)	Fusion 360 CAD (Autodesk, San Rafael, CA, USA), Ultimaker Cura 4.6.2 (Ultimaker B.V., Utrecht, The Netherlands), Graphical User Interface (GUI), Arduino IDE	Y	N
Zheng et al. (2023) [26]	3-layer insole-shaped sensing unit: The upper triboelectric layer is made of polydimethylsiloxane (PDMS) and has 44 hollow grooves. The middle layer consists of Cu electrodes and the bottom has polyimide layer	Triboelectric nanogenerators (TENG)	44-pixel TENG-based sensor array each	2x STM32F427	Self-powered using TENG	ADC, multichannel data acquisition (DAQ) board, SD card	ND *	N	N

Table 2. *Cont.*

Authors	Insole Manufacturing Materials	Types of Sensors	Number of Sensors	Microcontrollers (Hardware)	Power Supply (Hardware)	Communication/Transmission (Hardware)	Software Programs	Wireless (Yes (Y)/No (N))	AI/ML Techniques (Yes (Y)/No (N))
Crossland et al. (2023) [27]	Strain Analysis and Mapping of the Plantar Surface (STAMPS). 3 layers: a 5 mm thickness plasticine slab, a cross-patterned Nylon mesh, and an optimized computer-generated stochastic speckle (Correlated Solutions Speckle Generator, v1.0.5)	-	-	ND *	ND *	ND *	Digital Image Correlation (DIC), MATLAB (R2021b)	N	N
Gupta et al. (2023) [28]	3D-printed flexible insole (PLA), Velostat®, a polyethylene–carbon-black-infused composite material, Copper adhesive tape, Two layers of medium-density EVA (Ethylene–vinyl acetate), and high-quality leather	Piezoresistive pressure sensors made with Velostat®	5 each	ESP32	600 mAh lithium-ion battery	BLE	SolidWorks 2020, Mondopoint system, Arduino IDE, custom mobile app	Y	N
Fuchs et al. (2024) [29]	Pedar-X insole (Novel Electronics Inc., Saint Paul, MN, USA)	Capacitive sensors	99 each	ND *	ND *	ND *	Vicon Nexus 2.10.1 software, PASW Statistics 18.0 (SPSS Inc, Chicago, IL, USA) and Excel 365 (Microsoft Corporation, Redmond, WA, USA)	Y	N

Table 2. *Cont.*

Authors	Insole Manufacturing Materials	Types of Sensors	Number of Sensors	Microcontrollers (Hardware)	Power Supply (Hardware)	Communication/ Transmission (Hardware)	Software Programs	Wireless (Yes (Y)/No (N))	AI/ML Techniques (Yes (Y)/No (N))
Choi et al. (2024) [30]	F-Scan system and low-cost insole sensors	FSR sensors	6 each	ND *	ND *	ND *	ML models	Y	Y | Recurrent Neural Networks (RNNs), Long Short-Term Memory (LSTM)
Willemstein et al. (2023) [31]	3D-printed sensors: soft foam layer with the sensor, two insulating (top and bottom) layers, and a layer with grounding electrodes	Piezoresistive porous (foam-like) sensors	4 each	Arduino Uno (Arduino AG, Italy)	ND *	16-bit analog-to-digital converter ADS1115 (Texas Instruments, Dallas, TX, USA), breakout board (Adafruit, NY, USA) laptop	MATLAB (The MathWorks, Natick, MA USA) System Identification Toolbox	N	Y | Hammerstein-Wiener (HW) model

ND * = No data.

Some studies like the ones that are mentioned next do not provide much information about the materials used and procedures adopted. Crossland et al. [27] innovated a speckle-patterned system for strain analysis, utilizing digital image correlation; He et al. [22] created piezoresistive sensors using carbon nanotube conductive films and ethylene vinyl acetate (EVA) foam insoles; and Ciniglio et al. [14] and Fuchs et al. [29] employed commercial systems like Pedar-X and F-Scan for advanced analysis.

Regarding the smart insole prototypes with wireless connection, the 3D-printed capacitive insole [6] and the PDMS-based capacitive sensor insole [7] demonstrated superior performance, making them the most effective for high-precision plantar pressure and gait monitoring applications. The 3D-printed capacitive insole accomplished high scores in sensitivity, stability, and linearity, accurately distinguishing gait phases in real time. This device also offered strong reliability with low hysteresis and rapid response, making it well-suited for dynamic gait analysis. The PDMS-based capacitive sensor insole prototype presented a smooth composite material, a linear response, and precise and consistent pressure readings across different foot regions. Together, these two prototypes provide robust solutions for accurate real-time plantar pressure monitoring.

In recent investigations of prototypes with AI but no wireless capability, the focus was on studying ground reaction forces (GRF). Eguchi et al. [10] implemented the ML technique Gaussian Process Regression in an insole with pressure sensors. The results were accurate and appropriate for clinical trials and daily life. Willemstein et al. [31] presented a prototype with 3D-printed porous piezoresistive sensors in a commercial insole and with the employment of Hammerstein–Wiener models. According to the study, this device was a viable alternative compared to other commercial options due to its scores and cost.

The various studies incorporating wireless transmission and AI technology revealed significant improvements to diverse fields of knowledge and applications. For instance, the study on low-cost smart insoles employing recurrent neural networks enhanced GRF and CoP predictions by over 30% compared to traditional methods, showing promise for cost-effective gait analysis in clinical and day-to-day life [30]. The CapSoles system reached 95% accuracy in user identification and 82% in distinguishing floor types using capacitive sensing [17]. Another study involving a sensor insole designed for gait analysis achieved a remarkable 98.75% user recognition accuracy, underscoring its potential for sports and healthcare, where precision is critical [21].

An AI-powered smart insole developed for the construction industry [17] also demonstrated practical utility in reducing back pain risk among older workers by providing real-time feedback on lifting techniques. However, improvements in robustness are still needed for commercial purposes [18]. The smart insole timed up and go (SITUG) system, tailored explicitly for elderly users, refined the timed up and go test to operate in complex environments with a mean accuracy of 94.1% for gait feature extraction, making it a powerful tool for fall-risk assessment [19]. Meanwhile, the MovePort dataset introduced an extensive multimodal dataset that includes electromyography (EMG), IMU, and insole pressure sensors, enabling advanced analysis of abnormal movements in rehabilitation and setting a foundation for clinical research [24]. In the healthcare domain, a study leveraging pressure-map manifold learning achieved 86.6% accuracy in patient handling activity recognition, addressing injury risks for caregivers through robust task monitoring [16]. Lastly, a novel approach combined smart insoles with red–green–blue (RGB) cameras for identifying stationary individuals based on posture, presenting a feasible solution for IoT environments with controlled user monitoring [20]. Among these, the insole designed for gait analysis and the SITUG system obtained the best scores, presenting high accuracy and offering impactful resolutions in overcoming critical problems in healthcare and sports.

The studies presented demonstrate significant strides in integrating innovative materials, sensor designs, wireless technologies, and AI-driven frameworks. These advancements underscore the potential for developing advanced wearable technologies, which could enable real-time data collection and enhanced functionality across various applications.

4. Discussion

As expected, with the evolution of technology and the progress made in the smart insole field, it is essential to segment the final product to understand the components better while always understanding its purpose and future application. It is necessary to study the several manufacturing materials methodologies, the sorts of hardware, the database size, the numbers and types of sensors, and, ultimately, the integration of different ML techniques.

It is also known that smart insole devices evolve toward simplified designs, and the device's power consumption will be lower. Applications for daily use mainly focus on monitoring, detection, and interactions. It is desirable to have a streamlined design with the minimum number of sensors integrated, as the cost, power consumption, and robustness are the top considerations [2].

4.1. Insole Manufacturing Materials

Three very important features in this subchapter need to be addressed to better understand the procedures that come with it. First, different kinds of materials are already used in the production of insoles. For the manufacture of orthotics, various materials can be used, such as plastics (e.g., polypropylene), foams (e.g., polyurethanes and EVA), and carbon-based composite materials (e.g., carbon-fiber-reinforced plastics). It is crucial to carefully select the appropriate materials, considering that an excessively stiff insole will not provide adequate impact absorption, while an overly soft insole will lack sufficient support [32]. Among the materials mentioned, Valvez et al. [33] demonstrated that the mechanical properties of Poly(ethylene terephthalate)-glycol (PETG) can be significantly improved by optimizing printing parameters such as nozzle temperature, layer height, and infill density. Their findings reinforce the importance of process parameter tuning in additive manufacturing to ensure structural integrity and enhance mechanical performance [34]. In fact, in sensor applications, the structural integrity of the printed substrate is often a prerequisite for consistent and reliable signal response. Parameters such as layer thickness and infill density not only influence mechanical strength but also affect deformation behavior, dimensional stability, and interlayer bonding—all of which can impact the performance and durability of embedded or printed sensing elements [35]. Therefore, even when the final application is functional, structural optimization remains a critical foundation. Insoles with one single layer focus essentially on comfort, whereas insoles with two layers usually combine both soft and stiff materials to provide comfort and support to the user. Multi-layered insoles go one step further, incorporating various materials to balance support, control, and comfort. Thus, each structure and material choice affects the manufacturing process and therefore the insole performance in unique ways [36].

Second, it is also possible to choose a variety of manufacturing processes to fabricate a desired insole. There are various AM technologies such as selective laser sintering (SLS), selective laser melting (SLM), stereolithography (SLA) and fused filament fabrication (FFF) [36].

Finally, it is important to know the application or purpose of the equipment. As stated in Crabtree et al.'s [32] study, which focused on methodology of design and manufacture of sports insoles and referred to this niche as the symptom-specific sports (3S) insole, these processes must include the application or the specification of the sport to the equation. In that order, is important to prescribe the insole according to the geometry and biomechanical requirements, and to consider the specific sport (as well health problems, disabilities and disorders). This final point is important because the plantar pressures might differ and so the inherent sports movements can be characterized considering the loading patterns. Therefore, customizing insole design to match individual foot characteristics is essential to reduce the risk of pain and injury [36]. One task that can be implemented and could influence the outcome of the final product is the usage of three-dimensional foot-scanning equipment to reach a proper and accurate prescription of the foot insole [32]. In Figure 7, we show the synthetized information and the possible steps to obtain an adequate 3S insole.

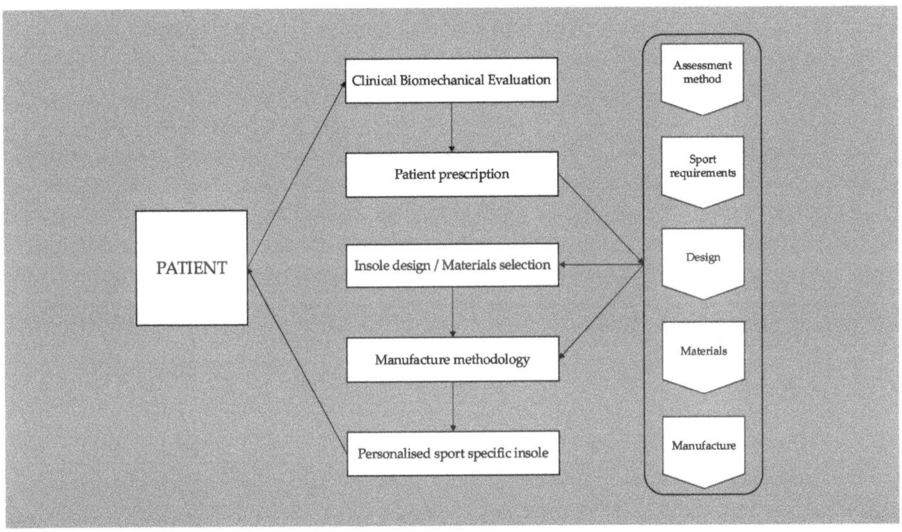

Figure 7. Adaptation of an image in Crabtree et al. [32] about a prescription methodology for a 3S insole.

4.2. *Significance of Hardware*

Choosing and integrating hardware components such as sensors, microcontrollers, batteries, and communication modules are crucial in the process of smart insole development. As is known, because of the inconvenience and impracticality of using force platforms as the main and preferable tool to measure GRF outside of laboratory settings, most studies have been constrained to indoor environments [3].

To overcome some obstacles, the precision and durability of the smart insole's elements determine the accuracy of data collection, which is essential in several applications. For example, high-quality sensors that can gather pressure points and force distribution allow users or healthcare providers to analyze gait patterns or detect early signs of foot-related health issues. Meanwhile, a compact but minimal, energy-efficient microcontroller is essential for processing data, ensuring the user's comfort.

Battery capacity, durability, and wireless communication capabilities are equally important; these specifications affect the range of the insole that can operate and how effectively it can transmit data to other devices, such as smartphones or computers. To transmit data to a computing platform in proximity, different technologies can be chosen and used, such as near-field communication (NFC), Bluetooth, Adaptive Network Topology (ANT) protocol, BLE, and ZigBee. Wireless communications, such as high-speed packet access (HSPA) or long-term evolution (LTE) services, can also be chosen [2].

4.3. Size and Type of Data

Full-body activities can be accurately recognized by smart insoles when compared to smartwatches and smartphones. Postures and movements like laying down, sitting, standing, walking, climbing staircases, running, cycling, and driving are possible to detect when using this type of technology. The non-obtrusive specification of the smart insole (if considered) permits the collection of different types of data, like the measurements of ground reaction force distribution, which can be valuable to disease detection and diagnosis [2]. Moreover, the integration of cycle time and activity repetition data could be instrumental in fatigue prevention initiatives [3].

Variables like the duration and the length of strides, steps, stances, swings, double supports, speeds, and cadences are also used to identify gait cycles. Gait spatiotemporal parameters represent the basics of gait analysis and are essential for balance, postural stability, and athletic performance evaluation. In conclusion, gait accuracy directly depends on spatiotemporal parameters and contributes to the overall gait detection accuracy [2].

One of the problems identified by most of the authors who were cited in this investigation is the lack of data diversity. For instance, some researchers have conducted pressure tests on diabetic foot insoles using the Pedar system [36] and others have recruited only college students [2]. Therefore, conducting a study with a diverse sample can be an asset, as it can help yield different conclusions and provide insight into the actual value of this technology. However, this can be an obstacle; there is evidence that models trained on small datasets, particularly in cases involving nonparametric regression, often perform best [37]. It is crucial to acknowledge that overtraining the data sample can contribute to this problem.

4.4. Types and Numbers of Sensors

It is known that among available sensors, each one has its own specifications, and considering the application and need, it is possible to choose the more appropriate one.

According to Razak et al. [38], to gather precise and accurate biomechanical data, it is first necessary to optimize the technology to the purpose, nature and application. Foot plantar sensors also need to fulfill some requirements such as mobility; limited wiring or wireless function; 15 sensors to cover most of the body weight changes; low cost; and low power consumption. To achieve good performance it is important to follow some key specifications, like hysteresis (the response when the sensor is loaded and unloaded), linearity (the response of the sensor to the applied pressure), sensing size, pressure range, temperature sensitivity, operation frequency/sampling and creep and repeatability (deformation of the material under high temperature and static stress and the ability to produce reliable results after a long period of time, respectively).

Recent studies concluded that IMU and pressure sensors are the most used in insoles for sports performance and rehabilitation [4,39]. In addition, there has been an increase in the integration of sensors into medical devices. With that said, it is also crucial to determine the adequate position and be precise in the sensor placement. Further calibrating it/them properly and proceeding to validation is essential [40].

Among the different types of sensors, triboelectrics can benefit from being self-powered, and according to a study by Fazio et al. [1] this technology is often applied to detect pressures from different body parts. Subsequently, in 2022, Almuteb et al. [2] stated that real-time pressure distribution monitoring insoles with piezoelectric nanogenerators combined with a triboelectric–electromagnetic nanogenerator and triboelectric nanogenerator-powered smart insole have the advantage of being self-charging or battery-free for daily use and the ability to use the energy generators as pressure sensors.

However, it is important to consider that integrating multiple sensors to gain the most critical information often results in redundant data, bulky designs, and uncomfortable equipment [2]. As another solution, integrating regression techniques may provide a viable alternative by allowing the use of a reduced number of sensors for estimating biomechanical time series and measuring all required independent variables [35].

4.5. Integration of ML Techniques and Others

ML is one AI subset. To better understand this, this technique uses statistical models and algorithms to enable computers to execute specific tasks without explicit programming, focusing primarily on making predictions or decisions based on data. Inside the world of ML, there exist great tools like Supervised Learning and Unsupervised Learning. In the first tool, the model is trained on labeled data to predict outcomes for new and unseen data. The other is used mainly to discover patterns in input data that are not labeled and is often used for clustering or association tasks [41].

The commonly used ML algorithms for daily activity recognition are Support Vector Machine (SVM); Decision Tree; Naïve Bayes; Nearest Neighbor (NN); Random Forest; and Deep Learning, such as Convolutional Neural Networks [2,4].

Regarding the recent curiosity about ML, fewer investigations into integrating this technology exist, but it has been proven that it has advantages. Incorporating a non-fixed machine learning algorithm is better than other algorithms [4]. They can handle large and complex datasets, "learn" from data, constantly improve their performance, and identify patterns and relationships in data that may not be readily noticeable with traditional methods. To better understand the integration of ML algorithms in devices, the process includes several steps: (1) collection of raw data; (2) filtering the noise/cleaning the data; (3) applying extraction techniques; (4) training the dataset; and (5) applying the model. The training process involves adjusting the model's parameters to reach a prediction with a higher accuracy score when compared to the actual output. Once the model has been trained with attention, it can predict unknown data [41].

In Abdollahi et al.'s [3] study, the authors reached the conclusion that the NN and SVM models were the most used but also showed consistently high performance when applied to smart insole data. These two models are more effective and suitable for smart insole applications. Employing force and pressure sensors with associated signal processing algorithms in smart insole devices has been well explored. The change in foot pressure when standing, walking, and exercising has proven valuable in gait studies. Also, some studies demonstrated that the integration of ML to classify and recognize activities achieved high accuracies [2]. In Yamaguchi et al.'s [42] study, a regression model was applied in a shoe sole sensor system to predict GRF. The results were compared with the values obtained in a force plate and the authors observed a 15% lower root-mean-square error (RMSE).

4.6. Limitations

This review synthesizes information from 39 studies focused on smart insole prototypes, with some of them integrating AI and wireless technologies. The potential impact of these studies on healthcare and other fields is significant. However, several limitations must be acknowledged. The level of detail provided by the original studies varies significantly, particularly in terms of dataset size, subject diversity, sensor calibration methods, and real-time validation. As a result, it was not always possible to provide a structured comparative analysis of all key parameters such as model performance metrics, dataset characteristics, or overfitting risks. Additionally, many studies were conducted in controlled environments, which may limit generalizability to real-world conditions. Finally, various prototypes were evaluated in research settings, and few have undergone clinical trials or large-scale user testing, which are critical steps toward commercialization and long-term adoption.

5. Future Directions

In a number of studies, significant issues have been identified, and connections have already been established. For instance, in 2012, Razak et al. [38] highlighted the necessity of a wireless system to deliver real-time and dependable foot plantar pressure data. The proposed solution to address the common issue was to minimize the hardware size and embed it within the shoe's insole along with the entire sensor. This detailed proposal called for a low-power, wearable wireless system, utilizing customized micro-electromechanical (MEMS) sensors and interfacing them with a data acquisition (DAQ) unit, to be integrated into the insole.

Another crucial specification in developing insoles is extending battery life; if this is achieved, it will enable continuous data collection for prolonged periods without requiring frequent recharging. To improve it, the focus on data storage constraints should be associated with the previous obstacle. The insoles should be able to stream data directly to the cloud during data collection [3]. In another study, some solutions were presented to improve these kinds of devices. An ideal smart insole includes a low-power consumption or self-sustainable system and presents biocompatibility [1]. For example, the most prominent issue in OpenGo and F-Scan models is their limited battery life, which allows for continuous data collection for only approximately two hours. As many wireless smart insoles require a connection to a device, such as a smartphone or PC, for data storage, some insoles, like OpenGo, can save data locally and transfer it later. However, they face the limitation of a determined local memory space, making continuous data collection difficult [3]. To solve this problem, AI algorithms can actively optimize the power consumption of wearable sensors. This can be achieved by reducing the amount of data transmitted and stored, as well as improving the efficiency of the algorithms that process the data [41].

Another significant concern in smart insole development is the need for a comprehensive and diverse dataset. For instance, to achieve 'high-accuracy' results, it is crucial to ensure that experiments are conducted in fully controlled environments and that subjects perform activities according to proper instructions. However, a significant issue is the limited number of subjects and their diversity. If all the above conditions are met, the ML models may be overfit to achieve high performance due to the limited data. On the other hand, when these models are applied to new subjects in uncontrolled environments such as daily life, accuracy may decrease abruptly [2]. To overcome this, simulating real-life scenarios in labs for semi-controlled and uncontrolled trials could be a valuable solution [2]. Therefore, it is essential to evaluate the effectiveness of a technique when deployed in a population that has not yet been validated. It is wrong to assume that a model trained and tested on impaired participants will have identical results as the same model trained and tested on healthy participants [37]. Gathering data from different regions, countries, industries, and activities to improve the accuracy, effectiveness, and robustness of models in smart insole applications is crucial for increasing the size and diversity of datasets [3]. However, if there is insufficient information, data augmentation can generate more data to overcome size limitations in the database without the need for additional experiments [2].

As smart insoles advance toward real-time monitoring and AI-driven analysis, ethical and data privacy concerns must be carefully considered. Continuous tracking of gait patterns, plantar pressure, and related biomechanical data constitutes sensitive personal information that can reveal health status, physical conditions, or behavioral patterns. Therefore, ensuring informed consent, transparency in data usage, and secure data transmission and storage is essential. Compliance with data protection regulations should be a baseline requirement in both research and commercial applications. Furthermore, future developments should explore privacy-preserving techniques, such as on-device processing, data anonymization, and federated learning, to minimize the risks associated with centralized data collection. Addressing these challenges proactively will help ensure the responsible and ethical deployment of AI-powered smart insole systems across clinical, occupational, and personal settings.

As previously stated, and supported by the references in this document, the integration of ML techniques holds significant potential to revolutionize wearable technology, particularly in smart insole devices. Fazio et al. [1] confirmed in their study the existence of an algorithm that automatically detects the sensor's position and calibrates it to achieve the best output. However, the deployment of many of these techniques is hindered by hardware limitations. Therefore, further research and investigation are necessary to explore other techniques, develop models that integrate multiple modalities, and determine their application and performance [3]. Another crucial aspect is the creation and application of specific subject models. This method suggests that current regression models may learn person-specific patterns rather than broader occurrences. This likely results from the small sample sizes used for model training in many of the reviewed studies [37]. Future work should emphasize the development of open-source algorithms that are validated in clinically impaired populations. This will likely enhance estimation performance, ensuring confidence in predictions and decisions, while reducing barriers and gaining trust from both medical professionals and patients [41]. Open-sourcing subject-general models will also allow the interest and integration of diverse professionals from other fields of study in a multidisciplinary research team. Open-source data, as well as open-source code, in future studies would help accelerate the development of these techniques [37].

Other aspects must be addressed and suggested to develop an ideal device as already explored by the research team [39]: (1) the possibility of manufacturing an invisible and comfortable insole with a thin, textile, and soft design, with high durability, bio-degradable material and appropriate to most applications; (2) user-friendliness with the integration of a more straightforward software and compatible with other devices; (3) availability for the everyday user, offering a reliable and low-cost device; (4) ability to provide analysis in real time with a proper time response; (5) using an adequate number of sensors with consideration of battery life; (6) a database with different feet (size number, type, and pathology) and the ability to analyze instantly and provide the position of each sensor; (7) selection of the best type of sensors with higher precision, efficiency, and durability and assembling it according to the application(s); (8) giving the user the possibility to remove and reuse the sensors in other shoe insole sizes.

6. Conclusions

Integrating AI, specifically using ML techniques and algorithms, in smart insole prototypes has shown promising results, significantly improving performance, reliability, and accuracy. This is concluded based on the results, both experimental (offline) and in real-time contexts, presented in the different works that were considered for this investigation. This systematic review provides a comprehensive overview of current methodologies, identifying key trends and future directions for research. The importance of continued innovation and rigorous testing cannot be overstated. It is these efforts that are essential to fully realize the potential of AI-driven smart insoles, which capture different biomechanical data crucial to enhancing foot health and overall well-being.

Author Contributions: V.M.S.: Conceptualization, methodology development, data collection, writing, analysis of the gathered data, review of the document. B.B.G.: Conceptualization, supervision and review. M.A.N.: Conceptualization, supervision and review. P.F.R.: Review of the document.

A.M.A.: Conceptualization, supervision and review. All authors have read and agreed to the published version of the manuscript.

Funding: Vítor Miguel Santos acknowledges Fundação para a Ciência e Tecnologia (FCT), Portugal, for the financial support through the PhD Grant 2024.03984.BD.

Acknowledgments: This research is sponsored by national funds through FCT—Fundação para a Ciência e a Tecnologia, under projects UID/00285—Centre for Mechanical Engineering, Materials and Processes and LA/P/0112/2020.

Conflicts of Interest: The authors declare no conflicts of interest.

Abbreviations

The following abbreviations are used in this manuscript:

AI	Artificial Intelligence
AML	Action Manifold Learning
ANT	Adaptive Network Topology
BLE	Bluetooth Low Energy
BN	Bayes Net
CAN	Controller Area Network
CDC	Capacitance-to-Digital Converter
CNT	Carbon Nanotube
CoP	Center of Pressure
DAQ	Data Acquisition (Board)
DIC	Digital Image Correlation
DT	Decision Tree
EMG	Electromyography
EVA	Ethylene–Vinyl Acetate
FSR	Force Sensing Resistor
GB	Gradient Boosting
GPS	Global Positioning System
GRF	Ground Reaction Forces
HSPA	High-Speed Packet Access
HW	Hammerstein–Wiener
IBK	Instance-Based Learner
IMU	Inertial Measurement Unit
K-NN	k-Nearest Neighbors
LDA	Linear Discriminant Analysis
Li-Po	Lithium Polymer
LSTM	Long Short-Term Memory
LTE	Long-Term Evolution
MCU	Microcontroller Unit
MEMS	Micro-Electromechanical Sensors
ML	Machine Learning
MTB	Microcontroller Tactile Board
MUX	Multiplexer
NB	Naive Bayes
NB-IoT	Narrowband Internet of Things
NFC	Near-Field Communication
NN	Nearest Neighbor
NNs	Neural Networks
PC	Personal Computer
PCB	Printed Circuit Board
PDMS	Polydimethylsiloxane

PETG	Poly(ethylene terephthalate)-glycol
PETG + CF	Poly(ethylene terephthalate)-glycol with Carbon Fibers
PETG + KF	Poly(ethylene terephthalate)-glycol with Aramid Fiber
PEVA	Ethylene-Vinyl Acetate Copolymer
PLA	Poly(lactic) Acid
PVC	Polyvinyl Chloride
RF	Random Forest
RGB	Red–green–blue
RNN	Recurrent Neural Networks
SITUG	Smart Insole Timed Up and Go (system)
SMO	Sequential Minimal Optimization
STAMPS	Strain Analysis and Mapping of the Plantar Surface
SVM	Support Vector Machine
TENG	Triboelectric Nanogenerators
TPU	Thermoplastic Polyurethane

References

1. De Fazio, R.; Mastronardi, V.M.; De Vittorio, M.; Visconti, P. Wearable Sensors and Smart Devices to Monitor Rehabilitation Parameters and Sports Performance: An Overview. *Sensors* **2023**, *23*, 1856. [CrossRef]
2. Almuteb, I.; Hua, R.; Wang, Y. Smart Insoles Review (2008–2021): Applications, Potentials, and Future. *Smart Health* **2022**, *25*, 100301. [CrossRef]
3. Abdollahi, M.; Zhou, Q.; Yuan, W. Smart Wearable Insoles in Industrial Environments: A Systematic Review. *Appl. Ergon.* **2024**, *118*, 104250. [CrossRef]
4. Wei, S.; Wu, Z. The Application of Wearable Sensors and Machine Learning Algorithms in Rehabilitation Training: A Systematic Review. *Sensors* **2023**, *23*, 7667. [CrossRef] [PubMed]
5. Page, M.J.; McKenzie, J.E.; Bossuyt, P.M.; Boutron, I.; Hoffmann, T.C.; Mulrow, C.D.; Shamseer, L.; Tetzlaff, J.M.; Akl, E.A.; Brennan, S.E.; et al. The PRISMA 2020 Statement: An Updated Guideline for Reporting Systematic Reviews. *BMJ* **2021**, *372*, n71. [CrossRef] [PubMed]
6. Samarentsis, A.G.; Makris, G.; Spinthaki, S.; Christodoulakis, G.; Tsiknakis, M.; Pantazis, A.K. A 3D-Printed Capacitive Smart Insole for Plantar Pressure Monitoring. *Sensors* **2022**, *22*, 9725. [CrossRef]
7. Luna-Perejón, F.; Salvador-Domínguez, B.; Perez-Peña, F.; Corral, J.M.R.; Escobar-Linero, E.; Morgado-Estévez, A. Smart Shoe Insole Based on Polydimethylsiloxane Composite Capacitive Sensors. *Sensors* **2023**, *23*, 1298. [CrossRef]
8. Sorrentino, I.; Andrade Chavez, F.J.; Latella, C.; Fiorio, L.; Traversaro, S.; Rapetti, L.; Tirupachuri, Y.; Guedelha, N.; Maggiali, M.; Dussoni, S.; et al. A Novel Sensorised Insole for Sensing Feet Pressure Distributions. *Sensors* **2020**, *20*, 747. [CrossRef]
9. Ho, J.-G.; Kim, Y.; Min, S.-D. Customized Textile Capacitive Insole Sensor for Center of Pressure Analysis. *Sensors* **2022**, *22*, 9390. [CrossRef]
10. Eguchi, R.; Takahashi, M. Estimation of Three-Dimensional Ground Reaction Forces During Walking and Turning Using Insole Pressure Sensors Based on Gait Pattern Recognition. *IEEE Sens. J.* **2023**, *23*, 31278–31286. [CrossRef]
11. Haron, A.H.; Li, L.; Shuang, J.; Lin, C.; Dawes, H.; Mansoubi, M.; Crosby, D.; Massey, G.; Reeves, N.; Bowling, F.; et al. In-Shoe Plantar Shear Stress Sensor Design, Calibration and Evaluation for the Diabetic Foot. *PLoS ONE* **2024**, *19*, e0309514. [CrossRef] [PubMed]
12. Peyer, K.E.; Brassey, C.A.; Rose, K.A.; Sellers, W.I. Locomotion Pattern and Foot Pressure Adjustments during Gentle Turns in Healthy Subjects. *J. Biomech.* **2017**, *60*, 65–71. [CrossRef] [PubMed]
13. Lakho, R.A.; Abro, Z.A.; Chen, J.; Min, R. Smart Insole Based on Flexi Force and Flex Sensor for Monitoring Different Body Postures. *Sensors* **2022**, *22*, 5469. [CrossRef] [PubMed]
14. Ciniglio, A.; Guiotto, A.; Spolaor, F.; Sawacha, Z. The Design and Simulation of a 16-Sensors Plantar Pressure Insole Layout for Different Applications: From Sports to Clinics, a Pilot Study. *Sensors* **2021**, *21*, 1450. [CrossRef]
15. Khandakar, A.; Mahmud, S.; Chowdhury, M.E.H.; Reaz, M.B.I.; Kiranyaz, S.; Mahbub, Z.B.; Ali, S.H.; Bakar, A.A.A.; Ayari, M.A.; Alhatou, M.; et al. Design and Implementation of a Smart Insole System to Measure Plantar Pressure and Temperature. *Sensors* **2022**, *22*, 7599. [CrossRef]
16. Lin, F.; Song, C.; Xu, X.; Cavuoto, L.; Xu, W. Patient Handling Activity Recognition through Pressure-Map Manifold Learning Using a Footwear Sensor. *Smart Health.* **2017**, *1–2*, 77–92. [CrossRef]

17. Matthies, D.J.C.; Roumen, T.; Kuijper, A.; Urban, B. CapSoles: Who is walking on what kind of floor? In Proceedings of the 19th International Conference on Human-Computer Interaction with Mobile Devices and Services, Vienna, Austria, 4–7 September 2017; ACM: New York, NY, USA, 2017; pp. 1–14.
18. Ishtiaque, T.A.; Cepuran, A.; Salaj, A.T.; Torp, O.; Diaconu, M.G. Developing an AI-Powered Smart Insole System to Reduce the Possibility of Back Pain among Older Workers: Lessons from the Norwegian Construction Industry. *IOP Conf. Ser. Earth Environ. Sci.* **2022**, *1101*, 032027. [CrossRef]
19. Yang, Z.; Song, C.; Lin, F.; Langan, J.; Xu, W. Empowering a Gait Feature-Rich Timed-Up-and-Go System for Complex Ecological Environments. In Proceedings of the 2017 IEEE/ACM International Conference on Connected Health: Applications, Systems and Engineering Technologies (CHASE), Philadelphia, PA, USA, 17–19 July 2017; IEEE: Piscataway, NJ, USA; pp. 340–347.
20. Poluan, S.E.R.; Chen, Y.-A. Using Smart Insoles and RGB Camera for Identifying Stationary Human Targets. In Proceedings of the 2019 IEEE 10th International Conference on Awareness Science and Technology, Morioka, Japan, 23–25 October 2019; Yuan Ze University: Taiwan, 2019.
21. Ivanov, K.; Mei, Z.; Lubich, L.; Guo, N.; Xile, D.; Zhao, Z.; Omisore, O.M.; Ho, D.; Wang, L. Design of a Sensor Insole for Gait Analysis. In Proceedings of the International Conference on Intelligent Robotics and Applications, Shenyang, China, 8–11 August 2019; pp. 433–444.
22. He, Y.; Lin, M.; Wang, X.; Liu, K.; Liu, H.; He, T.; Zhou, P. Textile-Film Sensors for a Comfortable Intelligent Pressure-Sensing Insole. *Measurement* **2021**, *184*, 109943. [CrossRef]
23. Burch, K.; Doshi, S.; Chaudhari, A.; Thostenson, E.; Higginson, J. Estimating Ground Reaction Force with Novel Carbon Nanotube-Based Textile Insole Pressure Sensors. *Wearable Technol.* **2023**, *4*, e8. [CrossRef]
24. Jiang, X.; Li, J.; Zhu, Z.; Liu, X.; Yuan, Y.; Chou, C.; Yan, S.; Dai, C.; Jia, F. MovePort: Multimodal Dataset of EMG, IMU, MoCap, and Insole Pressure for Analyzing Abnormal Movements and Postures in Rehabilitation Training. *IEEE Trans. Neural Syst. Rehabil. Eng.* **2024**, *32*, 2633–2643. [CrossRef]
25. Lueken, M.; Mueller, L.; Decker, M.G.; Bollheimer, C.; Leonhardt, S.; Ngo, C. Evaluation and Application of a Customizable Wireless Platform: A Body Sensor Network for Unobtrusive Gait Analysis in Everyday Life. *Sensors* **2020**, *20*, 7325. [CrossRef] [PubMed]
26. Zheng, Q.; Dai, X.; Wu, Y.; Liang, Q.; Wu, Y.; Yang, J.; Dong, B.; Gao, G.; Qin, Q.; Huang, L. Self-powered High-resolution Smart Insole System for Plantar Pressure Mapping. *BMEMat* **2023**, *1*, e12008. [CrossRef]
27. Crossland, S.R.; Siddle, H.J.; Brockett, C.L.; Culmer, P. Evaluating the Use of a Novel Low-Cost Measurement Insole to Characterise Plantar Foot Strain during Gait Loading Regimes. *Front. Bioeng. Biotechnol.* **2023**, *11*, 1187710. [CrossRef]
28. Gupta, S.; Jayaraman, R.; Sidhu, S.; Malviya, A.; Chatterjee, S.; Chhikara, K.; Singh, G.; Chanda, A. Diabot: Development of a Diabetic Foot Pressure Tracking Device. *J* **2023**, *6*, 32–47. [CrossRef]
29. Fuchs, P.X.; Chen, W.-H.; Shiang, T.-Y. Center of Pressure Measurement Accuracy via Insoles with a Reduced Pressure Sensor Number during Gaits. *Sensors* **2024**, *24*, 4918. [CrossRef] [PubMed]
30. Choi, H.S.; Yoon, S.; Kim, J.; Seo, H.; Choi, J.K. Calibrating Low-Cost Smart Insole Sensors with Recurrent Neural Networks for Accurate Prediction of Center of Pressure. *Sensors* **2024**, *24*, 4765. [CrossRef]
31. Willemstein, N.; Sridar, S.; Van Der Kooij, H.; Sadeghi, A. 3D Printed Graded Porous Sensors for Soft Sensorized Insoles with Gait Phase & Ground Reaction Forces Estimation. *arXiv* **2023**, arXiv:2303.04719. [CrossRef]
32. Crabtree, P.; Dhokia, V.G.; Newman, S.T.; Ansell, M.P. Manufacturing Methodology for Personalised Symptom-Specific Sports Insoles. *Robot. Comput. Integr. Manuf.* **2009**, *25*, 972–979. [CrossRef]
33. Valvez, S.; Silva, A.P.; Reis, P.N.B. Optimization of Printing Parameters to Maximize the Mechanical Properties of 3D-Printed PETG-Based Parts. *Polymers* **2022**, *14*, 2564. [CrossRef]
34. Rehman, R.U.; Zaman, U.K.u.; Aziz, S.; Jabbar, H.; Shujah, A.; Khaleequzzaman, S.; Hamza, A.; Qamar, U.; Jung, D.-W. Process Parameter Optimization of Additively Manufactured Parts Using Intelligent Manufacturing. *Sustainability* **2022**, *14*, 15475. [CrossRef]
35. Babich, N.; DeCohen, D.; Sun, H.; Faruk Emon, O. Effect of printing parameters on the sensing performance of a 3D printed elastomeric pressure sensor. *Manuf. Lett.* **2023**, *35*, 805–810. [CrossRef]
36. Ren, Y.; Wang, H.; Song, X.; Wu, Y.; Lyu, Y.; Zeng, W. Advancements in Diabetic Foot Insoles: A Comprehensive Review of Design, Manufacturing, and Performance Evaluation. *Front. Bioeng. Biotechnol.* **2024**, *12*, 1394758. [CrossRef] [PubMed]
37. Gurchiek, R.D.; Cheney, N.; McGinnis, R.S. Estimating Biomechanical Time-Series with Wearable Sensors: A Systematic Review of Machine Learning Techniques. *Sensors* **2019**, *19*, 5227. [CrossRef] [PubMed]
38. Abdul Razak, A.H.; Zayegh, A.; Begg, R.K.; Wahab, Y. Foot Plantar Pressure Measurement System: A Review. *Sensors* **2012**, *12*, 9884–9912. [CrossRef]
39. Santos, V.M.; Gomes, B.B.; Neto, M.A.; Amaro, A.M. A Systematic Review of Insole Sensor Technology: Recent Studies and Future Directions. *Appl. Sci.* **2024**, *14*, 6085. [CrossRef]

40. Resendes, T.; Freitas Rodrigues, P.; Cruz, F.; Gatões, D.; Santos, V.M.; Ramos, A.S.; Vieira, M.T. Advanced Medical Monitoring: 3D Printed Prosthetics with Integrated Strain Sensor. *Prog. Addit. Manuf.* **2024**, *10*, 219–229. [CrossRef]
41. Xiao, X.; Yin, J.; Xu, J.; Tat, T.; Chen, J. Advances in Machine Learning for Wearable Sensors. *ACS Nano* **2024**, *18*, 22734–22751. [CrossRef]
42. Yamaguchi, T.; Takahashi, Y.; Sasaki, Y. Prediction of Three-Directional Ground Reaction Forces during Walking Using a Shoe Sole Sensor System and Machine Learning. *Sensors* **2023**, *23*, 8985. [CrossRef]

actuators

MDPI

Article

A Lightweight, Simple-Structure, Low-Cost and Compliant Twisted String Actuator Featuring Continuously Variable Transmission

Chanchan Xu [1,2], Tong Liu [1,2], Shuai Dong [2], Yucheng Wang [1] and Xiaojie Wang [1,*]

[1] Institute of Intelligent Machines, Hefei Institutes of Physical Science, Chinese Academy of Sciences, Hefei 230031, China; xuchan@mail.ustc.edu.cn (C.X.); lt1997@mail.ustc.edu.cn (T.L.); ycwang@iamt.ac.cn (Y.W.)

[2] Department of Precision Machinery and Precision Instrumentation, University of Science and Technology of China, Hefei 230026, China; shuaizhi@ustc.edu.cn

* Correspondence: xjwang@iamt.ac.cn

Abstract: Twisted string actuators, which are an emerging artificial muscle, efficiently convert rotary motor motion into linear load movement, with advantages like high transmission ratio, compliance, simple structure, and long-distance power transmission. However, the limited range of transmission ratio adjustment remains a challenge. Thus, this paper introduces a novel twisted string actuator design that automatically and continuously adjusts its transmission ratio in response to external loads. Utilizing lightweight hyperelastic slender rods, the twisted string actuator with continuously variable transmission achieves a simple, compact, and cost-effective design. By manipulating the distance between two twisted strings through rod deformation, the transmission ratio continuously adapts to varying load conditions. Mathematical models of the twisted string actuator with continuously variable transmission are derived and experimentally validated, demonstrating a 2.1-fold transmission ratio variation from 0.1 kg to 1.5 kg loads. Application in an anthropomorphic robot finger showcases a 6.2-fold transmission ratio change between unloaded and loaded states. Our twisted string actuator with continuously variable transmission offers unparalleled advantages in weight, cost, simplicity, compliance, and continuous transmission ratio adjustability, making it highly suitable for robotic systems.

Keywords: twisted string actuator; continuously variable transmission; anthropomorphic robot finger

1. Introduction

In robotic systems, the need for versatility means that robots often operate under varying conditions. For instance, when lifting heavy objects, a robot requires greater torque output, while for rapid movement, a higher speed output is necessary. A standalone motor excels at high efficiency under high speeds with low-torque conditions, but its performance deteriorates when high torque is required at low speeds. To enhance the motor's load capacity while maintaining efficiency under high-torque conditions, it is common to pair the motor with a gearbox. However, this configuration often results in a significant reduction in speed under low-torque conditions. To simultaneously meet the demands for high speed and high torque, it is often necessary to use a motor with excessive power, which can lead to power redundancy and increased energy consumption. The development of continuously variable transmission (CVT) mechanisms is essential in enabling robotic systems to achieve both high-speed motion and substantial load capacity, thereby enhancing actuator efficiency [1]. Various CVT devices have been devised, encompassing wheel-type CVT, belt CVT, chain CVT, toroidal CVT, and spherical CVT, among others [2–5]. Each variant offers distinct approaches to achieving continuous variation in transmission ratio (TR), each with its own set of advantages and constraints.

Twisted string actuators (TSAs) represent a noteworthy soft actuator type, comprising one or more strings, an electric motor, and a load. They adeptly convert rotational motion from the motor into linear motion by winding the string(s). Owing to their straightforward design, high TR, substantial output force, and compliance, TSAs have found applications in a myriad of fields, including robotic hands, rehabilitation robots, and soft robots [6,7]. TSA effectively combines the functionalities of a motor and a gearbox, allowing it to operate efficiently under low speed with high-torque conditions; however, it operates within a relatively narrow efficient range. Thus, optimizing TSA performance necessitates integration with a CVT mechanism to tailor its TR in accordance with operational loads, typically achieved through modifications in string radius, string distance, or input torque.

Shin et al. [8] pioneered a passive dual-mode transmission mechanism rooted in TSA principles. In force mode, strings intertwine, while in speed mode, they rapidly coil around a thick shaft, effectively augmenting the string's radius. Despite offering a wide transmission ratio range, mode switching is beset by uncertainty owing to time-varying friction, leading to significant energy loss. To mitigate this challenge, the team proposed an active dual-mode variable transmission (VT) by incorporating a secondary motor [9]. However, this solution comes at the expense of compliance and introduces complexity. Furthermore, the dual-mode VT merely offers discrete TR levels, lacking continuous adjustment. Lu et al. [10] realized continuous automatic TR adjustment predicated on external load variations by compressing springs to varying extents, thereby inducing the string to wind around a shaft of variable radius. Nevertheless, their mechanism, characterized by a complex architecture and friction-induced energy losses, presents notable drawbacks. Additionally, Park's team [11] devised an elastomeric CVT-based TSA for automatic TR adjustment based on load size. Unlike Lu et al., they achieved continuous TR adjustment by modulating the input torque of the string according to the external load, accomplished by compressing the elastomer linked to the string's end. Nonetheless, the torque transmission pathway, from the motor through spur gears to a rigid lateral disc, reliant on friction between the disc and elastomer, harbors the risk of energy dissipation, silicon abrasion, and slippage. Consequently, its viability for large robot devices is compromised by the inherent friction limitations of the elastomer. Similarly, Shin et al. [12] engineered a CVT-based TSA by manipulating the input torque of the string in response to the external load. While their CVT delivers a twofold increase in TR compared to Park's team's, its highly intricate and bulky nature renders it unsuitable for lightweight robots.

In the pursuit of a lightweight and simplistic TSA with a CVT system, Ryu's team [13] introduced a passive TSA with CVT that adjusts TR by varying the distance between two strings using springs, contingent on external load fluctuations. Despite its relatively lightweight and straightforward design, the spring-based CVT still suffers from enlarged dimensions, rendering it less conducive for small robots. Moreover, to achieve continuous active TR adjustment, the team devised an active-type TSA with CVT employing planetary gears [14]. However, the incorporation of planetary gears and a secondary motor amplifies system weight, size, and cost, at the expense of compliance. Notwithstanding the plethora of TSA-based VTs developed, there persists a demand for a VT that is exceedingly lightweight, mechanically simple, compact, cost-effective, intrinsically compliant, continuous, and suitable for both small- and large-scale robotic applications.

To address this imperative, we present a novel TSA with CVT capable of continuously adjusting TR by modifying the distance between two twisted strings in response to external load variations. Our CVT comprises solely a pair of extremely lightweight hyperelastic slender rods, measuring a mere 1 mm × 1 mm × 32 mm, weighing 0.22 g, and costing only USD 0.041. Its TR undergoes a 2.1-fold variation as the load transitions from 0.1 kg to 1.5 kg. Comparative analysis against existing TSA-based VTs in Table 1 underscores our proposed TSA with CVT as the option that is lightest, simplest in structure, and most cost-effective. Furthermore, it boasts compliance and continuity, rendering it suitable for a plethora of robotic tasks across various scales. The proposed novel TSA with CVT system design significantly enhances the operational capabilities and broadens the potential applications of TSAs in diverse robotic contexts.

Table 1. The performance comparison with other existing TSA-based VTs.

TSA-Based VT Types	Weight of VT (g)	Size of VT (mm*mm *mm)	Complexity of VT's Structure	Cost of VT ($)	Compliant?	Continuous?	Range of TR Adjustment
Our CVT	0.22	1*1*30	Very simple	0.041	Yes	Yes	2.1 times
Dual-mode Passive [8]	>10	-	Moderate	>1	No	No	-
Dual-mode Active [9]	>29.6	69.2*19.5*15	Complex	>214	No	No	6.95 times
PVT [13]	80	>15*40*80	Simple	>1	Yes	Yes	<2 times
ElaCVT [11]	12	24*24*42	Complex	>10	Yes	Yes	2.31 times
Pat-CVT [12]	1400	85*120.5*115	Very complex	>10	Yes	Yes	4.6 times
Active CVT [14]	133.6	45*45*129	Complex	>232	No	yes	1.78 times
LAHM [10]	>10	14*14*67	Complex	>1	Yes	Yes	2.1 times

The subsequent sections are structured as follows: Section 2 elucidates the proposed design concept and operational mechanism of the TSA with CVT. Section 3 gives a theoretical analysis of the TSA with the CVT system. In Section 4, we delineate an experimental setup aimed at validating the efficacy of the proposed mechanism and the established theoretical model. Finally, in Section 5, we demonstrate the application of the proposed TSA with CVT to an anthropomorphic robot finger.

2. Concept of the Proposed TSA with CVT

Actuators employed in robotic systems are expected to exhibit a smaller TR for swift motion when operating without a load and a larger TR for delivering substantial supporting force at lower speeds when subjected to a load. To achieve this objective, we propose a TSA with CVT that automatically adjusts its TR according to the load weight. The structure and mechanism of the proposed TSA with CVT are illustrated in Figure 1. The proposed TSA with CVT consists of a pair of CVTs (CVT1 and CVT2) integrated within the TSA framework. CVT1 is linked to the motor's output shaft, while CVT2 is connected to the load. Both CVT1 and CVT2 comprise hyperelastic rods, interconnected by two strings. When CVT2 bears a lighter load, it undergoes minimal deformation. The distance between the strings is large, allowing rapid contraction, resulting in a small TR. Conversely, under heavier loads, CVT2 undergoes bending deformation, reducing the distance between the strings and slowing down the contraction speed, thereby increasing the TR and enhancing load capacity. As the load magnitude increases, the bending deformation of CVT2 intensifies, resulting in a smaller string distance, further increasing the TR and enhancing the load capacity.

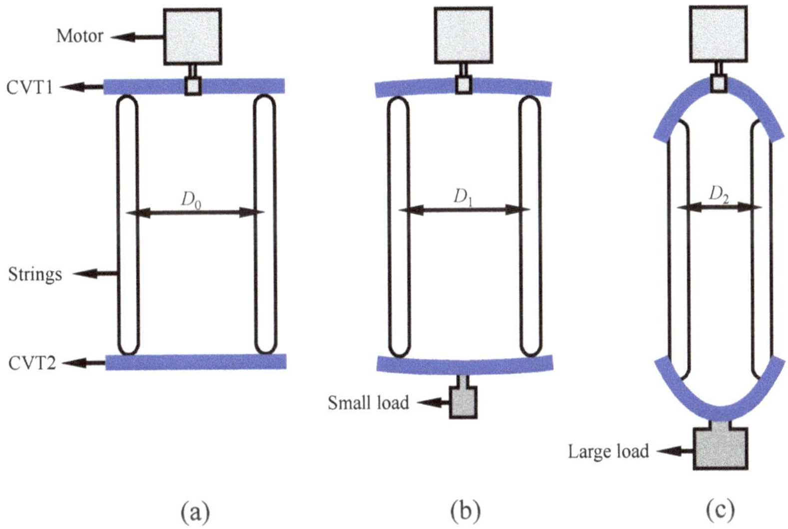

Figure 1. The TSA with a CVT mechanism. (**a**) The state without load. (**b**) The state with a small load. (**c**) The state with a large load.

3. Theoretical Analysis for the Proposed TSA with CVT

In this section, we develop a mathematical model for the proposed TSA with CVT that describes the relationship between input and output displacements, thereby characterizing the TR of the entire system. Additionally, we verify the effectiveness of the proposed TSA with CVT through theoretical computational results.

3.1. Mathematical Model

The proposed TSA with CVT is a combination of TSA and CVT. Notably, the conventional mathematical model for TSA initially assumes the strings to be rigid, overlooking the fact that the strings possess stretchable properties. While the conventional model simplifies the modeling process, it compromises the model's precision. Consequently, in our modeling approach, we consider the strings to be stretchable, with an associated stiffness coefficient denoted as K. Furthermore, we consider the CVT as a hyperelastic material, assuming its deformation to be within its elastic range and characterized by a constant elastic modulus E. To avoid an abrupt contraction of strings leading to an uncomfortable linear motion of load, the string is pre-twisted half a turn, as illustrated in Figure 2a. Once this pre-twisting is applied, we can calculate the initial distance, denoted as X_0, which represents the vertical distance between the connection points of the two CVTs and the strings on the same side, using the following formula:

$$X_0 = \sqrt{L_0^2 - D_0^2} \quad (1)$$

where L_0 represents the initial length of the string without load force and D_0 is the initial distance between the ends of the CVT and the string connection without load force. When a workload is attached to CVT2, as depicted in Figure 2b, the strings experience tension, causing them to stretch from L_0 to L_1. This simultaneously induces bending deformations in CVT1 and CVT2, leading to a reduction in the distance between the ends of the CVT and the string connection from D_0 to D_1 and an increase in the vertical distance between the connection points of the two CVTs and the string on the same side from X_0 to X_1.

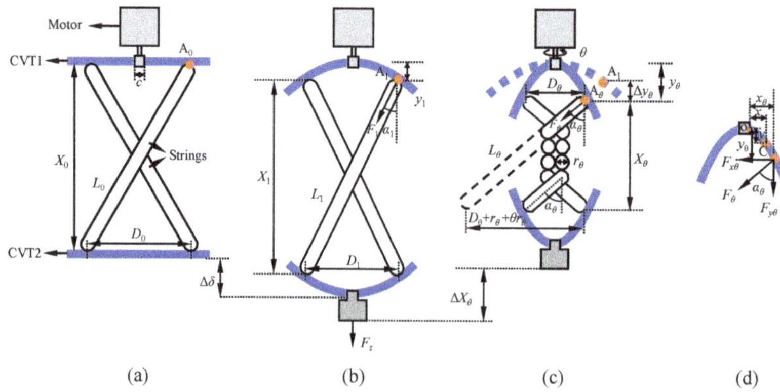

Figure 2. The proposed TSA with CVT, pre-twisted half a turn, for initial state without load (**a**) and with load (**b**). (**c**) The state with load after the motor makes a certain number of turns θ. (**d**) Force analysis for CVT1.

As the motor rotates by an angle of θ, the strings undergo a corresponding twisting motion, forming a cylindrical shape with a radius of r_θ in their midsection, as depicted in Figure 2c. This results in an increase in the tension experienced by the strings, causing their axial length to transition from L_1 to L_θ. Concurrently, CVT1 and CVT2 undergo further bending, resulting in a transition of the distance between the ends of the CVT and the string connection from D_1 to D_θ. Ultimately, this leads to a decrease in the vertical distance between the connection points of the two CVTs and the strings on the same side to X_θ, thereby inducing a linear motion of the load denoted by ΔX_θ. L_θ can be calculated by

$$L_\theta = L_0 + \frac{F_\theta}{K} \tag{2}$$

$$F_\theta = \frac{F_z}{2\cos\alpha_\theta} \tag{3}$$

where F_θ represents the axial tension of the string when it is twisted by an angle of θ, and F_z represents the load force. Additionally, α_θ is the angle between the string and the motor axis direction when it is twisted by an angle of θ. We assume that the helix angle of the formed cylinder is evenly distributed across the twists of the strings; thus, the helix angle equals the angle between the string and the motor axis direction, α_θ. Now suppose one of the strings is the unwound geometry of a cylinder, as shown in Figure 2c;

$$\cos\alpha_\theta = \frac{X_\theta}{\sqrt{X_\theta^2 + (D_\theta + r_\theta + \theta r_\theta)^2}} \tag{4}$$

$$X_\theta = L_\theta \cos\alpha_\theta = L_0 \cos\alpha_\theta + \frac{F_z}{2K} \tag{5}$$

The assumption of variable string radius was proved in [15]. However, given the elasticity of the strings, which experience deformation under tension, we have taken into account the impact of the applied load force on the radius. As a result, we approximate the radius r_θ of the cylindrical shape formed by the middle section of the strings using the following relationship:

$$r_\theta = \left(\frac{1}{k_1 F_z} + k_2 + k_3\sqrt{X_0/X_\theta}\right) r_0 \tag{6}$$

where k_1, k_2, and k_3 represent approximation coefficients that depend on the structure and mechanical properties of the strings, and r_0 stands for the diameter of the string when they are in an unloaded state.

To determine D_θ, we conduct a force analysis on CVT1, as depicted in Figure 2d. CVT1 is fixed to the motor's output shaft via a fixture of width c. The two ends of CVT1 experience bending deformation due to the tension in the two strings. Here, CVT is considered as a slender rod with a diameter d, which allows for the force and deformation characteristics of CVT1 to be likened to those of two symmetric cantilever beams, each subjected to concentrated loads at their endpoints. Therefore, we opt to analyze the force and deformation on the right half of CVT1. Select the center point of the right end face of the fixture as the coordinate origin O and assume that the entire assembly reaches equilibrium after the strings have twisted through an angle θ. We define the horizontal distance from any point C on the right half of CVT1 to O as x and the vertical distance to O as y. The resulting deformation of the bending curve satisfies the following flexural curve differential equation:

$$\frac{d^2y/d^2x}{(1+(dy/dx)^2)^{3/2}} = \frac{M}{EI} \tag{7}$$

where M represents the moment experienced at point C and I is the moment of inertia of the cross-section, where $I = \frac{\pi d^4}{64}$.

Point A_θ is the connection point between CVT1's right end and the string, bearing a concentrated load F_θ applied by the string. Let x_θ denote the horizontal distance from point A_θ to the origin O, and let y_θ represent the vertical distance from point A_θ to the origin O. Thus, the coordinates of point A_θ can be expressed as (x_θ, y_θ). By breaking down the tension F_θ at point A_θ into horizontal force $F_{x\theta}$ and vertical force $F_{x\theta}$, we have

$$\begin{cases} F_{y\theta} = F_z/2 \\ F_{x\theta} = F_z \tan\alpha_\theta/2 \end{cases} \tag{8}$$

Consequently, the moment M experienced at point C can be calculated as follows:

$$M = F_{y\theta}(x_\theta - x) + F_{x\theta}(y_\theta - y) \tag{9}$$

Assuming that the shape of the right half of CVT1 forms a quadratic curve after deformation, the relationship between y and x is as follows:

$$y = ax^2 \tag{10}$$

where a is the coefficient of the quadratic curve. Substituting (8)–(10) into (7), we obtain

$$\frac{2aEI}{(1+4a^2x^2)^{3/2}} = F_{y\theta}(x_\theta - x) + F_{x\theta}\left(ax_\theta^2 - ax^2\right) \tag{11}$$

Assuming that the length of the right half of CVT1 remains unchanged before and after deformation, we have

$$2\int_0^{x_\theta} \sqrt{1+(2ax)^2}dx + c = D_0 \tag{12}$$

Finally, the load contraction of the proposed TSA with CVT, represented by ΔX_θ, can be expressed as

$$\Delta X_\theta = X_1 - X_\theta - 2\Delta y_\theta = X_1 - X_\theta - 2(y_\theta - y_1) \tag{13}$$

where y_1 represents the vertical distance between point A_θ and point O when the strings with load are twisted by half a rotation, and Δy_θ is the distance by which point A_θ moves downward when the strings with load twist from half a rotation to θ. By solving Equations (5), (11) and (12) together, we can determine the values of X_θ and y_θ. Setting $\theta = \pi$ in these equations allows us to solve for X_1 and y_1. Further, we can resolve ΔX_θ. Thus, the TR of the proposed TSA with CVT is expressed as follows:

$$TR = \frac{1}{\partial(\Delta X_\theta)/\partial\theta} \tag{14}$$

3.2. Theoretical Results

We validate the effectiveness of the proposed TSA with CVT through theoretical analysis. Theoretical computational results depicting how the TR varies with the number of motor turns under various loads F_z are presented in Figure 3. The parameters utilized in the theoretical model are outlined in Table 2. It is observed that as the number of motor turns increases, the TR gradually decreases, consistent with the findings of Ryu's team [13]. Additionally, under the same number of motor turns, the TR increases with increasing workload, which is particularly noticeable with lower loads. Under the initial condition of half a turn pre-twist, the TR increases from 2.95 rad/mm to 4.83 rad/mm with the load ascending from 1 N to 15 N, marking a 1.6-fold increase. Moreover, with a pre-twist of 7.5 turns from the initial state, the TR surges from 0.59 rad/mm to 1.89 rad/mm, demonstrating a 3.2-fold rise under the corresponding load conditions. For lighter loads, the TSA with CVT demonstrates lower TR, resulting in faster contraction speed. Conversely, for heavier loads, the TSA with CVT shows higher TR, leading to slower contraction speed but enhanced load capacity. Thus, the effectiveness of the designed TSA with CVT is theoretically confirmed.

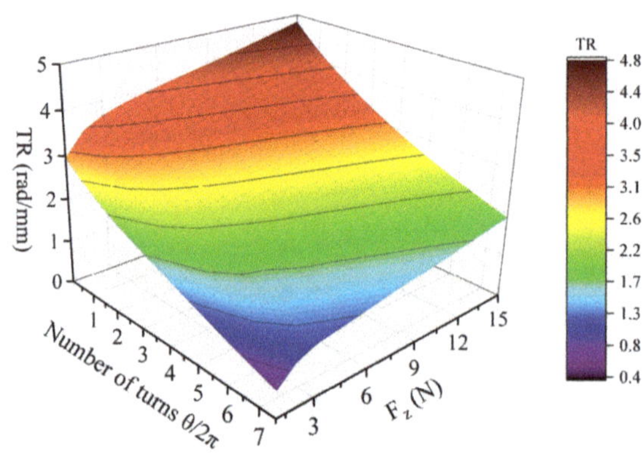

Figure 3. Theoretical TR of the proposed TSA with CVT varied with the number of turns of the motor and load F_z.

Table 2. The parameters used in theoretical model.

L_0 (m)	D_0 (m)	c (m)	r_0 (m)	d (m)	K (N/m)	E (GPa)	k_1 (1/N)	k_2	k_3
0.097	0.032	0.006	0.0015	0.001	1370	37	7.1	0.22	0.224

4. Performance Evaluation for the Proposed TSA with CVT

In this section, we develop an experimental setup to test the performance of the designed TSA with CVT and experimentally confirm the effectiveness of the proposed TSA with the CVT mechanism. Additionally, we validated the previously established theoretical model by comparing the experimental and theoretical values of TSA's contraction.

4.1. Experimental Setup

Figure 4 showcases the experimental setup utilized to assess the TR change capability of the TSA with CVT under varying loads. The key components of the setup included a direct current (DC) motor (Maxon DCX16L) controlled by an RMDS-405 motor driver, a pair of CVTs (CVT1 and CVT2), two strings, a load, a linear guideway (Hiwin, MG), and a linear variable differential transformer (LVDT, W-DC) with a stroke of 20 cm. The DC motor was horizontally mounted on a frame and operated via the RMDS-405 motor driver.

The load and LVDT were securely affixed to the linear guide. We conducted experiments using parameters consistent with those listed in Table 2. For the CVTs, two Ni-Ti memory alloy rods with a diameter of 1 mm, a length of 32 mm, and an elastic modulus of 37 Gpa, exhibiting superelastic characteristics, were utilized. CVT1 was connected to the output shaft of the motor using a 6 mm wide fixture, while CVT2, also with a 6 mm wide fixture, was positioned parallel to CVT1 on the linear guide. Two braided ultra-high-molecular-weight polyethylene (UHMWPE) fiber strings, each with a diameter of 1.5 mm, an initial length of 97 mm without external loads and a stiffness coefficient of 1370 N/m, were employed for the TSA. These strings were connected at either end of CVT1 and CVT2 in parallel to the motor's axis. To prevent sudden load movement, the strings were pre-twisted half a turn. Subsequently, the motor drove CVT1 to continuously rotate with a constant speed of 12.5 rpm/min for 7.5 turns under different loads.

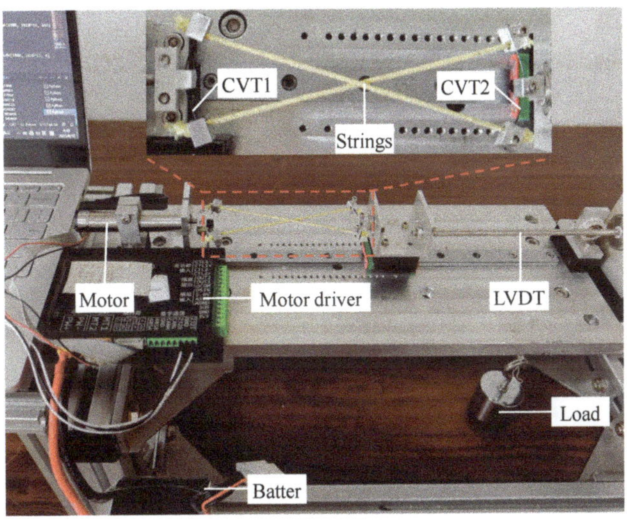

Figure 4. The test platform for the properties of TSA with CVT.

4.2. Results

The performance test results of the proposed TSA with CVT are illustrated in Figure 5. In Figure 5a, the experimental and theoretical load contractions for the proposed TSA with CVT are depicted, showcasing variations in the number of motor turns across different loads. The tension parameters used in our models were experimentally measured from the materials employed in the actuator, ensuring that they accurately reflect the material properties under operating conditions. For the deformation parameters in the theoretical modeling, we utilized a curve-fitting approach based on experimental data to derive these values. It is observed that as the number of motor turns increases, the load contraction gradually increases. Under the same number of motor turns, the load contraction decreases with an increasing workload, which is particularly evident with a higher number of turns. The TSA with CVT exhibits maximum contractions of 34.9 mm, 22.1 mm, and 16.64 mm when carrying workloads of 0.1 kg, 0.5 kg, and 1.5 kg, respectively. Comparing load contraction, there is a 36.7 % decrease from 0.1 kg to 0.5 kg a 24.7 % decrease from 0.5 kg to 1.5 kg, and a 52.3 % decrease from 0.1 kg to 1.5 kg. The TSA with CVT exhibits greater sensitivity to load variations between 0.1 kg and 0.5 kg compared to between 0.5 kg and 1.5 kg.

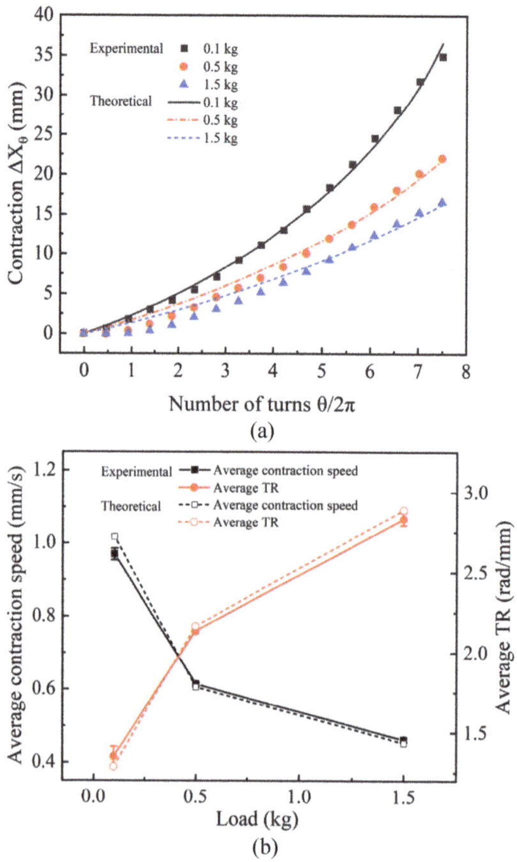

Figure 5. The performance test results of the proposed TSA with CVT. (**a**) The experimental and theoretical load contractions for the proposed TSA with CVT varied with the number of turns of the motor, maintaining a constant output speed of 12.5 rpm/min while carrying workloads of 0.1 kg, 0.5 kg, and 1.5 kg, respectively. (**b**) The experimental and theoretical average contraction speed of the load and average TR varied with the load, maintaining a constant motor output speed of 12.5 rpm/min.

Additionally, Figure 5b illustrates experimental and theoretical data concerning the average contraction speed of the load and the average TR as they vary with the load. The average contraction speed of the load represents the TSA's operational speed, and the operational speed of the TSA is inherently linked to the TR, which is the ratio of the motor speed to the load movement speed. When the motor operates at a constant speed, an increase in the TR results in a decrease in the TSA's operational speed. Firstly, we can find that the theoretical average contraction speed of the load and the average TR closely align with experimental results. Moreover, there is a decreasing trend in the average contraction speed of the load with increasing workload, while the average TR of the TSA with CVT exhibits a corresponding increase with workload. Under loads of 0.1 kg, 0.5 kg, and 1.5 kg, the TSA with CVT demonstrates experimental average contraction speeds of 0.97 mm/s, 0.61 mm/s, and 0.46 mm/s, respectively. Correspondingly, the experimental average TR varies at 1.35 rad/mm, 2.13 rad/mm, and 2.83 rad/mm for these respective loads. Notably, under a 1.5 kg load, the experimental contraction speed diminishes to 47.4% of that observed under a 0.1 kg load, while the average TR increases to 2.1 times. When the load is lighter, the TSA with CVT demonstrates reduced TR, facilitating quicker contraction rates. Conversely, under heavier loads, the TSA with CVT displays elevated TR, resulting in slower contraction rates but increased load capacity.

This phenomenon is demonstrated in Figure 6, which presents snapshots of the TSA with CVT carrying loads of 1 kg, 0.5 kg, and 1.5 kg in the initial state with half a turn of pre-twist with the load (Figure 6a) and subsequent to being twisted 7.5 turns from the initial state (Figure 6b). In Figure 6a, it is observed that when the load is 0.1 kg, the CVT undergoes minimal deformation in the initial state, with the distance between the ends of the CVT and the string connection (D) being 32 mm. With increasing turns, D remains almost unchanged, indicating that the CVT effectively operates as if under a fixed offset of 32 mm, resulting in relatively faster contraction speeds. When the load is 0.5 kg, there is a slight change in the CVT from 32 mm to 31 mm in the initial state, indicating a smaller change compared to the 0.1 kg load. However, as the turns increase, the tension on the string gradually increases, causing the CVT to bend and D to decrease while increasing the downward displacement (Δy_θ) of the ends of the CVT and the string connection. D decreases to 28.9 mm after 7.5 motor turns. The comprehensive effect of load contraction resulting from both the contraction of the string ($X_1 - X_\theta$) and the downward displacement of the CVT (Δy_θ) due to bending, is described by Equation (13). A larger ($X_1 - X_\theta$) leads to faster load contraction, while a larger Δy_θ results in slower load contraction. As D decreases with increasing turns, accompanied by an increase in Δy_θ, there is a dual reduction in the load contraction speed, which becomes more pronounced with higher load. When the load increases to 1.5 kg, the CVT exhibits significant initial bending, reducing D to 29.2 mm, which further decreases to 25.9 mm after 7.5 motor turns, resulting in even slower load contraction. Therefore, the experiments demonstrate that the proposed TSA with CVT can automatically adjust TR according to external load variations to meet the system's requirements.

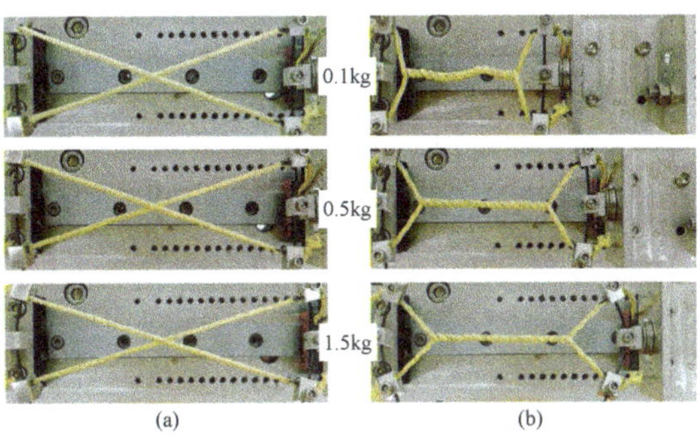

Figure 6. Snapshot of TSA with CVT carrying workloads of 0.1 kg, 0.5 kg, and 1.5 kg. (**a**) In the initial state and pre-twisted half a turn with the load. (**b**) After being twisted 7.5 turns from the initial state.

Moreover, from Figure 6a, it is evident that our established theoretical model aligns well with experimental results. While the theoretical model almost coincides with experimental data for a 0.1 kg load, there are slight deviations for 0.5 kg and 1.5 kg loads during the initial twisting stage, likely due to gaps within the braided string and the nonlinearity of the elastic modulus of the superelastic materials used in the CVT under a higher load. Overall, our established model effectively predicts the load contraction of TSA with CVT. This model serves as a valuable theoretical tool for analyzing and optimizing the performance of TSA with CVT.

5. Application for an Anthropomorphic Robot Finger

In this section, we apply the proposed TSA with CVT to an anthropomorphic robot finger and test its adaptability to the external load, further validating the practicality of the proposed TSA with CVT.

5.1. Design and Working Mechanism

Figure 7 illustrates the design principle of the proposed anthropomorphic robot finger with the proposed TSA with CVT. The anthropomorphic robot finger comprises a 3D-printed finger including four links and three rotating joints, and the TSA with the CVT system. Link 0 is fixed, while links 1, 2, and 3 can rotate around joints 1, 2, and 3 respectively, with all three joints equipped with elastic rings to provide resilience. To prevent rotation at the output end of the CVT system, only a superelastic rod is employed near the motor end, while a fixed offset is utilized near the load end. The output end of the CVT system is connected to the fingertip via a tendon. The finger's flexion is driven by the contraction force of the strings when the motor twists them, while the extension is achieved through the rebound force of the elastic rings at joints when the strings are untwisted.

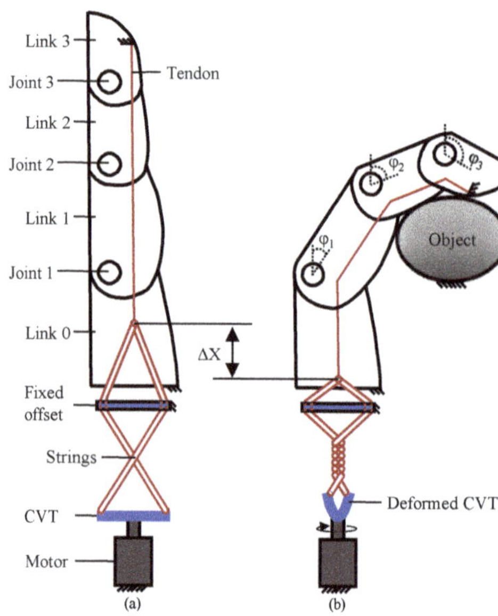

Figure 7. The design principle of the proposed anthropomorphic robot finger with the proposed TSA with CVT. (**a**) The initial state for pre-twisted half a turn without load. (**b**) The state after grasping the object.

The mechanism of the anthropomorphic robot hand with proposed TSA with CVT operates as follows: when the finger is not in contact with a load, the contraction force of the strings is minimal, and the CVT remains nearly unchanged. The distance between the strings is large, allowing rapid contraction, resulting in a small TR, thereby enabling the finger to quickly make contact with an object. Once the finger contacts an object, the contraction force of the strings increases, causing the CVT to bend and deform, reducing the distance between the strings and slowing down the speed of contraction. Consequently, the TR of the entire system increases, resulting in slower grasping speed but enhanced gripping capability. As the load increases, the degree of bending and deformation of the CVT becomes more pronounced, further reducing the distance between the strings and increasing the overall TR. Therefore, the anthropomorphic robot finger can adjust its TR according to the operational load conditions.

5.2. Performance Evaluation

To assess the TR change capability to a variable load of the proposed anthropomorphic robot finger, we designed an experimental setup as depicted in Figure 8. The motors (Maxon DCX16L) equipped with encoders (Maxon ENX16) were controlled using an RMDS-405 driver. The distance between the strings with a fixed offset is 30 mm. A superelastic Ni-Ti memory alloy rod with an initial length of 30 mm and a diameter of 1 mm was employed for CVT. The distance between the fixed offset and CVT is 86 mm. The string diameter is 1.5 mm. Given the varied motion trajectories inherent to underactuated robot fingers and the experimental objective of evaluating their capability to adapt to variable loads, joints 1 and 3 were immobilized, allowing only joint 2 to rotate freely. Consequently, an angle sensor (Inertial Measurement Units; IMUs) was affixed to link 3 to monitor the real-time flexion angle of link 2 around joint 2. A spring, connected at one end to the fingertip and at the other end to a tension sensor (SBT674 with a force of 50 N) secured on the frame, was utilized to provide the loading force. To prevent sudden finger movements, the string was pre-twisted half a turn. Subsequently, the motor drove the CVT to continuously rotate with a constant speed of 60 rpm/min for 8.5 turns. This caused link 2 to gradually rotate around joint 2 from a vertical position, gradually stretching the spring from its relaxed state. Both the IMU and tension sensor concurrently recorded data throughout the entire process.

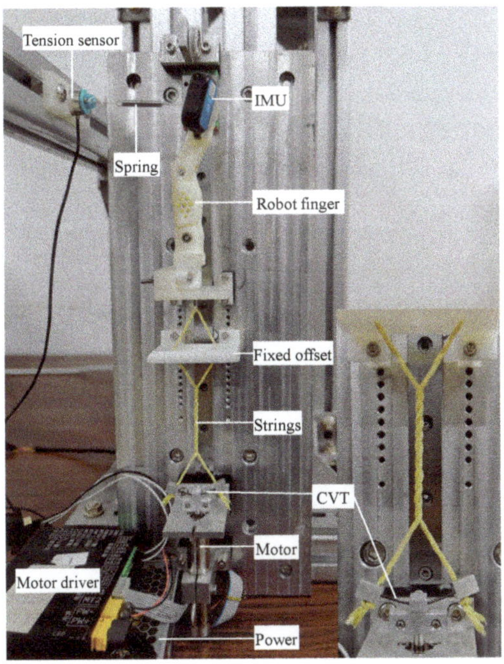

Figure 8. Experimental setup to evaluate the proposed anthropomorphic robot hand.

The findings, presented in Figure 9, encompass various facets of the experiment. Figure 9a, from top to bottom, shows the number of motor turns, the distance between the ends of the CVT and the string connection, the fingertip force provided by the spring, and the flexion angle of rotation of link 2 of the finger around joint 2 over time, respectively. Figure 9b showcases the sequential motion states of the proposed anthropomorphic robot hand at time intervals of 0 s, 3.8 s, and 8.6 s (further details in https://1drv.ms/v/s!AgaftTaT7Nn_gQKCWWQmciREtH1j?e=Z5P5xD (accessed on 25 September 2024)).

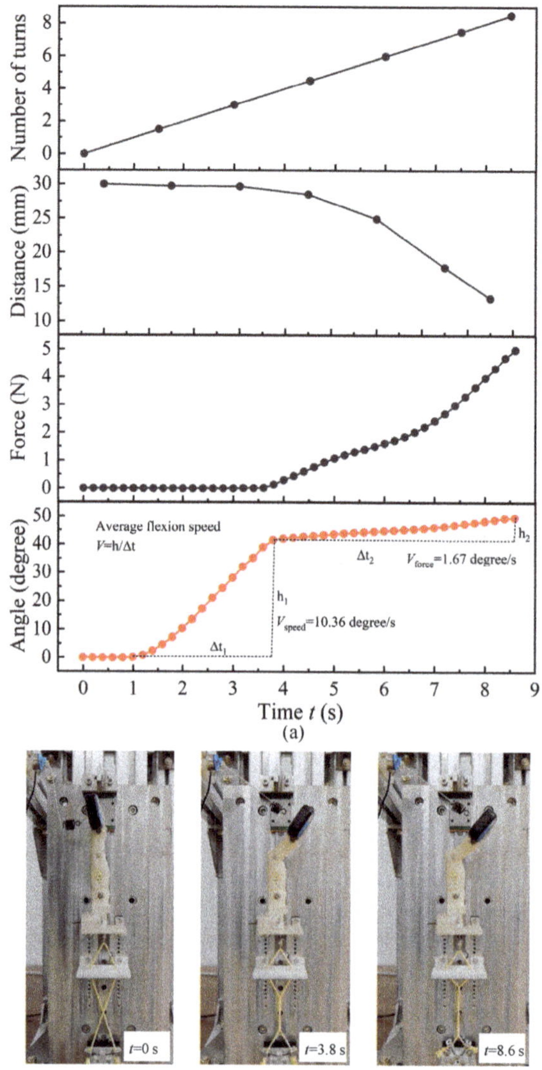

Figure 9. The results of the adaptability assessment of the proposed anthropomorphic robot hand. (**a**) illustrates the number of turns of the motor, the distance between the ends of the CVT and the string connection, the fingertip force provided by the spring, and the flexion angle of rotation of link 2 of the finger around joint 2 over time with a constant motor output speed of 60 rpm/min, respectively. (**b**) displays the motion states of the proposed anthropomorphic robot hand at time intervals of 0 s, 3.8 s, and 8.6 s.

It is evident that before t = 3.8 s, the spring is relaxed, providing zero force to the fingertip, and the distance between the ends of the CVT and the string connection remains almost constant. The average flexion speed of link 2 rotation around joint 2 is 10.36 degrees/s, except for the negligible motion because the initial twisting of the string does not significantly affect the movement of link 2. After $t > 3.8$ s, the spring gradually stretches, and the fingertip force increases from 0 N to 5 N. The distance between the ends of the CVT and the string connection is reduced to 12.4 mm, and the average flexion speed of the finger decreases to 1.67 degrees/s, indicating a notable 6.2-fold TR alteration between unloaded and loaded conditions. Additionally, from Figure 9b, it is apparent that before the stretching of the spring, the distance between the ends of the CVT and the string connection remains nearly constant. However, after the stretching of the spring, this distance noticeably decreases. The finger's movement speed is significantly faster before the spring stretching compared to after. Therefore, the proposed anthropomorphic robot hand demonstrates the TR change capability to adapt to variable loads. It also exhibits a considerable adjustment range and can withstand at least 5 N of fingertip force. This underscores the practicality of the proposed TSA with the CVT concept.

6. Conclusions and Future Work

In conclusion, this paper has introduced a novel TSA with CVT, offering automatic adjustment of TR in response to varying external loads. Through the utilization of lightweight hyperelastic slender rods, we have achieved a remarkably simple, compact, and cost-effective design. Our experimental results have validated the efficacy of the proposed TSA with CVT, demonstrating a notable 2.1-fold TR variation across load conditions ranging from 0.1 kg to 1.5 kg. Notably, the hyperelastic rods utilized in our CVT are exceptionally lightweight and cost-efficient, with dimensions of just 1 mm $\times$ 1 mm $\times$ 32 mm, weighing 0.22 g, and costing only USD 0.041. An application in an anthropomorphic robot finger has further underscored the adaptability and practicality of our TSA with CVT, revealing a significant 6.2-fold TR change between unloaded and loaded states.

This research represents a significant advancement in the field of robotic transmission systems, offering unparalleled benefits in terms of weight, cost-effectiveness, simplicity, compliance, and continuous TR adjustability. The proposed TSA with CVT holds great promise for a wide range of robotic applications, from small-scale robots to complex anthropomorphic robotic hands. Future work will focus on refining theoretical models to account for nonlinear material characteristics and the over-twist phase that occurs beyond 7.5 turns. Additionally, we aim to expand the application of our TSA with CVT to a fully anthropomorphic robotic hand.

Author Contributions: Conceptualization, C.X. and S.D.; methodology, C.X.; software, T.L.; validation, C.X. and Y.W.; formal analysis, T.L.; investigation, C.X.; resources, X.W.; data curation, X.W.; writing—original draft preparation, C.X.; writing—review and editing, C.X. and X.W.; visualization, C.X.; supervision, C.X.; project administration, S.D.; funding acquisition, Y.W. All authors have read and agreed to the published version of the manuscript.

Funding: This work is partially supported by the Anhui Provincial Key Research and Development Program No. 2022f04020008 and National Natural Science Foundation of China No. 62301522.

Data Availability Statement: Data are contained within the article.

Conflicts of Interest: The authors declare no conflicts of interest.

References

1. Yamada, H. A Radial Crank-type continuously variable transmission driven by two ball screws. In Proceedings of the 2012 IEEE International Conference on Robotics and Automation, Saint Paul, MN, USA, 14–18 May 2012; pp. 1982–1987.
2. Chen, X.; Hang, P.; Wang, W.; Li, Y. Design and analysis of a novel wheel type continuously variable transmission. *Mech. Mach. Theory* **2017**, *107*, 13–26. [CrossRef]
3. Srivastava, N.; Haque, I. A review on belt and chain continuously variable transmissions (CVT): Dynamics and control. *Mech. Mach. Theory* **2009**, *44*, 19–41. [CrossRef]
4. Verbelen, F.; Derammelaere, S.; Sergeant, P.; Stockman, K. A comparison of the full and half toroidal continuously variable transmissions in terms of dynamics of ratio variation and efficiency. *Mech. Mach. Theory* **2018**, *121*, 299–316. [CrossRef]
5. Capehart, T.; Moore, C.A., Jr. Variable Stiffness Mechanisms Using Spherical Continuously Variable Transmissions. In Proceedings of the ASME International Mechanical Engineering Congress and Exposition, American Society of Mechanical Engineers, Houston, TX, USA, 13–19 November 2015; Volume 57397, p. V04AT04A021.
6. Zhang, J.; Sheng, J.; O'Neill, C.T.; Walsh, C.J.; Wood, R.J.; Ryu, J.H.; Desai, J.P.; Yip, M.C. Robotic artificial muscles: Current progress and future perspectives. *IEEE Trans. Robot.* **2019**, *35*, 761–781. [CrossRef]

7. Bombara, D.; Coulter, R.; Konda, R.; Zhang, J. A twisted string actuator-driven soft robotic manipulator. *IFAC PapersOnLine* **2021**, *54*, 141–146. [CrossRef]
8. Shin, Y.J.; Rew, K.H.; Kim, K.S.; Kim, S. Development of anthropomorphic robot hand with dual-mode twisting actuation and electromagnetic joint locking mechanism. In Proceedings of the 2013 IEEE International Conference on Robotics and Automation, Karlsruhe, Germany, 6–10 May 2013; pp. 2759–2764.
9. Jeong, S.H.; Shin, Y.J.; Kim, K.S. Design and analysis of the active dual-mode twisting actuation mechanism. *IEEE/ASME Trans. Mechatron.* **2017**, *22*, 2790–2801. [CrossRef]
10. Lu, Z.; Wang, R.; Xiao, Y.; Liu, T.; Liu, C.; Zhao, H. A load-adaptive hoisting mechanism based on spring-loaded rope and variable radius reel. *Adv. Robot.* **2023**, *37*, 1520–1531. [CrossRef]
11. Kim, S.; Sim, J.; Park, J. Elastomeric continuously variable transmission combined with twisted string actuator. *IEEE Robot. Autom. Lett.* **2020**, *5*, 5477–5484. [CrossRef]
12. Shin, W.; Park, S.; Park, G.; Kim, J. A passively adaptable toroidal continuously variable transmission combined with twisted string actuator. In Proceedings of the 2022 International Conference on Robotics and Automation (ICRA), Philadelphia, PA, USA, 23–27 May 2022; pp. 11409–11415.
13. Singh, H.; Popov, D.; Gaponov, I.; Ryu, J.H. Twisted string-based passively variable transmission: Concept, model, and evaluation. *Mech. Mach. Theory* **2016**, *100*, 205–221. [CrossRef]
14. Jang, J.; Song, Y.U.; Ryu, J.H. Active-type continuously variable transmission system based on a twisted string actuator. *IEEE Robot. Autom. Lett.* **2022**, *7*, 2605–2612. [CrossRef]
15. Popov, D.; Gaponov, I.; Ryu, J.H. A study on twisted string actuation systems: Mathematical model and its experimental evaluation. In Proceedings of the 2012 IEEE/RSJ International Conference on Intelligent Robots and Systems, Algarve, Portugal, 7–12 October 2012; pp. 1245–1250.

actuators MDPI

Article

Optimal Realtime Toolpath Planning for Industrial Robots with Sparse Sensing

Enkhsaikhan Boldsaikhan * and Cole Birney

Industrial Systems and Manufacturing Engineering Department, Wichita State University, Wichita, KS 67260, USA; crbirney@shockers.wichita.edu
* Correspondence: enkhsaikhan.boldsaikhan@wichita.edu

Abstract: Non-contact surface processing does not involve direct contact between the tool and a worksurface. An industrial robot mostly uses preplanned toolpaths to perform non-contact surface processing. A preplanned toolpath may work well in repetitive conditions but may easily become inaccurate and unsafe if the tool needs to follow unknown worksurface variations. Many industrial processes, e.g., painting, coating, and sandblasting, typically involve worksurfaces with unknown variations. This study proposes an optimal toolpath planning method for an industrial robot equipped with end-of-arm distance sensors to automatically guide its tool motion along unknown worksurface variations. The distance sensors facilitate sparse sensing to acquire sparse data that is just enough for the quick and adequate perception of unknown worksurfaces by requiring fewer measurements and less computing. Optimization facilitates the optimality of multi-objective toolpath planning with a customizable value function, where the multiple objectives comprise adapting to unknown worksurface variations and traveling between known tool targets. To validate the proposed toolpath planning method, this study conducts a simulation experiment on a virtual robot with four end-of-arm distance sensors and a workpiece with unknown surface variations. The experimental results indicate that the proposed method is accurate and near-optimal even in the presence of sensor noises.

Keywords: industrial robotics; toolpath planning; sparse sensing; non-contact surface processing; optimization

1. Introduction

Industrial robots are emerging as major autonomous systems for handling various manufacturing tasks that are highly repetitive, difficult, and unsafe to humans [1–3]. The International Organization for Standardization (ISO) [4,5] identifies an industrial robot as an automatically controlled, reprogrammable, multipurpose, manipulative system with several reprogrammable axes, which may be either fixed in place or mobile for use in industrial automation applications. An industrial robot mostly uses preplanned toolpaths to perform non-contact surface processing. A preplanned toolpath may work well in repetitive conditions but may easily become inaccurate and unsafe if the tool needs to follow unknown worksurface variations. Many industrial processes, e.g., painting, coating, and sandblasting, typically involve worksurfaces with unknown variations.

Researchers have proposed theoretical methods for robotic toolpath planning [6–22] as well as their practical applications [6–9,13,14,16,18,23]. Navigation and the precision of toolpaths are practically important as an industrial robot may exhibit positioning errors [24,25] due to uncertain variations involved in the environment. Non-rigid robots with hyper-elastic, soft bodies experience even bigger challenges in terms of navigation, motion precision, stiffness, and compliance [26–29]. All these challenges in robotics indicate the importance of sensing, which can provide the needed perception of priorly unknown variations in environments, for better navigation and control of robots and autonomous systems.

For robotic non-contact surface processing, contactless sensing is always preferred as it allows a robot to sense surfaces from a distance. There are various contactless sensors (or non-contact sensors), such as vision cameras [30–39], laser trackers [24,40], Light Detection and Ranging (LiDAR) scanners [41,42], and so on, used in industrial applications. Vision cameras provide ample perception but require fast communication interfaces and extensive data processing. Laser sensors like range sensors, laser trackers, and LiDAR scanners can provide accurate spatial sensing but produce harmful laser beams. Ultrasonic distance sensors are simpler than machine vision cameras and safer than laser beams. They are not affected by the color or transparency of objects but have limitations in terms of acoustic disturbances and short detection ranges.

Researchers have documented various methods for the sensor-based routing and control of industrial robots [9,13,38,39,43,44], mobile robots [30,42,45,46], and aerial robots or drones [34,47,48]. However, these research outcomes mostly focus on solutions with resource-intensive sensing rather than generic solutions with less resource-intensive sensing for baseline perception. An example of resource-intensive sensing is machine vision, which can easily produce abundant data that is costly to compute in real time. In contrast, sparse sensing with simple distance sensors produces sparse data that is just enough for the quick, adequate perception of environments with less computing. In general, sparse sensing [49] is a sensing technique that leverages fewer measurements that are just enough to reconstruct full-field perception. It requires fewer hardware and software resources to provide quick, baseline perception.

There is a noticeable research gap in realtime toolpath planning with sparse sensing in industrial robotics. To bridge this gap, this study proposes an optimal path planning method with sparse sensing for an industrial robot equipped with end-of-arm distance sensors to automatically guide its tool motion along unknown worksurface variations. The end-of-arm distance sensors facilitate sparse sensing for quick, baseline perception. Optimization facilitates the optimality of multi-objective path planning with a customizable value function, where the multiple objectives of path planning comprise adapting to unknown worksurface variations and traveling between known toolpath targets.

To validate the proposed toolpath planning method with sparse sensing, this study conducted a simulation experiment on a virtual robot equipped with end-of-arm distance sensors and a workpiece with unknown surface variations. The simulation experiment systematically demonstrated the proposed method with and without sensor noises.

2. Materials and Methods

2.1. Problem Formulation

An end-effector is a tool that goes on the end of a robot arm. A tool pose defines the tool position and orientation in the reference frame. A robot manipulates its tool pose within the operational or task space of the robot. A robot tool must have a tool reference frame or simply a tool frame, and the tool frame origin is the tool center point or tool control point (TCP), which is the location where the robotic process takes place. A target is a tool pose, and a toolpath or path is a continuous stream of tool poses.

This problem formulation involves an industrial robot with an end-of-arm tool that has four distance sensors and a tool reference frame for non-contact surface processing as illustrated in Figure 1. Adaptive toolpath planning needs to guide the robot TCP along unknown worksurface variations by maintaining relative tool orientation and alignment, while the tool travels between known start and end tool poses. The end tool pose or target also serves as the frame of reference. A toolpath must adapt to unknown worksurface variations along the *XZ* plane of reference frame, which means an adapted toolpath must be constrained to the *XZ* plane of reference frame. The relative tool orientation depends on a manufacturing process, i.e., a welding process requires a certain attack angle between the welding tool and the toolpath. In this problem formulation, the robot tool axis, which is the z_T axis, needs to be orthogonal to the sensed worksurface for clarity and simplicity.

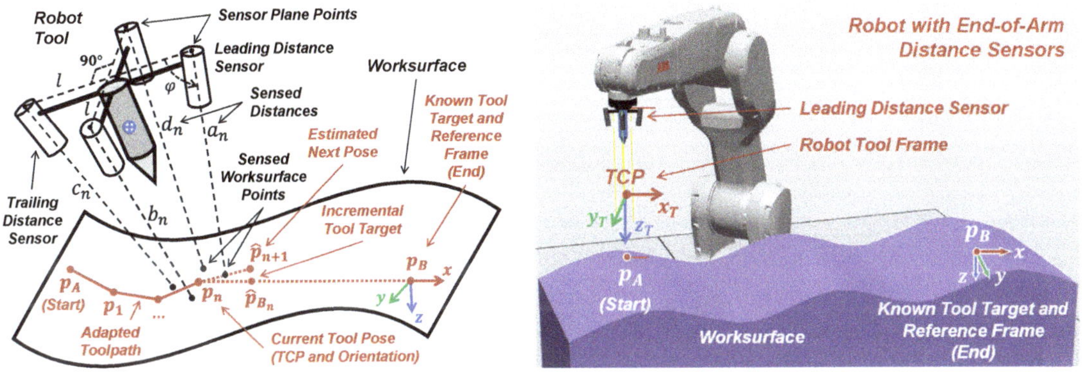

(**a**) (**b**)

Figure 1. This figure presents adaptive toolpath planning with sparse sensing for an industrial robot to perform non-contact surface processing: (**a**) Adaptive toolpath planning needs to guide the robot TCP along unknown worksurface variations by maintaining relative tool orientation and alignment, while the tool travels between known start and end tool poses. The end pose or target also serves as the frame of reference. A toolpath must adapt to unknown worksurface variations along the *XZ* plane of reference frame. l is the diagonal length of the sensor square arrangement, and φ is the sensor inclination angle. (**b**) A simulation setup involves an industrial robot equipped with end-of-arm distance sensors and a workpiece with unknown surface variations.

The four end-of-arm distance sensors have a square arrangement with a diagonal length of l on the sensor plane as shown in Figure 1. This square arrangement is centered at the tool axis, which is the z_T axis of the tool frame, and the tool axis is orthogonal to the sensor plane. Moreover, the two diagonals of the square arrangement are parallel to the tool $\boldsymbol{x_T}$ axis and the tool $\boldsymbol{y_T}$ axis, respectively. All the sensors have the same concentric inclination angle φ, which is assigned to 90 degrees to make each sensor axis orthogonal to the sensor plane for simplicity. Four distance sensors sense the distances—a_n, b_n, c_n, and d_n—between the sensor plane points and the sensed worksurface points, respectively, in each timestep $n \in \mathbb{N}$. The sensor plane is the interface plane between the robot end and the robot tool in Figure 1. The four distance sensors are associated with four customizable reference distances—a_{ref}, b_{ref}, c_{ref}, and d_{ref}—that signify a desired tool pose relative to a sensed worksurface. These four reference distances define a tool reference plane. In this formulation, let the *XY* plane of the robot tool frame be the tool reference plane for clarity and simplicity, which implies that all the reference distances are equal to the distance between the sensor plane and the *XY* plane of the robot tool frame. Note that the sensor plane is parallel to the *XY* plane of the robot tool frame.

The origin of the robot tool frame is tool center point (TCP), where a non-contact process takes place. The robot tool needs to follow sensed worksurface variations along the *XZ* plane of the reference frame ($\boldsymbol{p_B}$) by maintaining alignment between the tool reference plane (the XY plane of the robot tool frame) and the sensed worksurface, while traveling between known start and end tool poses.

Toolpath $\boldsymbol{P}(\boldsymbol{p_A}, \boldsymbol{p_B})$ in (1) is a series of adapted toolpath poses in between given or known start and end tool poses, $\boldsymbol{p_A} \in \mathbb{R}^6$ and $\boldsymbol{p_B} \in \mathbb{R}^6$, respectively.

$$\boldsymbol{P}(\boldsymbol{p_A}, \boldsymbol{p_B}) = \left\{ \boldsymbol{p_n} \in \mathbb{R}^6 \,\middle|\, \boldsymbol{p}_0 = \boldsymbol{p_A};\ \boldsymbol{p_N} = \boldsymbol{p_B};\ N \in \mathbb{N};\ n = 0, 1, \ldots, N \right\} \tag{1}$$

In (1), n and N are a timestep index and the finite number of timesteps, respectively. In each timestep n, realtime toolpath planning produces an adapted tool pose $\boldsymbol{p_n} \in \mathbb{R}^6$. Each tool pose or target $\boldsymbol{p_n}$ consists of the x, y, z coordinates of the TCP and the θ_x, θ_y, θ_z Euler angles with respect to the reference frame in Figure 1. In every timestep, the adaptive toolpath planner must estimate the next pose $\hat{\boldsymbol{p}}_{n+1}$ using the current tool pose $\boldsymbol{p_n}$ and a tool pose correction $\Delta\boldsymbol{p_n}$ as formulated in (2). The tool pose correction $\Delta\boldsymbol{p_n}$ in (2) consists of a correction $\Delta\boldsymbol{p}_{n|s}$ relative to the sensed worksurface and a correction $\Delta\boldsymbol{p}_{n|B_n}$ relative to an incremental tool target $\hat{\boldsymbol{p}}_{B_n}$. Let $\left\|\Delta\boldsymbol{p}_{n|B_n}\right\| = \Delta p$ be a known constant as a uniform increment in the direction of the final tool target $\boldsymbol{p_B}$. Figure 2 shows the geometry of an incremental tool target $\hat{\boldsymbol{p}}_{B_n} = \boldsymbol{p_n} + \Delta\boldsymbol{p}_{n|B_n}$. Moreover, Figure 2 illustrates the vector operations in (2) for estimating the next pose $\hat{\boldsymbol{p}}_{n+1}$.

$$\hat{\boldsymbol{p}}_{n+1} = \boldsymbol{p_n} + \Delta\boldsymbol{p_n} = \boldsymbol{p_n} + \Delta\boldsymbol{p}_{n|s} + \Delta\boldsymbol{p}_{n|B_n} \tag{2}$$

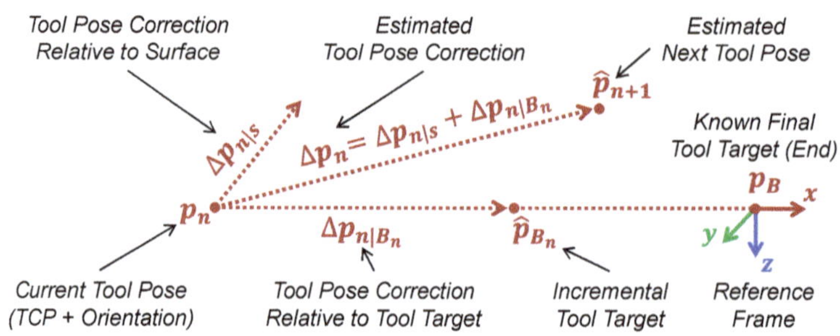

Figure 2. This figure illustrates the vector operations in (2). Note: $\left\|\Delta\boldsymbol{p}_{n|B_n}\right\| = \Delta p = constant$.

Finally, the problem formulation boils down to finding the tool pose correction $\Delta\boldsymbol{p_n}$ and its components, $\Delta\boldsymbol{p}_{n|s}$ and $\Delta\boldsymbol{p}_{n|B_n}$, to estimate the next tool pose $\hat{\boldsymbol{p}}_{n+1}$ based on the current tool pose $\boldsymbol{p_n}$ in every timestep. This study proposes a numerical solution to this problem by using realtime error tracking with linear quadratic regulation as outlined in Section 2.2.

2.2. Realtime Error Tracking with Linear Quadratic Regulation

The end-of-arm distance sensors in Figure 1 provide four distance sensor readings—a_n, b_n, c_n, and d_n—in every timestep n. The four distance sensors are associated with four customizable reference distances—a_{ref}, b_{ref}, c_{ref}, and d_{ref}—that signify a desired tool pose relative to a sensed worksurface. These four reference distances define a tool reference plane. In this study, the XY plane of the robot tool frame is the tool reference plane for clarity and simplicity, which implies that all the four reference distances are equal to the distance between the sensor plane and the XY plane of the robot tool frame. The deviations of four distance readings from their respective reference distances are $a_{n|ref}$, $b_{n|ref}$, $c_{n|ref}$, and $d_{n|ref}$ as derived in (3).

$$\begin{bmatrix} a_{n|ref} & b_{n|ref} & c_{n|ref} & d_{n|ref} \end{bmatrix}^T = \begin{bmatrix} a_n & b_n & c_n & d_n \end{bmatrix}^T - \begin{bmatrix} a_{ref} & b_{ref} & c_{ref} & d_{ref} \end{bmatrix}^T \quad (3)$$

Besides tracking worksurface variations, the robot tool must reach the known final tool target $\boldsymbol{p_B}$ in Figure 1. The final tool target $\boldsymbol{p_B}$ also serves as the frame of reference or reference frame for convenience. The TCP coordinates of the current tool pose $\boldsymbol{p_n}$ are x_n, y_n, and z_n. Since the final tool target $\boldsymbol{p_B}$ is located at a remote distance, an incremental tool target $\hat{\boldsymbol{p}}_{\boldsymbol{B_n}}$ is used at a uniform incremental distance from the current pose $\boldsymbol{p_n}$ in the direction of the final tool target $\boldsymbol{p_n}$ in each timestep as shown in Figure 2. The TCP coordinates of the current tool pose $\boldsymbol{p_n}$ that is relative to an incremental tool target $\hat{\boldsymbol{p}}_{\boldsymbol{B_n}}$ are $x_{n|B_n}$, $y_{n|B_n}$, and $z_{n|B_n}$, which are incremental tool target errors, according to (4).

$$\begin{bmatrix} x_{n|B_n} & y_{n|B_n} & z_{n|B_n} \end{bmatrix}^T = \begin{bmatrix} x_n & y_n & z_n \end{bmatrix}^T - \begin{bmatrix} x_{B_n} & y_{B_n} & z_{B_n} \end{bmatrix}^T \quad (4)$$

The sensed deviations $a_{n|ref}$, $b_{n|ref}$, $c_{n|ref}$, and $d_{n|ref}$ as well as the incremental tool target errors $x_{n|B_n}$, $y_{n|B_n}$, and $z_{n|B_n}$ constitute an error vector $\boldsymbol{e_n} \in \mathbb{R}^7$ in (5).

$$\begin{aligned} \boldsymbol{e_n} &= \begin{bmatrix} e_n\langle 1\rangle & e_n\langle 2\rangle & e_n\langle 3\rangle & e_n\langle 4\rangle & e_n\langle 5\rangle & e_n\langle 6\rangle & e_n\langle 7\rangle \end{bmatrix}^T = \\ &= \begin{bmatrix} a_{n|ref} & b_{n|ref} & c_{n|ref} & d_{n|ref} & x_{n|B_n} & y_{n|B_n} & z_{n|B_n} \end{bmatrix}^T \end{aligned} \quad (5)$$

Figure 3 demonstrates a tool θ_y angle as a pitch angle between the tool XY plane and a worksurface using the sensed deviations, $a_{n|ref}$ and $c_{n|ref}$. Likewise, a tool θ_x angle can also be identified as a roll angle between the tool *XY* plane and a worksurface using the sensed deviations, $b_{n|ref}$ and $d_{n|ref}$. Adaptive toolpath planning aims to maintain orientation alignment between the tool *XY* plane and a sensed worksurface using these two tool angles for tool pose corrections. The sensor inclination angle φ in Figure 1 is 90°. The sensitivity factor η is a heuristic scalar between 0 and 1, where 0 means no sensitivity, and 1 means full sensitivity. It can be used to lessen the amplifying effects of the two sensed deviations on the pivot point, which is the TCP. It can also be used to lessen the effects of sensor noises. The η sensitivity factor is 0.5 in this study.

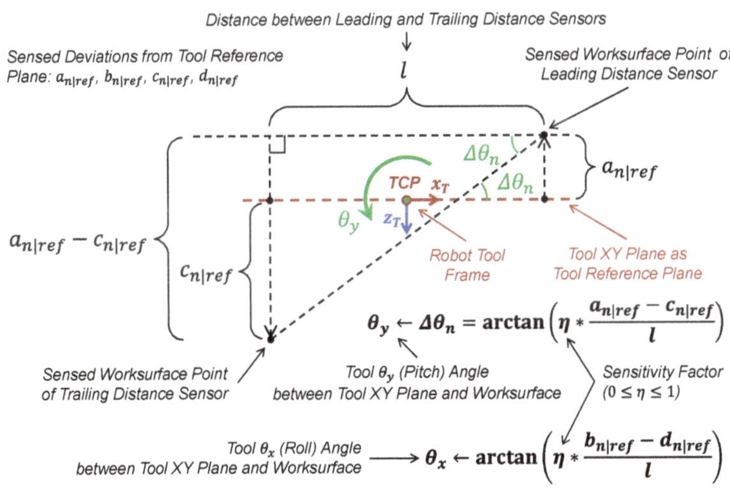

Figure 3. This figure demonstrates a tool θ_y angle as a pitch angle between the tool *XY* plane and a sensed worksurface. Likewise, a tool θ_x angle can also be identified as a roll angle between the tool *XY* plane and a sensed worksurface. The sensor inclination angle φ in Figure 1 is 90°.

The goal of adaptive toolpath planning is to minimize the error vector $\boldsymbol{e_n}$ with less effort. Therefore, the proposed approach involves a first-order error tracking model presented in (6).

$$\boldsymbol{e}_{n+1} = \boldsymbol{e}_n + \boldsymbol{u}_n \quad (6)$$

In (6), a control input vector $\boldsymbol{u_n} \in \mathbb{R}^7$ aims to minimize the error vector $\boldsymbol{e_n}$ in each timestep n. It is derived from the error vector $\boldsymbol{e_n}$ according to (7), (8), and (9) using a control gain matrix $\boldsymbol{K} \in \mathbb{R}^{7\times 7}$ that is a predefined diagonal matrix of the scalar gains K_s and K_B in (8) and (9). The gain K_s is used to regulate the deviations of the four sensor readings from their respective reference distances, and the gain K_B is used to guide the TCP motion in the direction of the known final tool target $\boldsymbol{p_B}$.

$$\boldsymbol{u}_n = -\boldsymbol{K}\boldsymbol{e}_n \quad (7)$$

$$\boldsymbol{K} = \boldsymbol{diag}(K_s, K_s, K_s, K_s, K_B, K_B, K_B) \in \mathbb{R}^{7\times 7} \quad (8)$$

$$\begin{aligned}\boldsymbol{u}_n &= \begin{bmatrix} u_n\langle 1\rangle & u_n\langle 2\rangle & u_n\langle 3\rangle & u_n\langle 4\rangle & u_n\langle 5\rangle & u_n\langle 6\rangle & u_n\langle 7\rangle \end{bmatrix}^T = \\ &= -\begin{bmatrix} K_s a_{n\,|\,ref} & K_s b_{n\,|\,ref} & K_s c_{n\,|\,ref} & K_s d_{n\,|\,ref} & K_B x_{n\,|\,B_n} & K_B y_{n\,|\,B_n} & K_B z_{n\,|\,B_n} \end{bmatrix}^T\end{aligned} \quad (9)$$

The tool pose correction $\Delta\boldsymbol{p}_n$ in (2) consists of a correction $\Delta\boldsymbol{p}_{n|s}$ relative to the sensed surface and a correction $\Delta\boldsymbol{p}_{n|B_n}$ relative to an incremental tool target $\hat{\boldsymbol{p}}_{\boldsymbol{B}_n}$. Hence, the formulas in (10) and (11), respectively, derive $\Delta\boldsymbol{p}_{n|s}$ and $\Delta\boldsymbol{p}_{n|B_n}$ from $\boldsymbol{u_n}$ in every timestep n. In (10), $\Delta p_{n|s}\langle z\rangle$ is the average of the tool–worksurface distance corrections, $u_n\langle 1\rangle$, $u_n\langle 2\rangle$, $u_n\langle 3\rangle$, and $u_n\langle 4\rangle$ to align the toolpath TCP with a sensed worksurface. The tool orientation corrections, $\Delta p_{n|s}\langle\theta_y\rangle$ and $\Delta p_{n|s}\langle\theta_x\rangle$, are derived from $\boldsymbol{u_n}$ by using the approach demonstrated in Figure 3, where the sensed deviations, $a_{n|ref}$, $b_{n|ref}$, $c_{n|ref}$, and $d_{n|ref}$, are replaced by the tool–worksurface distance corrections, $u_n\langle 1\rangle$, $u_n\langle 2\rangle$, $u_n\langle 3\rangle$, and $u_n\langle 4\rangle$, respectively, in (10) to align the tool reference plane with a sensed worksurface. The η sensitivity factor in (10) is 0.5 in this study. In (11), the TCP corrections, $u_n\langle 5\rangle$, $u_n\langle 6\rangle$, and $u_n\langle 7\rangle$, are applied to the tool pose correction $\Delta\boldsymbol{p}_{n|B_n}$ to reach the incremental tool target $\hat{\boldsymbol{p}}_{\boldsymbol{B}_n}$. Figure 4 provides an overall schematic of realtime error tracking with linear quadratic regulation for adaptive toolpath planning with sparse sensing, where $\dot{p}_{ref}$ is a reference tool speed that encompasses a predefined TCP speed and a predefined tool angular speed.

$$\Delta\boldsymbol{p}_{n|s} = \begin{bmatrix} \Delta p_{n|s}\langle x\rangle \\ \Delta p_{n|s}\langle y\rangle \\ \Delta p_{n|s}\langle z\rangle \\ \Delta p_{n|s}\langle \theta_x\rangle \\ \Delta p_{n|s}\langle \theta_y\rangle \\ \Delta p_{n|s}\langle \theta_z\rangle \end{bmatrix} = \begin{bmatrix} 0 \\ 0 \\ \frac{u_n\langle 1\rangle + u_n\langle 2\rangle + u_n\langle 3\rangle + u_n\langle 4\rangle}{4} \\ \tan^{-1}\left(\eta * \frac{u_n\langle 1\rangle - u_n\langle 3\rangle}{l}\right) \\ \tan^{-1}\left(\eta * \frac{u_n\langle 2\rangle - u_n\langle 4\rangle}{l}\right) \\ 0 \end{bmatrix} \quad (10)$$

$$\begin{aligned}\Delta\boldsymbol{p}_{n|B_n} &= \begin{bmatrix} \Delta p_{n\,|\,B_n}\langle x\rangle & \Delta p_{n\,|\,B_n}\langle y\rangle & \Delta p_{n\,|\,B_n}\langle z\rangle & \Delta p_{n\,|\,B_n}\langle \theta_x\rangle & \Delta p_{n\,|\,B_n}\langle \theta_y\rangle & \Delta p_{n\,|\,B_n}\langle \theta_z\rangle \end{bmatrix}^T \\ &= \begin{bmatrix} u_n\langle 5\rangle & u_n\langle 6\rangle & u_n\langle 7\rangle & 0 & 0 & 0 \end{bmatrix}^T\end{aligned} \quad (11)$$

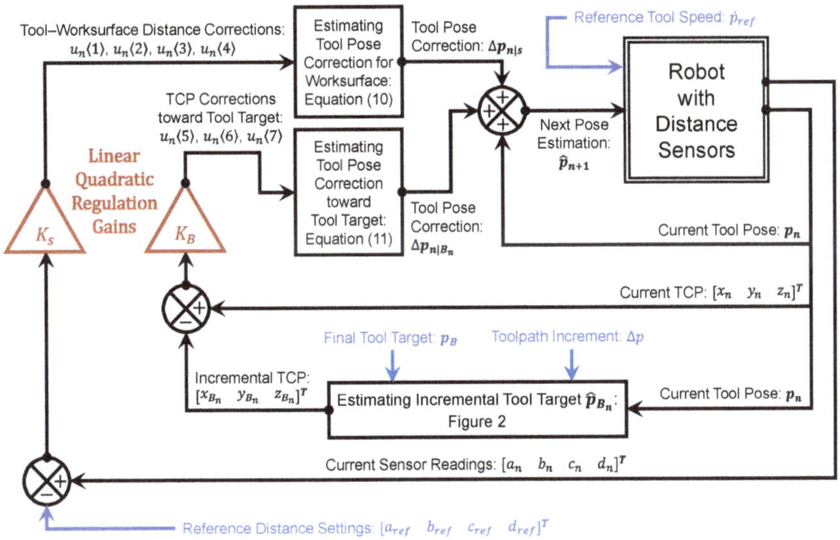

Figure 4. This figure shows realtime error tracking with linear quadratic regulation for realtime toolpath planning with sparse sensing. Note that $\dot{p}_{ref}$ is a reference tool speed. All reference inputs are in blue. The gains, K_s and K_B, are obtained via linear quadratic regulation.

The proposed realtime toolpath planning then employs linear quadratic regulation (LQR) to compute the gain matrix $\boldsymbol{K}$ in (8) using a customizable value function $\phi(\boldsymbol{e}_n, \pi)$ in (12), where π is a control sequence in (13) that steers the error tracking model in (6) from its current state $\boldsymbol{e}_{0|n}$ (or $\boldsymbol{e}_n$) to a final state $\boldsymbol{e}_{\infty|n}$ on an infinite time horizon.

$$\phi(\boldsymbol{e}_n, \pi) = \sum_{t=0}^{\infty} \left(\boldsymbol{e}_{t|n}^T \boldsymbol{Q} \boldsymbol{e}_{t|n} + \boldsymbol{u}_{t|n}^T \boldsymbol{R} \boldsymbol{u}_{t|n} \right); \ \boldsymbol{e}_{0|n} = \boldsymbol{e}_n, \boldsymbol{Q} = \boldsymbol{Q}^T \succcurlyeq 0, \ \boldsymbol{R} = \boldsymbol{R}^T \succ 0 \tag{12}$$

$$\pi = \left\{ \boldsymbol{u}_{t|n} \in \mathbb{R}^6 \ \middle| \ t = 0, 1, \ldots, \infty \right\} \tag{13}$$

In (12) and (13), $t \in \mathbb{N}$ is a timestep index on an infinite time horizon. $\boldsymbol{Q} \in \mathbb{R}^{7\times7}$ and $\boldsymbol{R} \in \mathbb{R}^{7\times7}$ are customizable state and control weight matrices, respectively. The goal of the optimization problem in (14), which is subject to (6) and (7), is to find an optimal control sequence π^* to steer the error tracking model in (6) from its current state $\boldsymbol{e}_{0|n}$ (or $\boldsymbol{e}_n$) to an optimal state $\boldsymbol{e}^*_{\infty|n}$ that converges to zero on an infinite time horizon.

$$\begin{cases} \phi(\boldsymbol{e}_n) = \min\limits_{\pi} \phi(\boldsymbol{e}_n, \pi) \\ \pi^* = \operatorname*{argmin}\limits_{\pi} \phi(\boldsymbol{e}_n, \pi) \end{cases} \tag{14}$$

To solve the optimization problem in (14) that is subject to (6) and (7), a cost-to-go matrix $\boldsymbol{P} \in \mathbb{R}^{7\times7}$ can be identified using the discrete algebraic Riccati equation (DARE) in (15) on an infinite time horizon.

$$\boldsymbol{P} = \boldsymbol{Q} + \boldsymbol{P} - \boldsymbol{P}(\boldsymbol{R} + \boldsymbol{P})^{-1}\boldsymbol{P}; \boldsymbol{P} = \boldsymbol{P}^T \succcurlyeq 0 \tag{15}$$

After solving (15) for $\boldsymbol{P}$, a time-invariant gain matrix of $\boldsymbol{K}$ can be found according to formula in (16) for the realtime error tracking model in (6) and its control input in (7).

$$\boldsymbol{K} = (\boldsymbol{R} + \boldsymbol{P})^{-1}\boldsymbol{P} \tag{16}$$

Many researchers including Al Tamimi et al. [50], Rawlings et al. [51], and Scokaert et al. [52] documented distinct variants of optimal control and linear quadratic regulation with in-depth mathematical derivations and proofs on stability and controllability. Their research papers can provide more theoretical details for further reading.

2.3. Experimental Settings

This study involves a simulation experiment on an industrial robot with end-of-arm distance sensors and a workpiece with unknown surface variations to demonstrate the proposed toolpath planning with sparse sensing using RobotStudio® 2024, which is an industrial robot programming and simulation program [53]. RobotStudio® 2024 has digital engineering capabilities of robot modeling, programming, and simulation with virtual sensors and devices. The simulation experiment employed a virtual ABB IRB-1200 robot with end-of-arm distance sensors and a workpiece with unknown surface variations as modeled in RobotStudio® 2024. Figure 1b portrays the setup of the simulation experiment. The robot tool frame with the tool center point (TCP) was arranged at 200 mm from the sensor plane along the tool z_T axis. The diagonal length l of the sensor square arrangement was 40 mm. The sensor inclination angle φ was 90 degrees for all four sensors. The robot was tasked to start from a known start pose and then reach a known final tool target by aligning the TCP with unknown worksurface variations and maintaining tool orthogonality relative to the sensed surface. It is assumed that the TCPs of both the start and the end poses are predefined worksurface points. The end pose served as the frame of reference. The $\dot{p}_{ref}$ reference tool speed of simulation comprised 100 mm/s as a TCP speed and 1000 deg./s as a tool angular speed. The Δp uniform increment of TCP was 10 mm. The timestep of the simulation was 0.5 s. The data sampling rate was 125 Hz.

This study used sensor noises to test the robustness of the proposed approach. Sensor noises can have various sources, such as vibrations, environmental disturbances, sensor calibration errors, etc. To apply noise to the sensor perception, each sensor reading was subjected to a random scalar from a 4 mm range between −2 mm and +2 mm.

To demonstrate customizable optimal toolpath planning, three optimal regulation cases were considered via customized weight matrices in Table 1, where $Q \in \mathbb{R}^{7\times7}$ and $R \in \mathbb{R}^{7\times7}$ are error state and control weight matrices, respectively. These weight matrices are used for solving the optimization problem in (14), which is subject to the error tracking model in (6) with the control input in (7). The customizable weights are based on a simple scoring range between 1 and 10, which serves as a handle on intuitively praising or penalizing certain error states and control actions for optimal toolpath planning.

Table 1. Experimental cases with customized weight matrices.

Cases	Error State Weight Matrix $Q \in \mathbb{R}^{7\times7}$	Control Weight Matrix $R \in \mathbb{R}^{7\times7}$
Case 1: Reaching the known toolpath target is priority.	$\boldsymbol{diag}(1,1,1,1,10,10,10)$	$\boldsymbol{diag}(10,10,10,10,1,1,1)$
Case 2: Balancing the Case 1 and Case 3 goals is priority.	$\boldsymbol{diag}(10,10,10,10,10,10,10)$	$\boldsymbol{diag}(10,10,10,10,10,10,10)$
Case 3: Following sensed surface variations is priority.	$\boldsymbol{diag}(10,10,10,10,1,1,1)$	$\boldsymbol{diag}(1,1,1,1,10,10,10)$

In Case 1, the weights for sensed deviations, which are the first four entries in the diagonal of the error state weight matrix Q, are 1s, while the remaining weights for toolpath target errors are 10s. This means toolpath target errors are more costly to endure than sensed deviations. Also, the weights for sensed deviation control, which are the first four entries in the diagonal of the control weight matrix R, are 10s, while the remaining weights for toolpath target control are 1s. It means sensed deviation control is more costly to execute than toolpath target control. Overall, the Case 1 weights imply that reaching the known toolpath target is the top priority.

In Case 2, the customized weights are all equal to 10s, which means reaching the known toolpath target and following sensed surface variations are equally important.

In Case 3, the weights for sensed deviations, which are the first four entries in the diagonal of the error state weight matrix Q, are 10s, while the remaining weights for toolpath target errors are 1s. It means sensed deviations are more costly to endure than toolpath target errors. Also, the weights for sensed deviation control, which are the first four entries in the diagonal of the control weight matrix $\boldsymbol{R}$, are 1s, while the remaining weights for toolpath target control are 10s. It means toolpath target control is more costly to execute than sensed deviation control. Overall, the Case 3 weights imply that surface tracking is the top priority.

A time-invariant gain matrix of $\boldsymbol{K}$ was computed according to (16) for each case using the weights in Table 1 and the linear quadratic optimization in (14) that is subject to (6) and (7). Table 2 presents the computed LQR gains. These LQR gains were used to systematically demonstrate optimal toolpath planning with and without sensor noises.

Table 2. Time-invariant gains obtained via linear quadratic regulation for the three cases in Table 1.

Cases	Gains for Surface Error Tracking: K_s	Gains for Toolpath Target Error Tracking: K_B
Case 1: Reaching the known toolpath target is priority.	0.27	0.92
Case 2: Balancing the Case 1 and Case 3 goals is priority.	0.62	0.62
Case 3: Following sensed surface variations is priority.	0.92	0.27

3. Results and Discussion

The simulation experiment involved three optimal realtime toolpath planning cases as specified in Tables 1 and 2 with and without sensor noises. To apply artificial noise to sensor perception, each sensor reading was subjected to a random number from a 4 mm range between −2 mm and +2 mm. In each case, an industrial robot with end-of-arm distance sensors was tasked to travel between known start and end tool poses by maintaining the tool orthogonality and the TCP alignment with the sensed worksurface as well as the ZX plane of the reference frame. The TCPs of both the start and the end tool poses were worksurface points. The end pose (the final tool target) also served as the reference frame. The $\dot{p}_{ref}$ reference tool speed of simulation comprised 100 mm/s as the TCP speed and 1000 deg./s as the tool angular speed. The Δp uniform increment of TCP was 10 mm, and the timestep of simulation was 0.5 s. The data sampling rate was 125 Hz.

Reaching the known toolpath target is the top priority in Case 1. Worksurface tracking is the top priority in Case 3. Balancing the Case 1 and Case 3 goals is the priority in Case 2. Figure 5 presents an image sequence of the Case 3 simulation.

The cycle times of Case 1, Case 2, and Case 3 without sensor noises were about 16 s, 18 s, and 23 s, respectively. The total lengths of the Case 1, Case 2, and Case 3 toolpaths without sensor noises were about 848 mm, 852 mm, and 866 mm, respectively. The cycle times of Case 1, Case 2, and Case 3 with sensor noises were about 16 s, 19 s, and 28 s, respectively. The lengths of the Case 1, Case 2, and Case 3 toolpaths with sensor noises were about 848 mm, 855 mm, and 908 mm, respectively. Sensor noises affected Case 3 more than any other cases since sensor-based surface tracking is the top priority in Case 3.

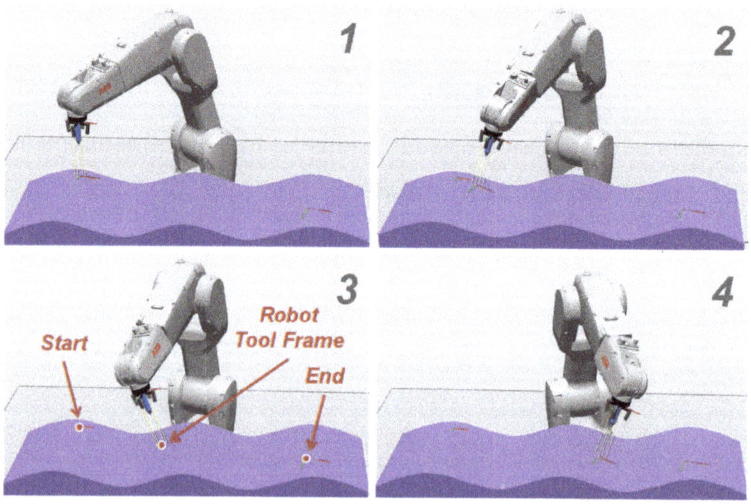

Figure 5. This figure presents an image sequence of the Case 3 simulation: (1) The robot tool is at the start pose; (2) The robot tool departs from the start pose; (3) The robot tool follows the sensed surface and advances toward the end target; (4) The robot tool keeps following the sensed surface and advancing toward the end target.

Figures 6–8 present comparisons between the toolpath orientation angles of an ideal case and the three cases with and without sensor noises. The ideal case involves ideal tool poses to achieve along an ideal toolpath defined by the intersection between the worksurface and the *XZ* plane of the reference frame. Figures 9 and 10 present comparisons between the ideal toolpath and the toolpaths of the three cases with and without sensor noises. Figure 11 presents the toolpath TCP errors of the three cases with and without sensor noises. A positive error in Figure 11 means the TCP is above the worksurface, and a negative error means the TCP is below the worksurface.

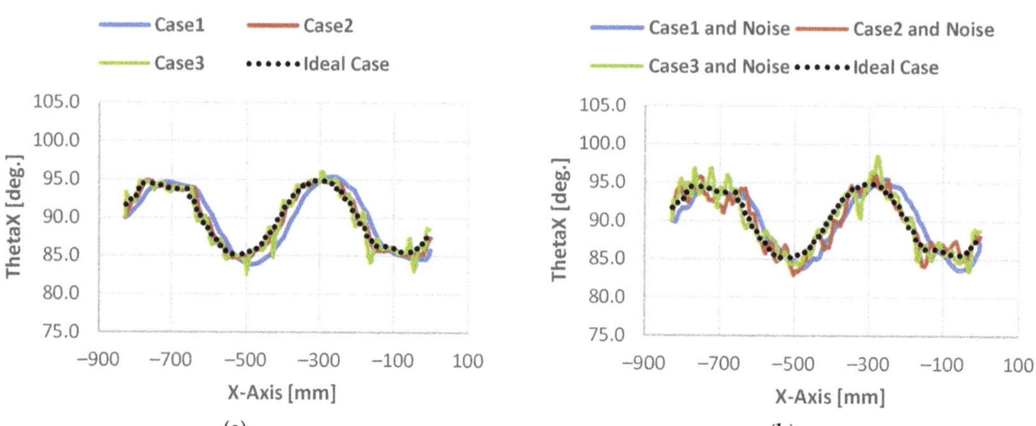

Figure 6. Toolpath θ_x angles: (**a**) no sensor noise; (**b**) added sensor noise.

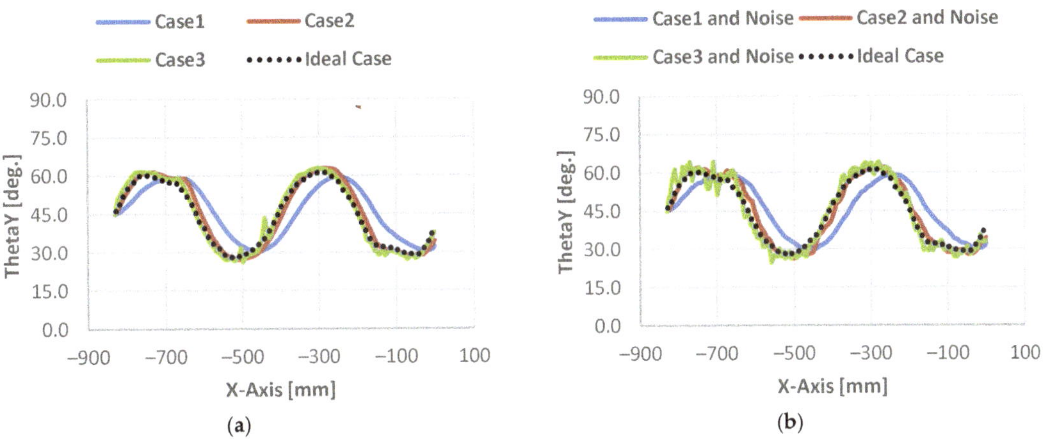

Figure 7. Toolpath θ_y angles: (**a**) no sensor noise; (**b**) added sensor noise.

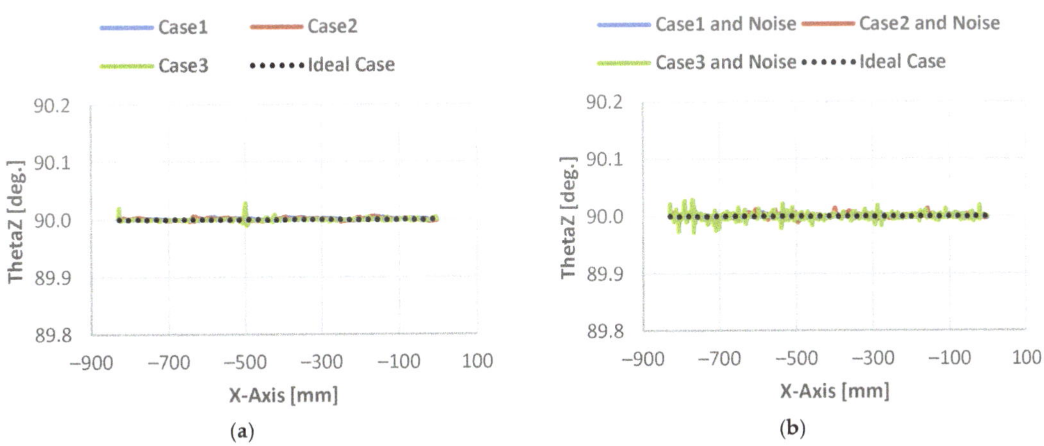

Figure 8. Toolpath θ_z angles: (**a**) no sensor noise; (**b**) added sensor noise.

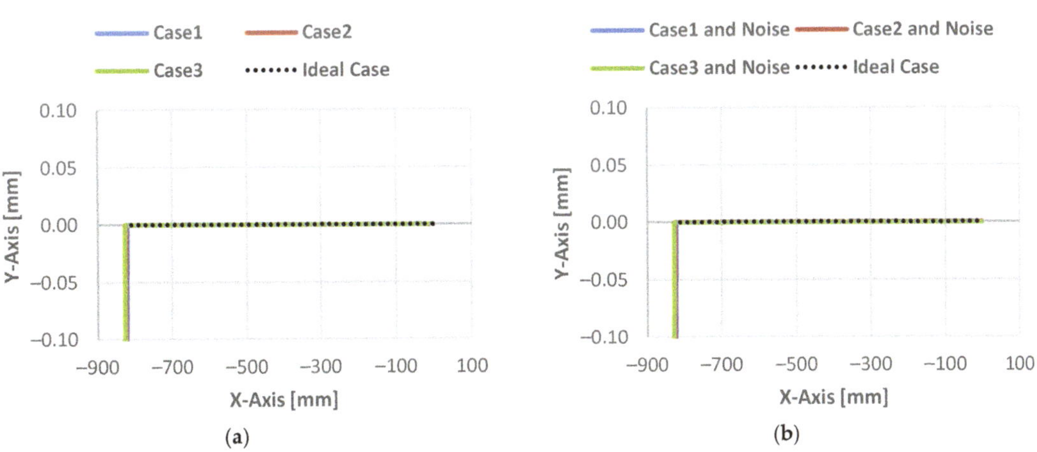

Figure 9. Top views of toolpaths: (**a**) no sensor noise; (**b**) added sensor noise.

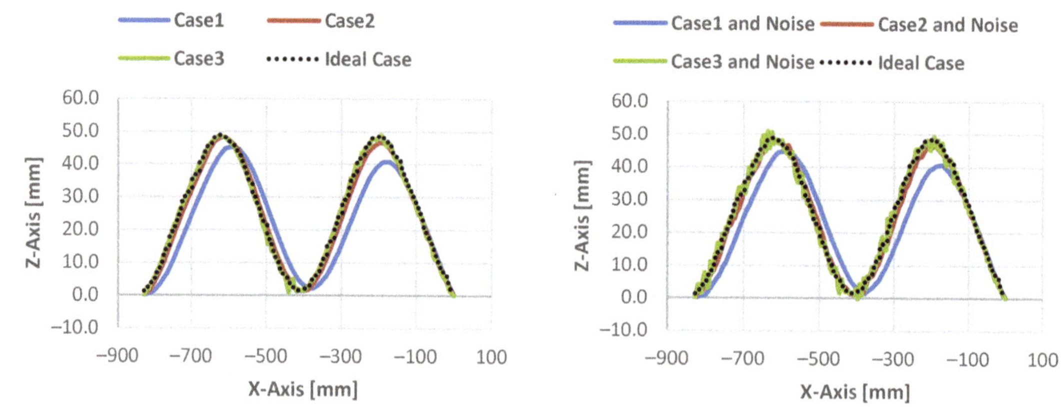

(a) (b)

Figure 10. Side views of toolpaths: (**a**) no sensor noise; (**b**) added sensor noise.

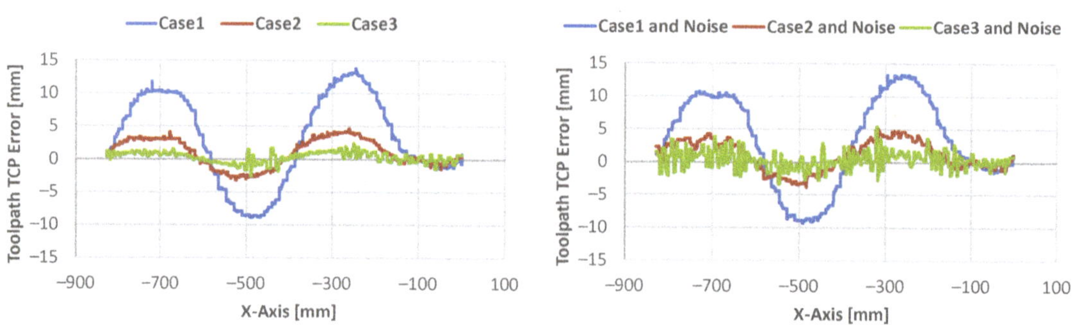

(a) (b)

Figure 11. Toolpath TCP errors along the tool z_T axis: (**a**) no sensor noise; (**b**) added sensor noise. A positive error means the TCP is above the worksurface, and a negative error means the TCP is below the worksurface.

The results in Figures 6–8 indicate that each experimental case was able to accomplish the anticipated optimal path planning goals in accordance with the customized weights specified in Table 1 and the LQR gains in Table 2. The tool θ_z angle is nicely maintained at 90 degrees in each of the three cases even in the presence of sensor noises. In comparison with the ideal case and Case 1, both Case 2 and Case 3 produced fairly accurate toolpath orientation angles but longer cycle times since tracking sensed surface variations requires more time to accomplish, especially in Case 3. In contrast, Case 1 produced less-accurate toolpath orientation angles but the shortest cycle time since reaching the known tool target is the top priority of Case 1. This urgency to reach the end target causes insufficient time to accurately track sensed surface variations. Added sensor noises barely affected the performances of Case 1 and Case 2 but noticeably affected the Case 3 performances. Since sensor-based surface tracking is the top priority in Case 3, sensor readings play essential roles in Case 3, which makes it more sensitive to sensor noises. This sensitivity can be reduced by reducing the η sensitivity factor in (10). Note that this study used a sensitivity factor of $\eta = 0.5$. The results in Figures 6–8 are summarized in Figure 12, which presents the Root Mean Square (RMS) errors [23] of the tool orientation for the three cases with and without sensor noises. Case 1 produced the largest angular RMS errors in the tool θ_x and θ_y

angles in comparison with the ideal toolpath orientation angles shown in Figures 6 and 7, while Case 3 and Case 2 produced much lower angular RMS errors.

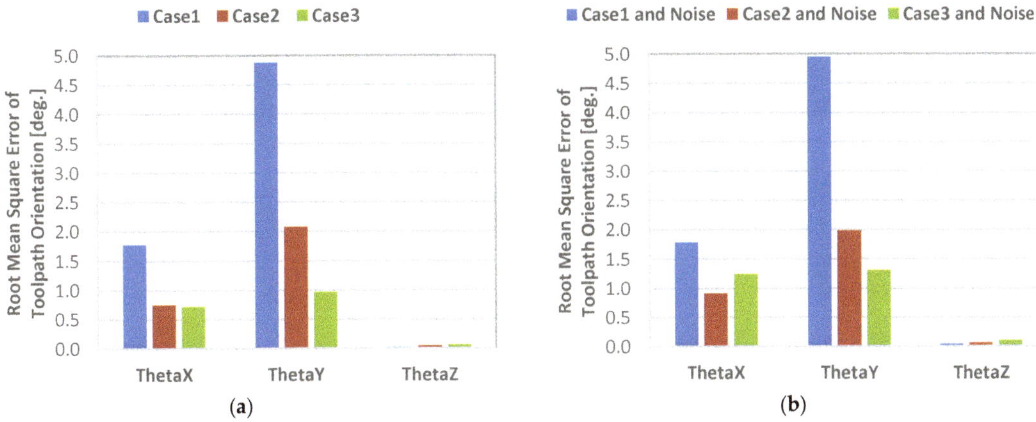

Figure 12. RMS errors of toolpath orientation errors: (**a**) no sensor noise; (**b**) added sensor noise.

According to Figures 9–11, the toolpath TCP coordinates of the three cases eventually converge at zero when the robot tool reaches the final tool target at the end of each process cycle. Note that the final tool target is also the reference frame. Figure 9 shows that the TCP alignment with the *XZ* plane of the reference frame looks flawless in each of the three cases with and without sensor noises as expected. Figure 10 presents comparisons between the ideal toolpath and the toolpaths of the three cases with and without noises. The results in Figures 9–11 are summarized in Figure 13. In terms of tracking unknown surface variations, the toolpaths of Case 3 and Case 2 are the most accurate ones with RMS errors of less than 2.5 mm even in the presence of sensor noises. The Case 3 toolpath has some fluctuations in Figures 10 and 11, especially in the presence of sensor noises. Such unwanted fluctuations can be reduced by reducing the η sensitivity factor in (10). The Case 1 toolpaths with and without sensor noises exhibit noticeable RMS errors of over 7 mm, which is expected since worksurface tracking has a lower priority in Case 1.

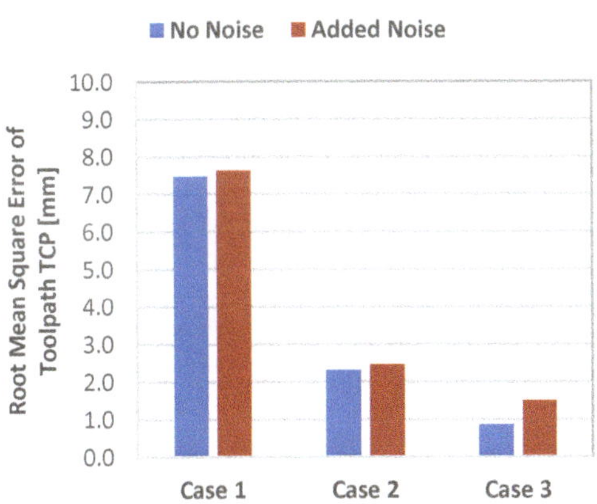

Figure 13. RMS errors of toolpath TCP for the Figure 11 data.

Figure 14 presents box-and-whisker plots of the actual toolpath speeds. Each box-and-whisker plot depicts the upper and lower extremes, the upper and lower quartiles, and the median of each dataset. Since the main emphasis of the proposed method is on enhancing the tool pose accuracy of the toolpaths, the proposed method does not regulate tool speeds by letting the robotic system control its tool speeds based on the commanded or reference tool speeds and kinematic constraints. The commanded or reference tool speeds of the simulation experiment were 100 mm/s as a linear speed of the toolpath TCP and 1000 deg/s as an angular speed of toolpath orientation. The actual toolpath speeds of the simulation experiment were lower than the reference tool speeds as the robot tool needed to slow down to apply new sensor-based changes to the tool pose in every timestep of 0.5 s. This limitation can be resolved by tool speed planning with speed interpolation as well as a specific sensor arrangement that allows the robot to sense the worksurface ahead of the tool along the toolpath. Moreover, Model Predictive Control (MPC) can be instrumental in achieving proper tool speed planning. Needless to say, both LQR and MPC are specific variants of optimal control.

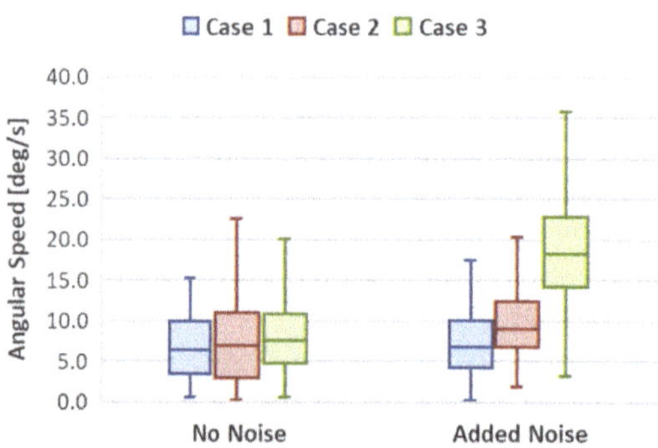

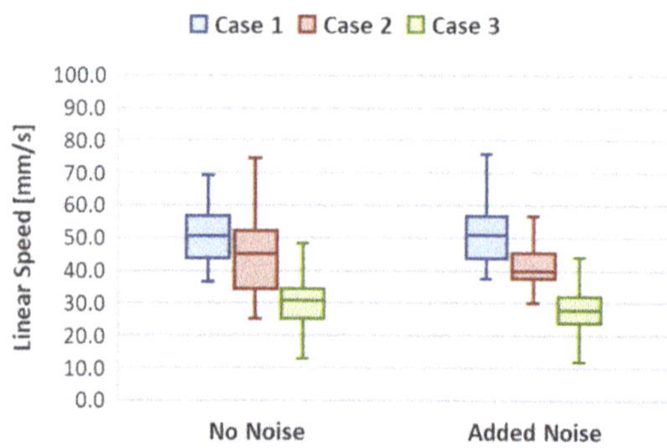

Figure 14. This figure presents box-and-whisker plots of actual toolpath speeds: (**a**) angular speeds of toolpath reorientation; (**b**) linear speeds of toolpath TCP. Each box-and-whisker plot depicts the upper and lower extremes, the upper and lower quartiles, and the median of each dataset.

The simulation experiment successfully demonstrated the predefined optimal path planning strategies with and without noises. These findings indicate that the proposed toolpath planning method with sparse sensing has the potential to automate non-contact surface processing. In this study, sparse sensing required fewer measurements and less computing than that for standard machine vision. The computational time complexity of sparse sensing was $O(n)$, which was much lower than $O(n \times m)$ of a standard image convolution algorithm typically used in machine vision, where n and m are the total number of timesteps and the total number of image pixels, respectively. The big O notation is a math notation that describes the limiting behavior of an algorithm or a function when the input argument tends toward infinity.

A limitation of the proposed path planning approach is an inability to deal with worksurface discontinuities, such as edges, holes, pockets, steep protruding surfaces, or sharp ridges. A possible solution to this problem can be smart sensor arrangements along with an intelligent algorithm that can detect surface discontinuities by examining distance sensor readings. Another limitation of the proposed toolpath planning approach is that it regulates the tool pose but not the tool speed. This limitation can be resolved by adding toolpath speed planning with speed interpolation. The future work will focus on resolving these limitations, establishing benchmark datasets for comparisons with other methods, and implementing artificial intelligence for better robot tool motion planning and control.

4. Conclusions

This study proposes an optimal toolpath planning method for an industrial robot equipped with end-of-arm distance sensors to automatically guide its tool motion along unknown worksurface variations. The end-of-arm distance sensors facilitate sparse sensing to acquire sparse data that is just enough for the quick, adequate perception of unknown worksurfaces. Optimization facilitates the optimality of multi-objective toolpath planning with a customizable value function, where the multiple objectives comprise adapting to unknown worksurface variations and traveling between known tool targets. To validate the proposed toolpath planning method with sparse sensing, this study conducted a simulation experiment on a virtual robot equipped with end-of-arm distance sensors and a workpiece with unknown surface variations. Sparse sensing required fewer measurements and less computing than that for conventional machine vision. The simulation experiment successfully demonstrated the predefined optimal strategies with and without sensor noises. These findings indicate that the proposed toolpath planning method with sparse sensing has the potential to automate non-contact surface processing. The future work will focus on adding tool speed planning, establishing benchmark datasets for comparisons with other methods, and implementing artificial intelligence for better robot tool motion planning and control.

Author Contributions: Conceptualization, E.B.; methodology, E.B.; software, E.B.; validation, E.B. and C.B.; formal analysis, E.B. and C.B.; investigation, E.B. and C.B.; resources, E.B.; data curation, E.B.; writing—original draft preparation, E.B.; writing—review and editing, E.B.; visualization, E.B.; supervision, E.B.; project administration, E.B. All authors have read and agreed to the published version of the manuscript.

Funding: This research received no external funding.

Data Availability Statement: The data presented in this study will be available upon request.

Acknowledgments: The authors would like to express gratitude to the department of Industrial Systems and Manufacturing Engineering, Wichita State University, for the administrative support.

Conflicts of Interest: The authors declare no conflicts of interest.

Abbreviations

The following abbreviations are used in this manuscript:

TCP	Tool Center Point
LQR	Linear Quadratic Regulation
RMS	Root Mean Square
ThetaX	Tool θ_x Angle
ThetaY	Tool θ_y Angle
ThetaZ	Tool θ_z Angle
MPC	Model Predictive Control

References

1. Landscheidt, S.; Kans, M.; Winroth, M.; Wester, H. The future of industrial robot business: Product or performance based? *Procedia Manuf.* **2018**, *25*, 495–502. [CrossRef]
2. Sanneman, L.; Fourie, C.; Shah, J.A. The State of Industrial Robotics: Emerging Technologies, Challenges, and Key Research Directions. *Found. Trends® Robot.* **2021**, *8*, 225–306. [CrossRef]
3. Bartoš, M.; Bulej, V.; Bohušík, M.; Stanček, J.; Ivanov, V.; Macek, P. An overview of robot applications in automotive industry. *Transp. Res. Procedia* **2021**, *55*, 837–844. [CrossRef]
4. *ISO 8373:2012*; Robots and Robotic Devices—Vocabulary. International Organization for Standardization: Geneva, Switzerland, 2012.

5. *ISO/TS 15066:2016*; Robots and Robotic Devices—Collaborative Robots. International Organization for Standardization: Geneva, Switzerland, 2016.
6. Ajoudani, A.; Tsagarakis, N.G.; Bicchi, A. Choosing Poses for Force and Stiffness Control. *IEEE Trans. Robot.* **2017**, *33*, 1483–1490. [CrossRef]
7. Dachang, Z.; Baolin, D.; Puchen, Z.; Shouyan, C. Constant Force PID Control for Robotic Manipulator Based on Fuzzy Neural Network Algorithm. *Complexity* **2020**, *2020*, 3491845. [CrossRef]
8. Shen, H.; Xie, W.F.; Tang, J.; Zhou, T. Adaptive Manipulability-Based Path Planning Strategy for Industrial Robot Manipulators. *IEEE/ASME Trans. Mechatron.* **2023**, *28*, 1742–1753. [CrossRef]
9. Wang, G.; Li, W.; Jiang, C.; Zhu, D.; Li, Z.; Xu, W.; Zhao, H.; Ding, H. Trajectory Planning and Optimization for Robotic Machining Based On Measured Point Cloud. *IEEE Trans. Robot.* **2022**, *38*, 1621–1637. [CrossRef]
10. Xie, H.; Wang, Y.; Zhang, H.; Tang, Y. Iterative Fine Point-Set Matching of a Free-Form Surface Based on the Point-to-Sphere Distance. *IEEE/ASME Trans. Mechatron.* **2022**, *27*, 5668–5678. [CrossRef]
11. Yao, H.; Laha, R.; Figueredo, L.F.C.; Haddadin, S. Enhanced Dexterity Maps (EDM): A New Map for Manipulator Capability Analysis. *IEEE Robot. Autom. Lett.* **2024**, *9*, 1628–1635. [CrossRef]
12. Angerer, A.; Hoffmann, A.; Larsen, L.; Vistein, M.; Kim, J.; Kupke, M.; Reif, W. Planning and Execution of Collision-free Multi-robot Trajectories in Industrial Applications. In Proceedings of the ISR 2016: 47st International Symposium on Robotics, Munich, Germany, 21–22 June 2016; pp. 1–7.
13. Ayyad, A.; Halwani, M.; Swart, D.; Muthusamy, R.; Almaskari, F.; Zweiri, Y. Neuromorphic vision based control for the precise positioning of robotic drilling systems. *Robot. Comput.-Integr. Manuf.* **2023**, *79*, 102419. [CrossRef]
14. Chen, C.; Peng, F.; Yan, R.; Li, Y.; Wei, D.; Fan, Z.; Tang, X.; Zhu, Z. Stiffness performance index based posture and feed orientation optimization in robotic milling process. *Robot. Comput.-Integr. Manuf.* **2019**, *55*, 29–40. [CrossRef]
15. Erdős, G.; Kovács, A.; Váncza, J. Optimized joint motion planning for redundant industrial robots. *CIRP Ann.* **2016**, *65*, 451–454. [CrossRef]
16. Heping, C.; Fuhlbrigge, T.; Xiongzi, L. Automated industrial robot path planning for spray painting process: A review. In Proceedings of the 2008 IEEE International Conference on Automation Science and Engineering, Washington, DC, USA, 23–26 August 2008; pp. 522–527.
17. Larsen, L.; Kim, J.; Kupke, M.; Schuster, A. Automatic Path Planning of Industrial Robots Comparing Sampling-based and Computational Intelligence Methods. *Procedia Manuf.* **2017**, *11*, 241–248. [CrossRef]
18. Lin, Y.; Zhao, H.; Ding, H. Posture optimization methodology of 6R industrial robots for machining using performance evaluation indexes. *Robot. Comput.-Integr. Manuf.* **2017**, *48*, 59–72. [CrossRef]
19. Lu, Y.-A.; Tang, K.; Wang, C.-Y. Collision-free and smooth joint motion planning for six-axis industrial robots by redundancy optimization. *Robot. Comput.-Integr. Manuf.* **2021**, *68*, 102091. [CrossRef]
20. Veeraraghavan, V.K. Auto-Normalization of Robot End-Effector. Master's Thesis, Industrial Systems and Manufacturing Engineering, Wichita State University, Wichita, KS, USA, 2021.
21. Reinhold, J.; Heilemann, L.L.; Lippross, S.; Meurer, T. Optimal path, orientation and trajectory planning along arbitrarily shaped surfaces for image-based robot-automated medical procedures. *Control Eng. Pract.* **2023**, *139*, 105656. [CrossRef]
22. Zanchettin, A.M.; Rocco, P. Motion planning for robotic manipulators using robust constrained control. *Control Eng. Pract.* **2017**, *59*, 127–136. [CrossRef]
23. Boldsaikhan, E.; Milhon, M.; Fukada, S.; Fujimoto, M.; Kamimuki, K. Metrology of Sheet Metal Distortion and Effects of Spot-Welding Sequences on Sheet Metal Distortion. *J. Manuf. Mater. Process.* **2023**, *7*, 109. [CrossRef]
24. Boldsaikhan, E. Measuring and Estimating Rotary Joint Axes of an Articulated Robot. *IEEE Trans. Instrum. Meas.* **2020**, *69*, 8279–8287. [CrossRef]
25. Ambaye, G.; Boldsaikhan, E.; Krishnan, K. Robot arm damage detection using vibration data and deep learning. *Neural Comput. Appl.* **2024**, *36*, 1727–1739. [CrossRef]
26. Ambaye, G.; Boldsaikhan, E.; Krishnan, K. Soft Robot Design, Manufacturing, and Operation Challenges: A Review. *J. Manuf. Mater. Process.* **2024**, *8*, 79. [CrossRef]
27. Moll, M.; Kavraki, L.E. Path planning for deformable linear objects. *IEEE Trans. Robot.* **2006**, *22*, 625–636. [CrossRef]
28. Papageorgiou, D.; Sigurðardóttir, G.Þ.; Falotico, E.; Tolu, S. Sliding-mode control of a soft robot based on data-driven sparse identification. *Control Eng. Pract.* **2024**, *144*, 105836. [CrossRef]
29. Ambaye, G.; Boldsaikhan, E.; Krishnan, K. Soft Robot Workspace Estimation via Finite Element Analysis and Machine Learning. *Actuators* **2025**, *14*, 110. [CrossRef]
30. Desouza, G.N.; Kak, A.C. Vision for mobile robot navigation: A survey. *IEEE Trans. Pattern Anal. Mach. Intell.* **2002**, *24*, 237–267. [CrossRef]
31. Răileanu, S.; Borangiu, T.; Anton, F.; Anton, S. Open source machine vision platform for manufacturing and robotics. *IFAC-Pap.* **2021**, *54*, 522–527. [CrossRef]

32. Zhu, K.; Zhang, T. Deep reinforcement learning based mobile robot navigation: A review. *Tsinghua Sci. Technol.* **2021**, *26*, 674–691. [CrossRef]
33. Ab Rashid, M.Z.; Mohd Yakub, M.F.; Zaki bin Shaikh Salim, S.A.; Mamat, N.; Syed Mohd Putra, S.M.; Roslan, S.A. Modeling of the in-pipe inspection robot: A comprehensive review. *Ocean Eng.* **2020**, *203*, 107206. [CrossRef]
34. Hossain, S.; Lee, D.-j. Deep Learning-Based Real-Time Multiple-Object Detection and Tracking from Aerial Imagery via a Flying Robot with GPU-Based Embedded Devices. *Sensors* **2019**, *19*, 3371. [CrossRef]
35. Mišeikis, J.; Brijačak, I.; Yahyanejad, S.; Glette, K.; Elle, O.J.; Torresen, J. Two-Stage Transfer Learning for Heterogeneous Robot Detection and 3D Joint Position Estimation in a 2D Camera Image Using CNN. In Proceedings of the 2019 International Conference on Robotics and Automation (ICRA), Montreal, QC, Canada, 20–24 May 2019; pp. 8883–8889.
36. Palleschi, A.; Hamad, M.; Abdolshah, S.; Garabini, M.; Haddadin, S.; Pallottino, L. Fast and Safe Trajectory Planning: Solving the Cobot Performance/Safety Trade-Off in Human-Robot Shared Environments. *IEEE Robot. Autom. Lett.* **2021**, *6*, 5445–5452. [CrossRef]
37. Liu, Q.; Liu, Z.; Xu, W.; Tang, Q.; Zhou, Z.; Pham, D.T. Human-robot collaboration in disassembly for sustainable manufacturing. *Int. J. Prod. Res.* **2019**, *57*, 4027–4044. [CrossRef]
38. Zacharia, P.; Aspragathos, N.; Mariolis, I.; Dermatas, E. A robotic system based on fuzzy visual servoing for handling flexible sheets lying on a table. *Ind. Robot: Int. J.* **2009**, *36*, 489–496. [CrossRef]
39. Zhu, J.; Navarro-Alarcon, D.; Passama, R.; Cherubini, A. Vision-based manipulation of deformable and rigid objects using subspace projections of 2D contours. *Robot. Auton. Syst.* **2021**, *142*, 103798. [CrossRef]
40. Slater, P.A.; Downey, J.M.; Piasecki, M.T.; Schoenholz, B.L. Portable Laser Guided Robotic Metrology System. In Proceedings of the 2019 Antenna Measurement Techniques Association Symposium (AMTA), San Diego, CA, USA, 6–11 October 2019; pp. 1–6.
41. Gul, F.; Rahiman, W.; Nazli Alhady, S.S. A comprehensive study for robot navigation techniques. *Cogent Eng.* **2019**, *6*, 1632046. [CrossRef]
42. Liu, B.; Xiao, X.; Stone, P. A Lifelong Learning Approach to Mobile Robot Navigation. *IEEE Robot. Autom. Lett.* **2021**, *6*, 1090–1096. [CrossRef]
43. Yang, Y.; Cai, Y.; Jung Yoon, Y.; Zhao, H.; Gupta, S.K. Sensor-Based Planning and Control for Conformal Deposition on a Deformable Surface Using an Articulated Industrial Robot. *J. Manuf. Sci. Eng.* **2023**, *146*, 011008. [CrossRef]
44. Huang, S.; Bergström, N.; Yamakawa, Y.; Senoo, T.; Ishikawa, M. High-performance robotic contour tracking based on the dynamic compensation concept. In Proceedings of the 2016 IEEE International Conference on Robotics and Automation (ICRA), Stockholm, Sweden, 16–21 May 2016; pp. 3886–3893.
45. Ravankar, A.; Ravankar, A.A.; Kobayashi, Y.; Hoshino, Y.; Peng, C.-C. Path Smoothing Techniques in Robot Navigation: State-of-the-Art, Current and Future Challenges. *Sensors* **2018**, *18*, 3170. [CrossRef]
46. Yang, H.; Qi, J.; Miao, Y.; Sun, H.; Li, J. A New Robot Navigation Algorithm Based on a Double-Layer Ant Algorithm and Trajectory Optimization. *IEEE Trans. Ind. Electron.* **2019**, *66*, 8557–8566. [CrossRef]
47. Ollero, A.; Heredia, G.; Franchi, A.; Antonelli, G.; Kondak, K.; Sanfeliu, A.; Viguria, A.; Dios, J.R.M.-D.; Pierri, F.; Cortes, J.; et al. The AEROARMS Project: Aerial Robots with Advanced Manipulation Capabilities for Inspection and Maintenance. *IEEE Robot. Autom. Mag.* **2018**, *25*, 12–23. [CrossRef]
48. Eskandaripour, H.; Boldsaikhan, E. Last-Mile Drone Delivery: Past, Present, and Future. *Drones* **2023**, *7*, 77. [CrossRef]
49. Yuan, J.; Szydlowski, M.; Wang, X. An optimal sparse sensing approach for scanning point selection and response reconstruction in full-field structural vibration testing. *Mech. Syst. Signal Process.* **2024**, *212*, 111298. [CrossRef]
50. Al-Tamimi, A.; Lewis, F.L.; Abu-Khalaf, M. Discrete-Time Nonlinear HJB Solution Using Approximate Dynamic Programming: Convergence Proof. *IEEE Trans. Syst. Man Cybern. Part B* **2008**, *38*, 943–949. [CrossRef] [PubMed]
51. Rawlings, J.B.; Muske, K.R. The stability of constrained receding horizon control. *IEEE Trans. Autom. Control* **1993**, *38*, 1512–1516. [CrossRef]
52. Scokaert, P.O.M.; Rawlings, J.B. Constrained linear quadratic regulation. *IEEE Trans. Autom. Control* **1998**, *43*, 1163–1169. [CrossRef]
53. ABB. ABB RobotStudio Software. Available online: https://new.abb.com/products/robotics/robotstudio (accessed on 10 May 2024).

actuators

MDPI

Article

Actuation and Control of Railcar-Mounted Sensor Systems

Caroline Craig and Mehdi Ahmadian *

Railway Technologies Laboratory (RTL), Center for Vehicle Systems and Safety (CVeSS), Virginia Polytechnic Institute and State University, Blacksburg, VA 24060, USA; ccaroline20@vt.edu
* Correspondence: ahmadian@vt.edu

Abstract: This study provides the design, analysis, and prototype fabrication of a remotely controlled actuation system for railcar-mounted sensors. Frequent railway inspections are essential for detecting and preventing major defects that could lead to train derailments or accidents. Integrating supplemental automated inspection systems into existing trains can aid inspection crews without interfering with standard railway operations. However, many sensors and cameras require protection during transit, motivating the need for a deployable mounting assembly. The feasibility of a deployable sensor system was successfully assessed by creating and demonstrating a functional prototype mounting assembly that can be used with future automated inspection systems. Typical loads and accelerations experienced by a train were used to design a lead screw and stepper motor system capable of working within desired tolerances. Optimized inputs controlling this motion with an Arduino Uno were found through the iterative testing of digital signals and direct port manipulation. Further research testing in a field-like environment is suggested.

Keywords: actuation system; control; railway inspection; sensor deployment; railcar integration; lead screw actuation; Arduino

1. Introduction

The rail network in the United States is a crucial infrastructure that supports national logistics and transit, often out of the public's eyes. With an extensive freight rail network covering approximately 225,000 km of track across the continental U.S., this industry is critical for moving and delivering goods [1]. In addition to freight rail, passenger trains in the U.S. operate alongside freight trains. In the 2024 fiscal year, Amtrak reported an all-time ridership record of more than 32.8 million passengers [2]. Inspections and maintenance of the existing infrastructure are essential to meet the demands of this ever-growing industry upon which millions rely yearly. In recent years, Class 1 railroads in the U.S. have focused on improving safety, upgrading rail networks, and increasing reliability. In 2023, they collectively reinvested USD 26.8 billion toward these goals and have made many technological improvements to the network [3].

With such a complex system of railways, track crossings, and signaling equipment, railroad workers must spend significant time, manpower, and resources properly inspecting everything by hand. As such, railroad companies have been pursuing ways to add routine automated inspection technology to their maintenance procedures. In many cases, automated systems can find defects and problems earlier, allowing workers to respond to the issue before it impacts safety or performance. Early detection and prevention are fundamental to lowering the number of derailments and accidents that occur each year.

Some recent equipment that has been successfully implemented into existing railways include train inspection portals (TIPs), autonomous track inspection (ATI) systems, and

https://doi.org/10.3390/act14060289

wayside train monitoring systems (WTMS) [4]. TIPs implement a system of high-speed cameras, machine vision, and artificial intelligence to inspect railcars as they pass through a tunnel at speed [5]. Similarly, many ATI systems use comparable camera, sensor, and machine learning technologies, but these typically operate mounted to a railcar or Hyrail truck [6]. WTMS are stationary systems that are mounted alongside a track bed to monitor train health, such as wheel integrity, as it passes by the monitoring location [4].

Many of the newer inspection technologies focus on keeping trains moving and reducing idle time, which benefits consumers, companies, and the environment. These advantages include decreased fuel consumption, fewer delays, extended locomotive life, and lower emissions, all of which improve operational efficiency and environmental sustainability [7]. Prioritizing the ratio of track time for trains over crews of workers to keep freight and passengers moving makes sense economically but can lead to problems when inspections must be completed on a schedule. If some of the sensors and equipment normally operated by crews of workers could be integrated into existing trains, this would allow data collection to occur without slowing down the rail network.

As train-mounted data collection systems have become more widespread, challenges have arisen from the harsh and unpredictable environments in which trains operate. Some sensors, cameras, and other systems are not made to operate in such places. Thus, there is an occasional need to detach or cover the equipment to prevent damage to sensitive components. However, requiring manual setup and takedown of this equipment can diminish the advantages of train-mounted automated systems over Hyrail trucks. Manual intervention can also become problematic in emergency scenarios. For example, suppose a train traveling at high speeds encounters sudden inclement weather. In that case, it may not be able to stop in time for personnel to safely retract or secure vulnerable equipment.

This study aims to design, test, and validate an automated and retractable platform capable of deploying track inspection sensors during train motion and stowing them in a protected storage space to protect sensitive components when not in use.

2. System Requirements and Design Considerations

In this section, the system requirements and design considerations are outlined based on the integration, environmental, and functional constraints.

2.1. Railcar Integration

In recent years, a common trend in the railroad industry has been to convert old passenger coach cars into towable cars equipped with ATI systems [8]. Using this concept as a guide, the system in this study was designed considering the structural geometry and space limitations of a passenger railcar. To achieve a full field of view of the environment around a track, the railroads often select the rear of a test platform as the place to deploy scanning equipment such as Lidars. Since the coaches used for testing (geometry or metrology cars) are often towed at the end of a train, this ensures that other railcars will not interfere with or block the field of view of mounted sensors. Given that many coach cars include a rear platform, this is the guiding geometry used for the specification and constraints of the assembly designed, tested, and validated in this work.

2.2. Environmental Constraints and Assumptions

To meet the design objective of environmental protection, it was necessary to determine where in the railcar the system could be safely and effectively housed. First, the system must be fully enclosed, eliminating the possibility of directly mounting an assembly to the open rear platform. This location would be easy for operators to access but offers little protection. Second, mounting equipment underneath the train introduces a greater risk of damage due to increased exposure to debris and railbed impact. This area also experiences higher vibrations and less damping than the suspended passenger section, making it less ideal for sensitive components. Considering these assumptions, we determined that the safest and most protected area towards the rear of a coach car is the ceiling space above the platform. This space is already built to be weather-protected and has few systems critical to the railcar's function. The low-vibration environment and minimal functional interference also make the rear ceiling space ideal for integrating additional equipment. The prototype developed in this study was not designed to meet a specific ingress protection (IP) rating, given that environmental impacts could not be studied in laboratory tests. However, future iterations of the system intended for field deployment will require proper sealing and protection.

2.3. Project Specifications

With a design space selected for the system, there are several design constraints that affect how the retraction and stowing can be accomplished. Table 1 summarizes the essential space constraints and vibration requirements. The size of the system needs to be kept as small as possible while maintaining strength to allow it to be installed into the tight areas of a railcar. For this study, the maximum height and width were selected as 558.8 (22 inches) and 685.8 (27 inches) millimeters, respectively. These dimensions keep the overall system compact while leaving a reasonable space for mounting sensors and accompanying electronics. Since one of the overall goals of this project is to quickly deploy and retract sensors for data collection, the ideal actuation time was set to 30 s, which is a reasonable time for rapid deployment without risking damage to the mounted sensors due to uncontrollable or sudden movements.

Table 1. Design specifications.

Parameters	Target Value
Height	558.8 mm (22 inches)
Width	685.8 mm (27 inches)
Actuation Speed	30 s
Deployment Tolerances	3.175 mm (0.125 inches)
User Interface	Passive monitoring
Lateral Force Resistance	0.25 G
Vertical Force Resistance	0.25 G

Positional accuracy is essential for calibrating sensors when deployed to collect data. If the system does not return to the same position behind the railcar, it causes errors between data collected at different times. Applying this constraint, the system needs to be able to actuate repeatably within a 3.175 mm (0.125-inch) tolerance. Another important factor to consider in the design is how the operator will interact with the equipment. At a minimum, the operator must be able to monitor the positional status of the actuated platform passively. Hence, the interface should indicate whether the system is extended or retracted, without the need to provide an intermediate location between the two ends of travel.

The final three design specifications outline the types of forces that a train-mounted assembly could experience while traveling down the railway. Using the Federal Railroad Association (FRA) regulations for passenger cars, the lateral and vertical accelerations experienced must not exceed 0.15 g [9]. Using a higher limit of 0.25 g for this design guarantees that the system can withstand all forces generated in high-speed turns. The longitudinal force or the force in line with the rail could be higher if the railcar is hit hard during coupling. To account for such an impact, a conservative acceleration of 2.5 g was selected to account for forces in the longitudinal direction [10]. This value ensures that large acceleration spikes and impulses that can be applied to a passenger car during coupling are accounted for in calculations.

3. Actuation Mechanism and Control System

A reliable, precise, and durable actuation subsystem must be implemented into the assembly to meet the project specifications outlined in the previous section. Given that the overall shape of the system needs to be shelf-like to accommodate sensor mounting, the actuation subsystem must consist of a guideway and drive system, as shown in the preliminary system layout in Figure 1. It is also important to note that a door or hatch subsystem is also required to protect the equipment from environmental conditions but depends highly on the specific geometry and constraints when installed in a passenger car, which is beyond the scope of this study.

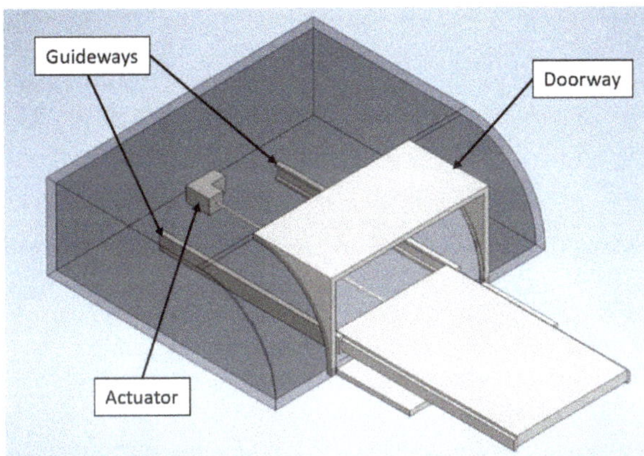

Figure 1. Preliminary system layout.

3.1. Drive System Concepts

There are several different driving methods that can be applied to this type of assembly, each with unique advantages and disadvantages. The five versions applicable for this type of assembly include the following:

- Electric cylinders;
- Pneumatic cylinders;
- Hydraulic cylinders;
- Belt-driven motors;
- Motorized lead screws.

After evaluating the five concepts, the one that was determined to best meet the overall system requirements is the motorized lead screw. This drive mechanism can be designed to be integrated into a more compact area and provides adequate stability and rigidity. In comparison, the belt-driven motor concept lacks rigidity depending on the tensioning of the belt and could experience more wear in service than a lead screw. The three-cylinder actuators were not selected mainly because of their higher costs and potentially more complex control requirements. Hydraulic and pneumatic cylinders require pumps and compressors to actuate them, which would occupy valuable space when integrating them into the ceiling of a passenger railcar.

3.2. Linear Guideway Concepts

Five different guideway concepts were also considered during the design phase to fully stabilize and actuate the system. The possibilities included the following:

- T-slot extrusion linear bearings;
- Linear rails with blocks;
- Ball-bearing drawer slides;
- Linear rods;
- Dovetail slides.

While each has certain advantages, based on cost, versatility, and strength in the load directions, we determined that the best option was the t-slot extrusion linear bearings. Dovetail slides are highly precise and strong but often expensive to manufacture and implement. The drawer slides and linear rods may not be best suited to the cantilever and compression load scenarios. These concepts could also be prone to buckling or bending under the stresses caused during operation. Ultimately, the decision was made to proceed using the t-slot extrusion linear bearings for the prototype assembly manufactured in this study with the understanding that upgrading to a stronger linear bearing and rail could be accomplished if there is a need for even more strength in the future.

3.3. Structural Design

With the drive and guideway subsystem selected, the deployable shelf was designed using 6061 aluminum for a high strength-to-weight ratio, as shown in Figure 2. This setup is designed to have a versatile and customizable top to allow different types of equipment and sensors to be installed. The linear bearings are 3-slotted slides moving along a 1-inch by 3-inch extruded aluminum beam with T-slots to guide the bearings [11]. The bearings have t-shaped pads made of ultra-high-molecular-weight polyethylene (UHMW) seated in the guideways. UHMW is a tough, wear-resistant plastic that is ideal for this application. Using the material data, we determined a static coefficient of friction of 0.20, which we used for the load calculations and sizing of the electric drive motor [12]. This assembly offers 609.6 mm (24 inches) of travel for the shelf while moving from fully stowed to fully extended, with a total mass of 24.78 kg, including an added margin for sensor equipment. The structural design also accounted for the anticipated lateral and vertical force resistances of 0.25 g, as detailed earlier.

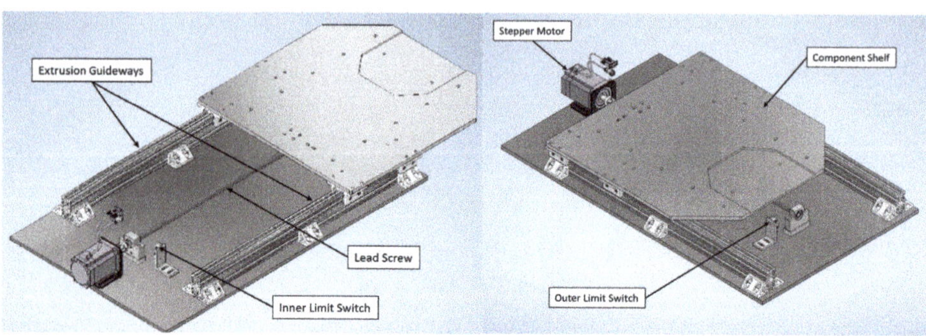

(a) (b)

Figure 2. (a) Fully extended deployable shelf assembly; (b) fully retracted deployable shelf assembly.

3.4. Lead Screw Sizing and Load Calculations

Sizing the lead screw acting in a horizontal configuration with an uncentered CG (center of gravity) from the lead screw nut required calculating the bending stress acting on the screw and comparing it with the material's yield strength. Using a 20 mm lead screw with a 4 mm pitch made from 1045 carbon steel, the yield strength is approximately 310 MPa [13]. Applying the known load from the mass, 243.1 N, at a known offset of approximately 500 mm gives an applied moment of 121.5 Nm, which can be used in Equation (1) for calculating the bending stress.

$$\sigma = \frac{M \cdot c}{I} = \frac{M \cdot (d/2)}{(\pi \cdot d^4)/64} = \frac{32 \cdot M}{\pi \cdot d^3} = \frac{32 \cdot 121.5 \text{ Nm}}{\pi \cdot (0.020)^3 \text{m}^3} = 154.8 \text{ MPa} \tag{1}$$

The above represents the worst-case static moment when the platform is fully extended. Comparing the bending stress against the material's known yield strength gives a factor of safety (FOS) of approximately 2.0 against yielding.

3.5. Motor Selection and Torque Requirements

The lead screw and applied loading can be used to calculate the required torque needed to actuate the system. To account for the worst-case scenario, the longitudinal acceleration will be applied to the total forces acting on the lead screw. The total applied force is the sum of the frictional force and coupling acceleration force [14]:

$$F_{total} = F_{friction} + F_{longitudinal} = (0.20 \cdot 243.1)\text{N} + (2.5 \cdot 234.1)\text{N} = 656.4 \text{ N} \tag{2}$$

From the lead screw used earlier, the efficiency is approximately 0.20 according to industry standards [15]. To determine an acceptable torque range for the motor, two FOS values of 2.0 and 4.0 are applied to Equation (3) [15].

$$T = \frac{FOS \cdot F_{total} \cdot p}{2\pi \cdot \eta} = \frac{(2.0 \ or \ 4.0) \cdot 656.4 \text{ N} \cdot (0.004) \text{ m}}{2\pi \cdot 0.20} = 4.2 \ to \ 8.4 \text{ Nm} \tag{3}$$

A NEMA 34 stepper motor was selected to actuate the system using this range of required torque values. This motor has a holding torque of 8.5 Nm, which falls in the upper range of the torque values for a high FOS on the motor [16]. A stepper motor is a reliable, strong, and implementable driver for this application. The required RPMs for the actuation to occur within the time constraints defined earlier can be found using Equation (4) with a travel speed of 20.3 mm/s from the 609.6 mm (24 inches) of travel divided by the actuation time of 30 s.

$$RPM = \frac{v \cdot 60}{p} = \frac{(20.3 \text{ mm/s}) \cdot (60 \text{ s/min})}{(4 \text{ mm/rev})} = 304.8 \frac{\text{rev}}{\text{min}} \tag{4}$$

The NEMA 34 stepper motor can achieve the 304.8 RPMs needed to move the platform within the required distance and time. This RPM is one of the key parameters used to develop the controller.

3.6. Control System Architecture

Among various options considered, we selected an Arduino Uno board as the controller due to its relatively low cost, ease of use, intuitive interface, and effectiveness in the prototyping phase of design. As shown in the control block diagram in Figure 3, two different voltages are required for the assembly. The Arduino and control signals are controlled by 5 volts, and the stepper motor and driver need 24 volts. Given that the system only needs to be in retracted or fully extended modes for sensor deployment, limit switches can detect these positions. The limit switches ensure the positioning is sufficiently accurate and the 0.125-inch positional accuracy is met. Simultaneously, wiring LED lights to indicate to the operator that the system has finished traveling in one direction can fulfill the specification of passive monitoring. A rotary switch is implemented to control the starting and stopping of the motor and its direction rotation, which controls the platform's direction of travel.

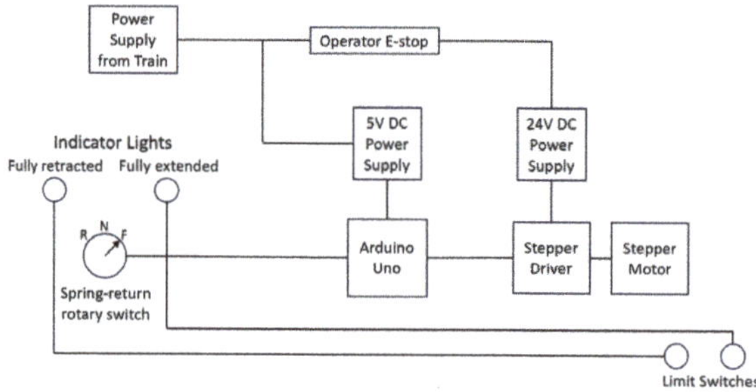

Figure 3. System control block diagram.

4. Prototype Development and Testing

Using the developed actuation and control system described earlier, this section presents a physical prototype assembly able to be built and tested in a controlled laboratory setting. This system provides validation of the performance of the actuation concepts before integrating into a railcar.

4.1. Mechanical Assembly

To fully test and validate the theoretical concepts presented by this study, a prototype assembly was designed and fabricated to conduct actuation tests on the control and actuation assemblies. A simple wooden platform, shown in Figure 4, is used for the laboratory tests to emulate mounting to a railcar. The platform shelf, spacers, and alignment blocks were manufactured from 6061 aluminum. The alignment blocks are critical for ensuring that the stepper motor output shaft is in-line with the lead screw. When a stepper motor and lead screw are not properly aligned, this can lead to extra stress on components, premature wear on the threads, and lower efficiency while operating [17]. In addition to making sure the axes of both the stepper motor output shaft and the lead screw are properly aligned, a flexible shaft coupling is used to connect them. A flexible coupling is used to compensate for shaft misalignment while still allowing effective torque transmission and accounting for slight axial movement by either shaft [18].

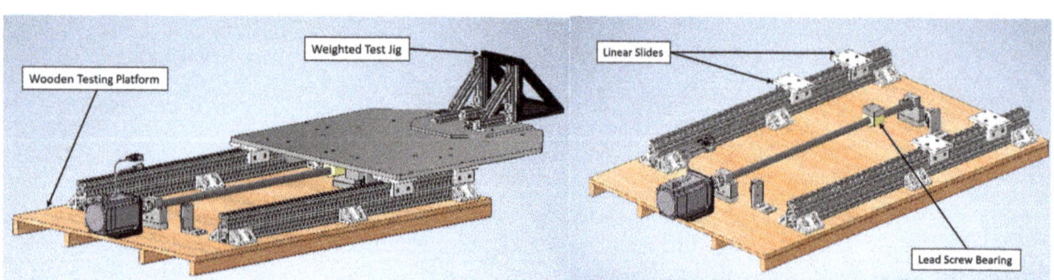

(a) (b)

Figure 4. (**a**) Prototype assembly with equipment shelf; (**b**) Prototype assembly with actuation system shown.

All components were mounted to a wooden test platform using accompanying brackets and hardware. The wooden platform was selected for assembly and laboratory testing. In practice, it will be replaced by the railcar base to which the actuation system is mounted.

The fully assembled prototype assembly shown in Figure 5 adequately represented the actual assembly on board a railcar while allowing the ease of adjustability needed for the laboratory tests. Microswitches were properly positioned at the limits of travel to prevent excessive motion in either direction. Once the platform contacts the limit switches, the motor's power is disabled, and the actuation stops. The extrusion rails that allow the bearings to slide from the retracted to extended positions in the final assembly have a travel distance of 615.95 mm (24.25 inches), allowing for adequate testing and performance assessment.

Figure 5. Fully assembled prototype.

4.2. Wiring and Electrical Integration

The controls and wiring system were implemented according to the wiring diagram in Figure 6. An Arduino Uno board was used to send control signals to the stepper driver using 5 V power. The limit switches, rotary knob, and LED indicator lights were wired to the board using pins 2 through 9. The wiring was set up so the limit switches could read them as normally closed (NC). With this arrangement, the circuit through the switch is complete when the limit switch is unpressed. The circuit opens when the switch is pressed, and the Arduino reads no current flow through those pins, preventing any more motion in that direction. The advantage of wiring a limit switch in an NC configuration is that if a wire or switch breaks, the Arduino will cease any travel in that direction, acting as an additional safety measure. The rotary knob is a center spring return switch that allows motion in either direction based on how it is turned. When held, the Arduino reads the circuit and actuates the stepper motor clockwise or counterclockwise. Once the switch is released, it returns to the middle, disconnecting the circuit and ceasing movement. The LED indicator lights are each wired and programmed to turn on when a limit switch is pressed. The indicator light provides a visual alert to the operator that the system can no longer travel in that direction. A 24 V power supply unit (PSU) was connected to the driver through an emergency stop button to provide power to the stepper motor. It is worth noting that the NEMA 34 stepper motor used in this study is also equipped with an encoder, even though it was not used in our design. The simplified control approach we adopted, where only the platform position at the two limits is needed, did not require using the stepper motor's embedded encoder.

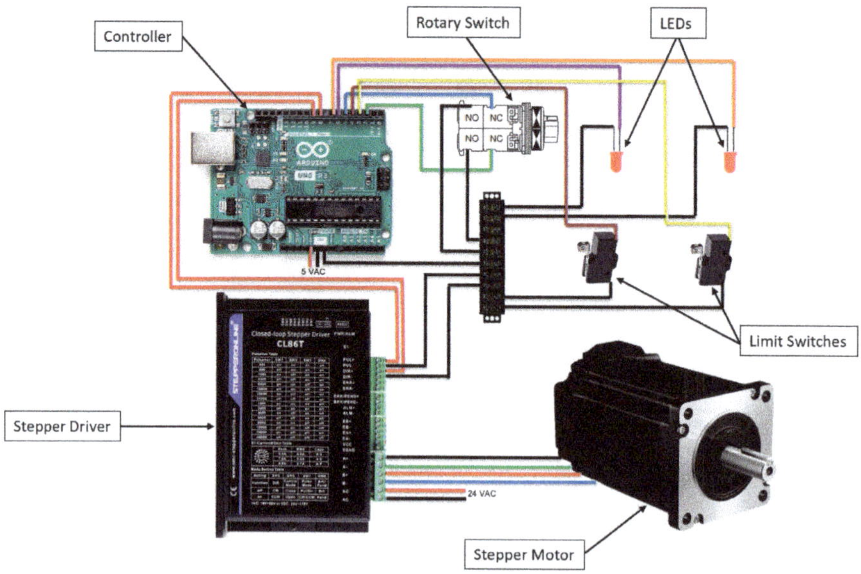

Figure 6. Control system wiring diagram dependent on components used in assembly.

5. Performance Evaluation and Results

Testing of the prototype assembly was completed in the laboratory using 24 V PSU and 5 V DC power supplies that emulate the onboard power of a railcar. Various actuation modes were evaluated to test the system responses to different controlled inputs. Significant testing was conducted to refine the Arduino code and stepper motor controller. After issues involving signal discrepancies were resolved, the prototype underwent a series of tests with weights to mimic the loaded configuration in practice. The tests included extending and retracting. The time to complete each movement was recorded and then used to calculate actual speeds.

5.1. Pulse Delay Analysis

The rotational speed of the stepper motor was calculated by using the timing delay of pulses sent from the Arduino to the stepper driver. These pulses trigger the motor to move one micro-step at a time, controlling the number of pulses per revolution. For the NEMA 34 stepper motor used in this study, the micro-step is adjustable from 1 through 256, which can be used to refine the pulses per revolution needed from 200 to 51,200. The number of pulses per revolution is calculated by multiplying the micro-step by the motor default setting of 200 steps per revolution. These pulses are sent from the Arduino by inputting a pulse delay in microseconds. A smaller delay sends faster pulses to the stepper motor, moving the platform faster. Several test runs were conducted on the system to determine the optimal delay and micro-step setting needed to meet the system requirements mentioned earlier.

Applying these values to Equation (5) allows for calculating the theoretical RPM of the motor [19].

$$RPM = \frac{60 \cdot 10^6 (\mu s/\min)}{200\ (\mu s) \cdot (microstep\ 8) \cdot (200\ \text{steps/rev})} = 187.5\ \frac{\text{rev}}{\text{min}} \tag{5}$$

Combining the calculated RPM with the travel distance of 615.95 mm (24.25 inches) and the 4 mm pitch of the lead screw provides a predicted actuation time:

$$t = \frac{60 \cdot L}{RPM \cdot p} = \frac{60(\mathrm{s/\ min}) \cdot 615.95\ \mathrm{mm}}{187.5\ (\mathrm{rev/\ min}) \cdot 4\ (\mathrm{mm/\ rev})} = 49.28\ \mathrm{s} \quad (6)$$

Conducting preliminary testing using a micro-step of 8 and various pulse delays, a discrepancy was found between the theoretical and observed actuation times. As shown in Table 2, the observed time was consistently more than 2 times higher than the predicted time based on the input delay. Using the error, the actual speed of the motor and pulse delay received by the stepper motor could be calculated by reversing Equations (5) and (6).

Table 2. Initial speed testing; Microstep:8.

Delay (μs)	RPM	Predicted Time (s)	Observed Time (s)	Error (%)	Measured Delay (μs)
100	375	24.64	59.90	143.12	243.12
200	187.5	49.28	109.67	122.56	445.13
300	125	73.91	159.20	115.39	646.17
400	93.8	98.55	208.76	111.83	847.32

As the motor's speed increases, the error also increases. Other trials were performed by varying the micro-steps; however, the same error percentages were still observed. By eliminating potential factors such as this, it was concluded that the system's error was related to the input delay. The signal being sent by the Arduino was two times slower than what the stepper motor was receiving.

5.2. Control Signal Optimization

Initially, the code sent pulses to the stepper motor and read the limit switches, rotary switch, and LED statuses using many digitalWrite() and digitalRead() commands in loops. While it takes the Arduino only microseconds to complete each function, it requires the loops to run thousands of times, causing processing delays. These extra operating delays in the Arduino code added an unacceptable amount of delay over time. To help mitigate this error, the code was converted to Direct Port Manipulation (DPM). This method directly sets and clears pins by using bits and is roughly 30 times faster than the digital commands [20].

After implementing DPM into the code, the system response became much faster; however, this revealed a second flaw in the code. Since the Arduino was sending very fast signals, the stepper motor did not always correctly receive them at the start of an actuation, causing it to skip steps and not actuate properly. The implementation of DPM yielded an 87.19% decrease in error between the predicted and observed times, as included in Table 3. However, the missed steps at the beginning of travel still contributed to a 35.37% error.

Table 3. Digital vs. DPM—RPM: 187.5; predicted time: 49.28 s; delay: 200 μs.

Pin Method	Observed Time (s)	Error (%)
Digital	109.67	122.56
DPM	66.71	35.71

5.3. Acceleration Implementation

The best way to mitigate the misread signal problem at the start of actuation was to add a ramp-up acceleration into the signal pulse. The ramp-up allows the stepper motor to start motion at a slower speed to overcome the initial static friction's slowing effect. Then, it can speed up to a faster RPM without missing steps. Keeping the same target RPM of 187.5 and an input pulse delay of 200 μs, a new predicted time for the total actuation can be estimated by accounting for the time for acceleration. The minimum pulse of 200 μs controls the speed for most of the actuation. The maximum pulse of 5000 μs is the starting speed, which is set to be much slower than the target speed. For every step that is completed, the pulse delay being sent to the motor is decreased by the acceleration rate of 10 μs/step. This acceleration rate was determined through preliminary testing to optimize the overall system response to prevent misread signals by the stepper motor during deployment. Equation (7) uses these values to solve the number of steps needed to accelerate to the maximum speed.

$$N_{accel} = \frac{maxPulse - minPulse}{acceleration} = \frac{5000\ \mu s - 200 \mu s}{10\ (\mu s/\text{step})} = 480 \text{ steps} \quad (7)$$

Since every step during the acceleration phase is faster than the step before, an acceleration time can be estimated using Equation (8). Assuming the acceleration is linear, the time to accelerate is found by multiplying the number of acceleration steps by the average acceleration pulse per step.

$$\begin{aligned} t_{accel} &= N_{accel} \cdot \frac{(maxPulse + minPulse)}{2} \\ &= 480 \text{ steps} \cdot \frac{(5000+200)}{2} \frac{\mu s}{\text{step}} \cdot 10^{-6} \frac{s}{\mu s} = 1.25 \text{ s} \end{aligned} \quad (8)$$

Factoring in the acceleration time with the remaining travel time at full speed, the new predicted time to complete one actuation is 50.52 s. Table 4 compares the improved system code to the previous versions, showing the error decreased by 29.64% after the implementation of acceleration time.

Table 4. Comparison of error—RPM: 187.5; minPulse: 200 μs; maxPulse: 5000 μs.

Pin Method	Predicted Time (s)	Observed Time (s)	Error (%)
Digital	49.28	109.67	122.56
DPM	49.28	66.71	35.71
DPM with acceleration	50.52	53.59	6.07

The actuation speed testing was redone using the optimized code to compare the observed times to the predicted times. Additionally, to aid in the stepper motor's ability to ready the signals, the micro-step setting was reduced to 4, equating to 800 pulses per revolution. Reducing the micro-step setting decreases the precision of the motor, but it can increase output torque and allow longer pulse delays to be used to achieve the same speeds as before [21]. Since the motor was occasionally struggling with faster pulses, this change was able to help it process the signals it was receiving. Figure 7 demonstrates the overall performance improvements after implementing acceleration into the code. As expected, the error between the predicted time and the observed time still increases as maximum speed increases; however, the target time of 30 s was met by running the stepper motor at 375 RPM and higher. It was observed that at RPMs higher than 428.6, the motor struggled to overcome the initial resistance again. One can reason that further adjustments of the acceleration time and micro-step settings may allow for even higher RPMs to be achieved. However, such adjustments were unnecessary because the target actuation time was already met.

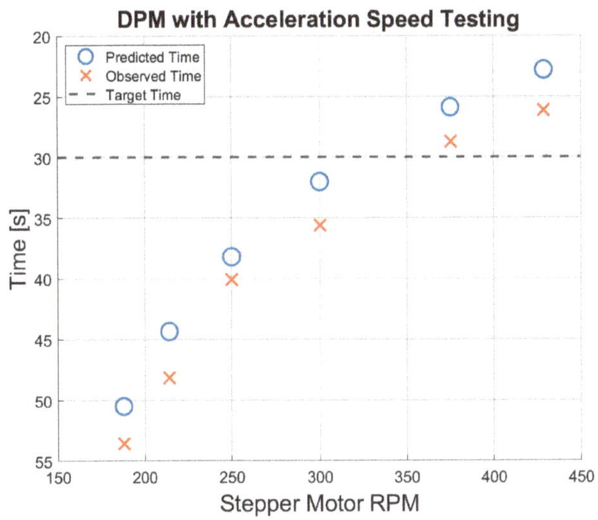

Figure 7. Actuation time vs. motor RPM.

5.4. Weighted Testing

Additional testing was conducted by adding extra weight as an overhanging load to emulate the response of the actuation mechanism in maximum loading conditions. As shown in Figure 8, various weighted plates were added to hang off the end of the shelf to simulate the worst case of loading that could be applied to the stepper motor and lead screw.

Figure 8. Weighted testing.

The same system inputs were given to the stepper motor as in the fastest unweighted test from before. A minimum pulse of 175 μs, a maximum pulse of 5000 μs, an acceleration rate of 10 μs/step, and a micro-step setting of 4. Actuations from 0 to 30.86 lb. shown in Figure 9a revealed a consistent error percentage of around 14.94% between the observed time and the predicted time of 22.81 s until around 20 lbs. At this point, the stepper motor began to experience some issues when retracting from the fully extended position but still performed mostly well once it overcame the initial resistance. An iterative process with different pulse delays and accelerations was conducted to assess the system performance in various actuation and loading conditions. To counteract the retracting problems, the minimum pulse was increased to 200 μs, and the acceleration rate was decreased to 5 μs/step. In doing so, the predicted time of actuation was lowered to 27.13 s, which is closer to the target time of 30 s but allows the motor to output more torque in extreme conditions. Running the refined test, the system could fully actuate without issues, starting from the fully extended position, as shown in Figure 9b. This more optimized version of step inputs reduced the error percentage between the predicted and observed time to drop from 14.94% to 8.70% while still meeting the actuation time requirement.

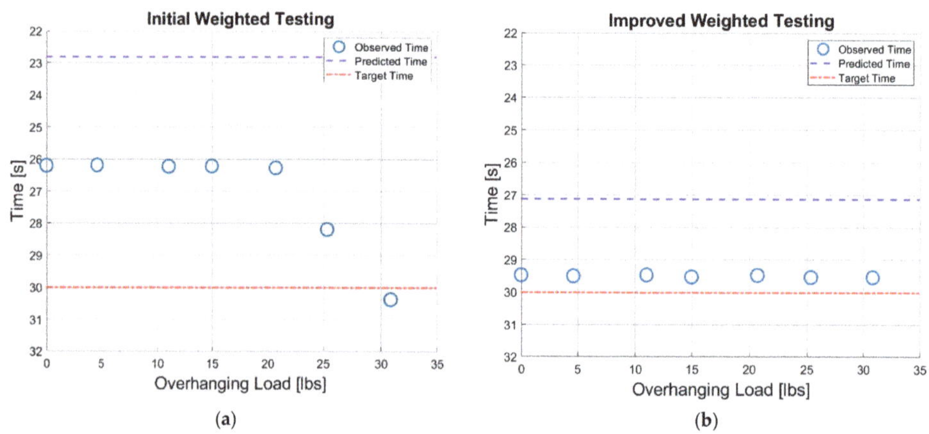

Figure 9. (**a**) Weighted testing with faster overall speed and higher acceleration; (**b**) Optimized speed and acceleration.

6. Results and Discussion

This section analyzes the performance results and key limitations of the prototype assembly that may affect the implementation of the system into a real-world environment.

6.1. Achieved Project Requirements

The prototype testing results validated the actuation and control concepts proposed in this study while highlighting opportunities for refinement in future studies. Through iterative design and testing, the prototype met the following project requirements:

- Designed within overall width and height size constraints;
- Achieved target actuation time of 30 s;
- Met deployment tolerance of 3.175 mm (0.125 inches);
- Facilitated passive monitoring with LED indicators;
- Designed to withstand lateral and vertical accelerations of 0.25 g;
- Designed to withstand longitudinal accelerations of 2.5 g.

Overall, the prototype performance demonstrated a successful design that met the requirements in the mechanical, structural, and control system areas. The final width of the prototype was 673.1 mm (26.5 inches), within the specified 685.8 mm (27 inches). The height was able to be minimized from the specified 558.8 mm (22 inches) to a low profile of 152.4 mm (6 inches), leaving adequate room on top of the shelf for mounting a variety of sensor equipment. Using the t-slot extrusion rails and bearings, the prototype achieved a travel distance of 615.95 mm (24.25 inches), mimicking a real-world deployment scenario.

Using microswitches to define travel limits, the assembly could return to the same position each time it was actuated and trigger the indicator LED lights for the operator to passively monitor the system position with intuitive manual control. Testing the system with a high, overhanging load ensured that the torque capability of the lead screw and stepper motor was suitable for anticipated loading conditions. The Arduino control system allowed for quick and effective design iterations to be implemented, particularly the shift to DPM and the addition of a ramp acceleration to the system motion, which met the target actuation time of 30 s.

6.2. Current System Limitations

As a prototype control system, the Arduino performed as required for this study; however, there are areas that could be improved upon to give more control and feedback on the state of the system. The current setup of limit switches provides no feedback on the intermediate position of the platform. If such information is needed, one can use the motor's encoder to provide the position of the platform at all states, not just to travel limits. The open-loop type control system limitations were highlighted during the assembly testing. When the motor struggled to move by misreading the control signals or having insufficient torque, there was no additional indication to the operator to release the rotary switch other than auditory and visual cues from the motor. Since the testing took place in a controlled lab, the environmental conditions in this study, such as vibrations, dirt, and temperature, are not comparable to those faced by a train in the field. These environmental factors were unable to be tested in this study but should be considered in the future to assess their influence on motor performance and system response. Another small limitation observed during testing was a slight kickback by the stepper motor during motion due to the more flexible mounting on the wooden surface. This motion could potentially cause unusual wear on the lead screw and stepper motor as well as cause small alignment issues. The effects of the kickback appeared negligible; however, they could become an issue as the assembly performs more actuation cycles. Long-term wear testing on the lead screw and bearings was not able to be performed in this study but would provide valuable estimations on the life of the system. With the current benchtop prototype, the lateral and vertical force resistances were not able to be validated experimentally; however, the system is designed to withstand the expected forces.

7. Summary and Future Work

The design and implementation of the actuation system for a railcar-mounted sensor platform were described in detail. The platform is intended for retracting and stowing sensitive inspection systems used onboard test cars for railroad inspection. The platform includes a lead screw mechanism operated by a stepper motor. The controller uses an Arduino Uno that allows ease of control and implementation.

The strength requirements for the drive mechanisms and structural components were calculated to achieve a design factor of two or higher. Once built, the actuation platform was tested under various conditions to ensure it could operate at the required retraction and stowing of onboard sensors. The tests were designed to emulate the actual use of the platform in practice, including in loaded configurations. The tests were valuable in highlighting deficiencies and implementing improvements to resolve them. One such deficiency proved to be the slower-than-expected actuation speed, which was caused by errors in generating the stepper motor pulse delay. Once corrected, the actuation system operated according to the design specifications.

The study identified a few areas for future improvements. Since the stepper motor is equipped with a built-in encoder, it can provide real-time position feedback to the operator, which could improve the platform's operation. Another improvement that could be made to the control system is to experiment with PID control or a closed-loop feedback setup. Such control enhancements would strengthen the overall system architecture and performance.

The developed prototype system offers insight into how this can be implemented in practice. It provides a framework for enhancing onboard automated track inspection equipment. With further refinement, the platform could be deployed on existing trains to increase the data collection frequency, ultimately contributing to a safer rail network.

The next phase of testing can also be set up to emulate the dynamic impacts, vibrations, and environmental conditions experienced in a railcar. Such tests can be carried out onboard railroad test cars or Hyrail trucks, which we could not access during this study.

Author Contributions: Design, C.C.; building, C.C.; testing, C.C.; analysis, C.C.; writing, C.C.; review and editing, C.C. and M.A.; supervision, M.A.; project administration, M.A. All authors have read and agreed to the published version of the manuscript.

Funding: This research received no external funding.

Data Availability Statement: The datasets presented in this article are not readily available because the data are part of an ongoing study. Requests to access the datasets should be directed to ccaroline20@vt.edu.

Conflicts of Interest: The authors declare no conflicts of interest.

Nomenclature

The following parameters are used equations in this manuscript:

σ	Bending stress
M	Applied moment
c	Distance to axis of bending
I	Moment of inertia
d	Lead screw diameter
FOS	Factor of safety
F_{total}	Total applied force
$F_{friction}$	Frictional force between slides and rails
$F_{longitudinal}$	Coupling force (in line with rail)
T	Torque required by motor
p	Pitch of lead screw
η	Efficiency of lead screw
RPM	Revolutions per minute
v	Travel velocity
t	Actuation time
L	Length of travel
N_{accel}	Number of acceleration steps
maxPulse	Slowest travel velocity pulse speed
minPulse	Faster travel velocity pulse speed
acceleration	Acceleration rate
t_{accel}	Acceleration time

References

1. Overview of America's Freight Railroads. 2020. Available online: https://www.aar.org/wp-content/uploads/2018/08/Overview-of-Americas-Freight-RRs.pdf (accessed on 16 April 2025).
2. Abrams, J. FY24 Year End Ridership. Amtrak Media. 2024. Available online: https://media.amtrak.com/2024/12/fy24-year-end-ridership/#:~:text=Amtrak%20achieved%20an%20all-time,in%20markets%20across%20the%20nation (accessed on 16 April 2025).

3. Association of American Railroads. Railroad 101. 2025. Available online: https://www.aar.org/rail-facts/#! (accessed on 16 April 2025).
4. Association of American Railroads. How Technology Enhances Rail Safety & Efficiency. 2025. Available online: https://www.aar.org/freight-rail-tech/ (accessed on 16 April 2025).
5. Toon, J. Digital Inspection Portal Uses AI and Machine Vision to Examine Moving Trains. GTRI. 2024. Available online: https://www.gtri.gatech.edu/newsroom/digital-inspection-portal-uses-ai-and-machine-vision-examine-moving-trains (accessed on 16 April 2025).
6. Association of American Railroads. Automated Rail Inspections: AAR. 2025. Available online: https://www.aar.org/issue/automated-track-inspections/ (accessed on 16 April 2025).
7. EPA. Learn About Idling Reduction for Locomotives. 2024. Available online: https://www.epa.gov/verified-diesel-tech/learn-about-idling-reduction-locomotives (accessed on 16 April 2025).
8. High Speed Research Car T-16 | FRA. High Speed Research Car T-16. 2019. Available online: https://railroads.dot.gov/program-areas/testing-facilities-equipment/high-speed-research-car-t-16#:~:text=T-16%20also%20assists%20Amtrak,removable%20media%20for%20offline%20viewing (accessed on 16 April 2025).
9. Electronic Code of Federal Regulations (eCFR). Code of Federal Regulations, Title 49, Subtitle B, Chapter II, Part 299, Subpart C, Section 299.315—Vehicle Structure. Available online: https://www.ecfr.gov/current/title-49/subtitle-B/chapter-II/part-299/subpart-C/section-299.315 (accessed on 16 April 2025).
10. Cole, C. Longitudinal Train Dynamics and Vehicle Stability in Train Operations. In *Handbook of Railway Vehicle Dynamics*, 2nd ed.; Iwnicki, S., Spiryagin, M., Cole, C., McSweeney, T., Eds.; CRC Press: Boca Raton, FL, USA, 2019; pp. 457–517. [CrossRef]
11. 80/20 Inc. A Guide to 80/20 T-Slot Linear Bearings. 2024. Available online: https://8020.net/blog/a-guide-to-8020-t-slot-linear-bearings.html (accessed on 16 April 2025).
12. Redwood Plastics and Rubber. Virgin UHMW. 2024. Available online: https://www.redwoodplastics.com/materials/uhmw-polyethylene/virgin-uhmw-polyethylene/ (accessed on 16 April 2025).
13. Corner, L. Aisi 1045 Medium Carbon Steel. AZO Materials. 2025. Available online: https://www.azom.com/article.aspx?ArticleID=6130 (accessed on 16 April 2025).
14. Oberg, E.; Jones, F.D.; Horton, H.L.; Ryffel, H.H.; McCauley, C.J. Ball and ACME Leadscrews. In *Machinery's Handbook*, 31st ed.; Brengelman, L., Ed.; Industrial Press, Inc.: South Norwalk, CT, USA, 2020; pp. 2640–2641.
15. Lipsett, R. Why Lead Screws are the Best Fit for Many Linear Motion Applications—And How to Rightly Apply Them. Thomson Linear. Available online: https://www.thomsonlinear.com/downloads/articles/Why_Lead_Screws_Best_Fit_Linear_Motion_Applications_taen.pdf (accessed on 16 April 2025).
16. Stepper Online. P Series NEMA 34 Closed Loop Stepper Motor 8.5nm(1203.7oz.in) with Encoder 1000PPR(4000CPR)—34E1K-85 | STEPPERONLINE. Available online: https://www.omc-stepperonline.com/p-series-nema-34-closed-loop-stepper-motor-8-5nm-1203-94oz-in-with-encoder-1000ppr-4000cpr-34e1k-85 (accessed on 16 April 2025).
17. Tao, L.; Zhang, L.; Wang, X.; Fan, X.; Shi, H. Dynamic screw lead accuracy measurement: Research on the compensation of the workpiece placed away from the ideal position. *Sensors* **2024**, *24*, 6829. [CrossRef] [PubMed]
18. Boyce, M. Ch. 18 Couplings and Alignment. In *Gas Turbine Engineering Handbook*; Gulf Professional Publishing: Houston, TX, USA, 2006; pp. 635–662.
19. GlobalSpec. Stepper Motor Calculations. Available online: https://www.globalspec.com/pfdetail/motors/stepper-motor-calculations (accessed on 16 April 2025).
20. Margolis, M.; Jepson, B.; Weldin, N.R. *Arduino Cookbook: Recipes to Begin, Expand, and Enhance Your Projects*; O'Reilly Media: Sebastopol, CA, USA, 2020.
21. Bednarski, B.; Jackiewicz, K.; Gałecki, A. Influence of microstepping signal shape on shaft movement precision and torque variation of the Stepper Motor. *Energies* **2021**, *14*, 6107. [CrossRef]

actuators

MDPI

Article

Minimizing Power Losses in BLDC Motor Drives Through Adaptive Flux Control: A Real-Time Experimental Study

Mohamed Fadi Kethiri [1], Omar Charrouf [1], Achour Betka [1], Muhammad Salman [2,*] and Chiara Boccaletti [2]

[1] Department of Electrical Engineering, University of Biskra, Biskra 07000, Algeria; fadi.kethiri@univ-biskra.dz (M.F.K.); omar.charrouf@univ-biskra.dz (O.C.); a.betka@univ-biskra.dz (A.B.)
[2] Department of Astronautical, Electrical and Energy Engineering, Sapienza University of Rome, 00184 Rome, Italy; chiara.boccaletti@uniroma1.it
* Correspondence: muhammad.salman@uniroma1.it

Abstract: This paper presents a novel methodology for minimizing power losses in brushless DC (BLDC) motors through the implementation of adaptive flux control techniques. Conventional motor control strategies, such as direct torque control (DTC), typically employ fixed flux values, which often result in suboptimal performance, particularly under dynamic load and speed variations. To mitigate this inherent limitation, two adaptive flux control methods are introduced: incremental conductance (IncCond) and fuzzy logic. These proposed strategies facilitate real-time dynamic adjustment of the stator flux, thereby optimizing motor performance and significantly enhancing system efficiency. Experimental validation confirms the efficacy of these adaptive techniques, demonstrating substantial improvements in power loss reduction and overall efficiency when compared to traditional fixed flux control strategies. Notably, the fuzzy logic control strategy achieves the highest efficiency, registering a system efficiency of 66.59%, which surpasses both the incremental conductance method and conventional fixed flux control. These findings underscore the considerable potential of adaptive flux control in applications where energy efficiency is paramount, including electric vehicles and renewable energy-driven systems.

Keywords: BLDC motors; adaptive flux control; incremental conductance; fuzzy logic; power loss minimization; system efficiency; direct torque control

1. Introduction

BLDC motors are widely used in modern drive systems due to their high efficiency, compact size, high torque-to-weight ratio, and low maintenance requirements. They play a critical role in a broad range of applications, including electric vehicles, robotics, industrial automation, HVAC systems, and renewable energy systems such as wind turbines and solar water pumps. Their inherent advantages over brushed DC and induction motors make them ideal for high-performance and energy-sensitive applications.

Although BLDC motors are widely recognized for their high efficiency, power losses resulting from non-optimal flux control remain a significant concern, particularly in real-time applications with stringent energy-saving requirements. To mitigate these losses and enhance drive performance, the implementation of adaptive flux control strategies is imperative, especially under varying load and speed conditions.

Minimizing power losses in BLDC motors remains a central focus of research, particularly through the use of adaptive flux control strategies such as fuzzy logic controllers (FLCs) and IncCond. FLCs have proven highly effective in dealing with nonlinearities and system uncertainties. Ref. [1] demonstrated that fuzzy-based speed controllers significantly reduce overshoot and torque ripple while halving the settling time compared to conventional PID methods. Similarly, ref. [2] confirmed that fuzzy logic, when combined with power factor correction, leads to reduced rise time and enhanced dynamic performance. Hybrid control approaches, such as the one proposed by [3], merge the benefits of digital adaptive PID with fuzzy logic to achieve superior torque regulation and disturbance rejection.

Incremental conductance, while traditionally applied in Maximum Power Point Tracking (MPPT) in photovoltaic systems, is emerging as a viable method for real-time flux adaptation in BLDC motors. Ref. [4] has successfully applied IncCond to BLDC pump systems, showing improved efficiency under dynamic operating conditions. An ANFIS-enhanced IncCond-based control scheme is explored in [5], reporting significant improvements in total harmonic distortion and system responsiveness.

T–S fuzzy PI controllers in buck-boost converters for enhanced performance are presented in [6], while studies such as [7,8] confirm the benefits of fuzzy-PID and sliding-mode methods in regulating BLDC motor dynamics. Furthermore, the use of IncCond in FPGA-based systems for renewable energy-driven BLDC control has been investigated by [9].

In the area of sensorless estimation and control, a recent study detailed advancements using sliding-mode observers, Kalman filters, and deep learning-based zero-shot estimation for flux and speed control [10]. These developments align with the growing need for robust and intelligent motor control systems in embedded and real-time environments.

A commutation point optimization method is presented for sensorless BLDC motors that reduces commutation error and improves efficiency by compensating for the phase difference between back EMF and phase current. Ref. [11] proposes a commutation point optimization method for sensorless BLDC motors that accurately compensates for commutation errors by analyzing the phase difference between back EMF and phase current, improving efficiency. The study presents a new switching technique to minimize losses in a trapezoidal brushless DC motor drive, demonstrating superior performance in terms of torque ripple, current ripple, and overall efficiency [12]. A study compares two PWM techniques for BLDC motor control, with one technique showing superior performance in minimizing torque ripple, DC current ripple, and losses [13]. Machine learning models for sensorless estimation of BLDC motor winding temperature are proposed in [14], achieving high accuracy with a mean absolute percentage error below 4.5% and demonstrating the viability of eliminating physical temperature sensors. A study in [15] presents an energy-efficient BLDC motor optimized for a centrifugal monoset pump using Grey Wolf Optimization (GWO), demonstrating a 6% increase in efficiency validated through 3D finite element analysis.

Applying a tilt integral derivative (TID) controller to BLDC motors significantly enhances real-time speed control, torque and current stability, and overall system efficiency in electric vehicle applications [16]. An improved robust deadbeat predictive sensorless current control scheme is proposed for permanent magnet synchronous motors (PMSMs), employing an enhanced super-twisting sliding-mode observer to simultaneously estimate back electromotive force (EMF), inductance, and flux linkage. This unified observer effectively mitigates the impact of parameter mismatches, enhancing overall system robustness [17]. An improved adaptive fractional-order sliding mode control (AFOSMC) system is proposed for PMSMs, integrating a fractional-order sliding mode disturbance observer to enhance robustness, anti-interference capabilities, and overall dynamic and steady-state performance [18]. A novel four-point trajectory tracking control method for PMSMs is proposed to enhance dynamic response speed and steady-state efficiency across the full-speed range by adaptively adjusting current trajectories [19].

A study introduced a novel adaptive flux observer for direct torque and flux control (DTFC) of permanent magnet synchronous motor (PMSM) drives, employing an online loss minimization algorithm to estimate the air-gap flux for optimal efficiency across various operating conditions [20]. A multi-objective optimization approach for PI controllers in BLDC motor speed control for electric vehicles is developed, integrating constrained swarm-based algorithms and real-time road gradient data to significantly improve speed control precision and reduce energy consumption [21]. An adaptive PID controller integrated with direct torque control (DTC) for fuel cell-fed BLDC motor drives is proposed, demonstrating effective regulation of torque and speed while addressing nonlinearity, parameter fluctuations, and load transient issues in industrial applications [22].

Despite these advancements, direct experimental comparisons between fuzzy logic and IncCond methods for flux control in BLDC motors, specifically with respect to real-time power loss minimization, remain limited. This paper addresses this gap by providing a comparative experimental study using a Hardware-in-the-Loop (HIL) setup, demonstrating the superior efficiency and adaptability of fuzzy control, while also validating the promising performance of the IncCond method.

This paper investigates two adaptive flux control strategies for BLDC motors, each aiming to overcome the limitations of fixed flux control by dynamically adjusting the stator flux reference in real time. The first strategy applies the incremental conductance (IncCond) method, originally developed for photovoltaic MPPT, to refine the flux reference according to instantaneous system operating conditions. The second strategy employs a fuzzy logic controller, specifically a Mamdani-type fuzzy inference system, which incorporates expert knowledge to manage system nonlinearities and uncertainties by adjusting the stator flux based on motor speed and load torque. Each method is implemented and evaluated independently to assess its effectiveness in improving torque performance, reducing power losses, and enhancing overall system efficiency, particularly in applications such as electric vehicles and renewable energy systems.

This study employs a 100 W BLDC motor with well-characterized internal parameters, enabling detailed analysis and accurate validation of the proposed adaptive flux control strategies. Although the absolute power losses in such low-power systems are relatively small, their proportionate impact on overall efficiency is considerable, making the setup well-suited for comparative evaluation. Moreover, the control methods developed and tested in this work are designed to be scalable and applicable to higher-power systems, including those used in electric vehicles and industrial motor drives, where efficiency optimization remains a critical objective.

The paper is structured as follows: Section 2 introduces the mathematical model of the BLDC motor. Then, the proposed adaptation flux control technique is discussed in Section 3, while Section 4 presents the power losses. Section 5 discusses the experimental results. Finally, Section 6 concludes the paper by summarizing the key findings and future directions.

2. Mathematical Model of the BLDC Motor

The state-space approach is a fundamental tool in modern control theory for modeling and analyzing dynamic systems [23]. This method provides a compact mathematical framework applicable to both linear and nonlinear control strategies, making it essential for designing high-performance motor controllers.

This study considers a three-phase BLDC motor controlled via a full-bridge inverter operating in sinusoidal conduction mode, where at any instant, all three phases conduct current simultaneously with sinusoidal waveforms synchronized to the rotor position. To simplify the analysis, the following assumptions are made: the magnetic circuit is unsaturated, the phase resistances remain constant and are unaffected by temperature variations, and the self-inductances do not vary with rotor position. This modeling approach enhances efficiency, ensures smoother torque, and simplifies commutation. The system dynamics are represented in state-space form as:

$$\begin{bmatrix} \dot{i}_a \\ \dot{i}_b \\ \dot{\omega} \\ \dot{\theta}_s \end{bmatrix} = \begin{bmatrix} -\frac{R}{L} & 0 & 0 & 0 \\ 0 & -\frac{R}{L} & 0 & 0 \\ 0 & 0 & -\frac{B}{J} & 0 \\ 0 & 0 & 1 & 0 \end{bmatrix} \begin{bmatrix} i_a \\ i_b \\ \omega \\ \theta_s \end{bmatrix} + \begin{bmatrix} \frac{2}{3L} & \frac{1}{3L} & 0 \\ -\frac{1}{3L} & \frac{1}{3L} & 0 \\ 0 & 0 & \frac{1}{J} \\ 0 & 0 & 0 \end{bmatrix} \begin{bmatrix} u_{ab} - e_{ab} \\ u_{bc} - e_{bc} \\ T_e - T_L \end{bmatrix} \tag{1}$$

where u_{ab}, u_{bc} are the line voltages, e_{ab}, e_{bc} the back-EMFs, R the phase resistance, L the phase inductance, J the rotor moment of inertia, B the friction coefficient, T_e the electromagnetic torque, T_L the load torque, ω the rotor speed, θ_s the rotor position, and i_a, i_b the phase currents.

To facilitate control implementation, the Clarke transformation maps the three-phase (a, b, c) voltages, currents, and back-EMFs into a two-phase (α, β) stationary reference frame:

$$\begin{bmatrix} v_\alpha \\ v_\beta \end{bmatrix} = \begin{bmatrix} \frac{2}{3} & -\frac{1}{3} & -\frac{1}{3} \\ 0 & \frac{\sqrt{3}}{3} & -\frac{\sqrt{3}}{3} \end{bmatrix} \begin{bmatrix} v_a \\ v_b \\ v_c \end{bmatrix},$$

$$\begin{bmatrix} i_\alpha \\ i_\beta \end{bmatrix} = \begin{bmatrix} \frac{2}{3} & -\frac{1}{3} & -\frac{1}{3} \\ 0 & \frac{\sqrt{3}}{3} & -\frac{\sqrt{3}}{3} \end{bmatrix} \begin{bmatrix} i_a \\ i_b \\ i_c \end{bmatrix}, \tag{2}$$

$$\begin{bmatrix} e_\alpha \\ e_\beta \end{bmatrix} = \begin{bmatrix} \frac{2}{3} & -\frac{1}{3} & -\frac{1}{3} \\ 0 & \frac{\sqrt{3}}{3} & -\frac{\sqrt{3}}{3} \end{bmatrix} \begin{bmatrix} e_a \\ e_b \\ e_c \end{bmatrix}$$

The flux linkages in the $\alpha\beta$ frame are given by:

$$\lambda_\alpha = \int (V_\alpha - Ri_\alpha)\, dt; \quad \lambda_\beta = \int (V_\beta - Ri_\beta)\, dt \tag{3}$$

with the estimated flux magnitude:

$$\phi_s = \sqrt{\lambda_\alpha^2 + \lambda_\beta^2} \tag{4}$$

The electromagnetic torque is expressed as:

$$T_e = \frac{3}{2} P (\lambda_\alpha i_\beta - \lambda_\beta i_\alpha) \tag{5}$$

where P is the number of pole pairs. The motor parameters used in this study are listed in Table 1. These values are based on a standard BLDC motor model available in the MATLAB/Simulink library R2024b, commonly used for academic and validation purposes.

In this paper, a DTC strategy for the BLDC motor is implemented using a predefined switching pattern presented in Equation (6), consisting of six active voltage vectors. Figure 1a depicts the circular arrangement of voltage vectors within their corresponding sectors, which guide the switching decisions. Figure 1b shows the resulting sinusoidal variation of phase currents over a 60° electrical cycle. Switching times are determined based on these patterns, enabling accurate torque estimation. This approach allows for a comprehensive comparison of torque and flux responses for different voltage vectors, ensuring improved dynamic performance and control precision [24,25].

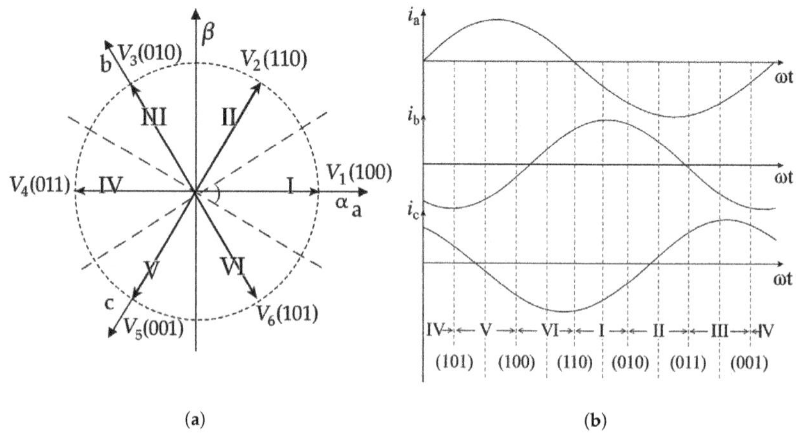

Figure 1. Illustration of the DTC switching principle: (**a**) Circular voltage sectors with labeled vectors. (**b**) Sinusoidal phase currents over one electrical cycle.

Table 1. BLDC motor parameters.

Parameter	Value	Unit
Electric power P_{elec}	100	W
Maximum speed ω	60	rad/s
Maximum torque T_L	1.5	Nm
Stator resistance R_s	0.18	Ω
Nominal flux ϕ_n	0.07145	Wb
Stator phase inductance L_s	8.5×10^{-3}	H
Number of pole pairs p	4	-
Moment of inertia J	0.00062	$kg \cdot m^2$
Damping coefficient B	0.0003035	Ns/m
Rated voltage V_{dc}	48	V
Hysteresis constant K_h	0.05	-
Eddy current constant K_{eddy}	0.001	-

$$(S_a, S_b, S_c) = \begin{cases} (1,0,0), & \text{if } V = V_1 \\ (1,1,0), & \text{if } V = V_2 \\ (0,1,0), & \text{if } V = V_3 \\ (0,1,1), & \text{if } V = V_4 \\ (0,0,1), & \text{if } V = V_5 \\ (1,0,1), & \text{if } V = V_6 \end{cases} \tag{6}$$

3. Adaptive Flux Control Techniques

Adaptive flux control represents a paradigm shift from traditional fixed-flux motor control strategies by enabling real-time adjustment of the stator flux reference in response to varying operating conditions [26]. Unlike conventional methods that maintain a constant flux regardless of load or speed changes, adaptive control dynamically adjusts the flux magnitude to minimize power losses while preserving torque performance [1]. This methodology leverages the fundamental observation that optimal flux levels are not fixed, but rather depend on the motor's instantaneous operating point, allowing for improved efficiency across a broad range of speed–torque combinations [27].

The theoretical foundation of this approach lies in the relationship between electromagnetic losses and flux magnitude: copper losses scale with the square of current, while core losses are influenced by flux density. By intelligently adapting the stator flux, both types of losses can be minimized simultaneously. This is particularly advantageous in applications with frequent variations in speed and load, such as electric vehicles and renewable energy systems, where efficiency directly impacts performance and operational costs [28].

In the following subsections, two adaptive flux control strategies are presented. The first is the incremental conductance (IncCond) method, which relies on power–flux characteristics to determine optimal flux levels. The second is a Fuzzy Logic-based controller, which adapts the flux reference using a rule-based inference system driven by real-time motor speed and load torque.

3.1. Adaptive Flux Control Using Incremental Conductance

The IncCond method is primarily utilized for MPPT in PV systems [29], rather than for direct flux control in motors. In traditional applications, flux control in motors, especially PMSMs, is often achieved through methods like flux-weakening control. For instance, a study by [30] presents a voltage-based flux-weakening strategy using a variable incremental controller to manage the flux component current in PMSMs. In this work, it is employed for adaptive flux control of the BLDC motor, aiming to optimize the electromagnetic performance while maintaining stability under varying operating conditions. The IncCond method leverages the instantaneous relationship between power and flux to achieve faster convergence and reduced oscillations.

The proposed Algorithm 1 dynamically adjusts the flux reference based on the incremental change in flux ($\Delta\phi$) and power (ΔP_{em}). The slope $\frac{\Delta P_{em}}{\Delta\phi}$ is computed, and its sign is used to adjust the flux:

$$\phi_{\text{ref}}(k+1) = \begin{cases} \phi_{\text{ref}}(k) + \Delta\phi, & \text{if } \frac{\Delta P_{em}}{\Delta\phi} > 0 \\ \phi_{\text{ref}}(k) - \Delta\phi, & \text{if } \frac{\Delta P_{em}}{\Delta\phi} < 0 \\ \phi_{\text{ref}}(k), & \text{if } \frac{\Delta P_{em}}{\Delta\phi} = 0 \end{cases} \quad (7)$$

This iterative process ensures that the flux converges to the point where $\frac{\partial P_{em}}{\partial\phi} = 0$, which corresponds to the maximum power point.

To maintain system stability and prevent excessive adjustments, the flux reference is limited within a predefined range:

$$\phi_{\text{ref}} = \min(\max(\phi_{\text{ref}}, \phi_{\min}), \phi_{\max}) \quad (8)$$

where $\phi_{\min}$ and $\phi_{\max}$ are the lower and upper bounds of flux, typically set as a percentage of nominal flux.

Algorithm 1. Incremental Conductance Adaptation Algorithm Steps

Input:

- Instantaneous stator voltages (e_α, e_β)
- Instantaneous stator currents (i_α, i_β)
- Estimated flux magnitude (ϕ_s)
- Rotor speed (ω)

Output:

- Adjusted flux reference (ϕ_{ref})

Initialization:

1. Define nominal optimal flux ϕ_{opt}.
2. Define flux saturation bounds: ϕ_{min} and ϕ_{max}.
3. Initialize current flux reference $\phi_{\text{ref}}(0)$ (e.g., to ϕ_{opt} or a predefined value).
4. Initialize previous flux reference $\phi_{\text{ref,old}} \leftarrow \phi_{\text{ref}}(0)$.
5. Initialize previous electromagnetic power $P_{\text{em,old}} \leftarrow 0$.
6. Define distance thresholds $d_{\text{large}}, d_{\text{medium}}$ and corresponding adaptation step sizes $\Lambda_{\text{large}}, \Lambda_{\text{medium}}, \Lambda_{\text{small}}$.

Loop (executed at each control cycle):

1. Calculate instantaneous electromagnetic power: $P_{\text{em}} = e_\alpha i_\alpha + e_\beta i_\beta$.
2. Compute the change in flux reference from the previous iteration: $\Delta\phi = \phi_{\text{ref}} - \phi_{\text{ref,old}}$.
3. Compute the power–flux gradient (slope):
    - If $\Delta\phi \neq 0$: gradient $= (P_{\text{em}} - P_{\text{em,old}})/\Delta\phi$.
    - Else $(\Delta\phi = 0)$: gradient $= 0$.
4. Compute the absolute distance from the nominal optimal flux: $d = |\phi_{\text{ref}} - \phi_{\text{opt}}|$.
5. Determine the adaptation step size Λ based on distance d:
    - If $d > d_{\text{large}}$: $\Lambda \leftarrow \Lambda_{\text{large}}$
    - Else if $d > d_{\text{medium}}$: $\Lambda \leftarrow \Lambda_{\text{medium}}$
    - Else: $\Lambda \leftarrow \Lambda_{\text{small}}$
6. Adjust flux reference ϕ_{ref}:
    - If gradient > 0: $\phi_{\text{ref}} \leftarrow \phi_{\text{ref}} + \Lambda$
    - Else if gradient < 0: $\phi_{\text{ref}} \leftarrow \phi_{\text{ref}} - \Lambda$
    - Else (gradient $= 0$): No change to ϕ_{ref}
7. Saturate ϕ_{ref} within defined bounds: $\phi_{\text{ref}} \leftarrow \max(\phi_{\text{min}}, \min(\phi_{\text{max}}, \phi_{\text{ref}}))$.
8. Update state variables for the next iteration:
    - $P_{\text{em,old}} \leftarrow P_{\text{em}}$
    - $\phi_{\text{ref,old}} \leftarrow \phi_{\text{ref}}$

3.2. System Integration and Control Architecture

Figure 2 illustrates the complete control architecture for the proposed BLDC motor drive system, integrating adaptive flux regulation with multi-level control strategies to ensure dynamic performance, torque accuracy, and energy-efficient operation.

The control loop operates as follows:

1. Measurement and Feedback Acquisition: The system continuously measures the stator current components and rotor speed, either via sensors or through sensorless estimation based on back-EMF. These signals form the foundation of the real-time feedback loop.
2. Clarke Transformation and Estimation Block: Acquired phase currents are transformed into the stationary $\alpha\beta$ reference frame. This enables the estimation of electromagnetic torque, stator flux magnitude, and rotor position—essential parameters for both control and observation tasks.
3. Adaptive Flux Regulation via Incremental Conductance: Leveraging the incremental conductance (IncCond) principle, the adaptive control module dynamically adjusts the

reference flux. By analyzing the power contribution in the $\alpha\beta$ frame ($P_{em} = e_\alpha i_\alpha + e_\beta i_\beta$), the controller tunes the flux reference with adaptive step sizes to maintain optimal operating conditions across varying load and speed profiles.

4. Speed and Torque Control using Optimized PI Loops: A cascade control structure is adopted, where an outer-loop PI controller regulates speed, and the inner-loop handles current (torque) control. The PI parameters are optimized to maintain closed-loop stability while incorporating the flux reference updates from the adaptive process.
5. Hysteresis Current Regulation: Two independent hysteresis regulators compare the measured and reference current components (i_α, i_β) and generate switching signals when deviations exceed a predefined threshold. This ensures fast current response and reduced torque ripple.
6. Switching Table Logic: Based on the outputs from the hysteresis controllers, the switching table determines the appropriate gating signals for the inverter. The commutation strategy adheres to the six-step method, ensuring proper sector alignment with minimal switching losses.
7. Inverter and Motor Actuation: The inverter translates control signals into high-speed switching patterns, delivering the regulated voltage and current to the BLDC motor. This final actuation ensures precise torque production and effective utilization of the adaptive flux strategy.

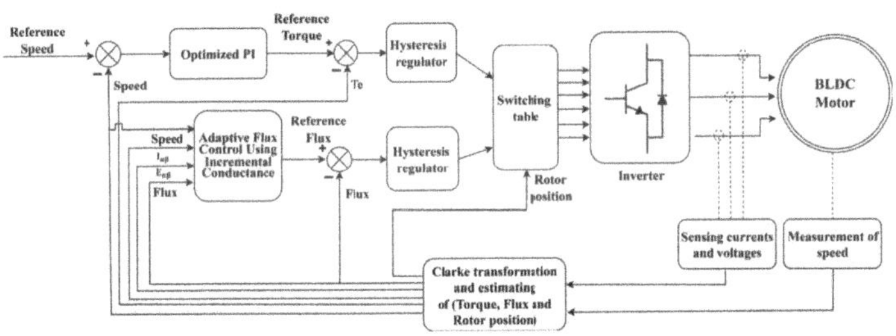

Figure 2. Control scheme integrating adaptive flux regulation with BLDC motor drive system.

This layered control structure effectively integrates adaptive flux regulation with conventional motor control techniques to enhance overall drive performance. One of its key advantages is the real-time optimization of stator flux using the IncCond method, which operates without the need for a detailed motor model. The control scheme also improves dynamic response through fast hysteresis-based current regulation, while robust speed and torque tracking are achieved via cascade-structured PI controllers. Additionally, the implementation of reliable switching logic ensures efficient commutation across all operating conditions.

Figure 3 illustrates the complete control architecture integrating adaptive fuzzy logic-based flux regulation into the DTC scheme for the BLDC motor drive.

The control system follows a cascaded and real-time feedback loop as described below:

1. Clarke Transformation and Estimation Block: The stator currents and voltages are processed through the Clarke transformation to compute the rotor position, electromagnetic torque, stator flux magnitude, and motor speed.
2. Adaptive Flux Regulation via Fuzzy Logic: Based on real-time speed and torque feedback, the fuzzy logic controller adaptively updates the reference flux ϕ^* using a predefined rule base and membership functions.
3. Speed and Torque Control Using Optimized PI Loop: A speed controller generates the torque reference, which is compared to the actual torque via a hysteresis regulator.
4. Hysteresis Regulation: Flux and torque regulators track their respective references and trigger control actions when deviations exceed the hysteresis bounds.

5. Switching Table Logic and Inverter Operation: The outputs from the hysteresis regulators and rotor position are used by the switching table to select the appropriate voltage vector for inverter control.

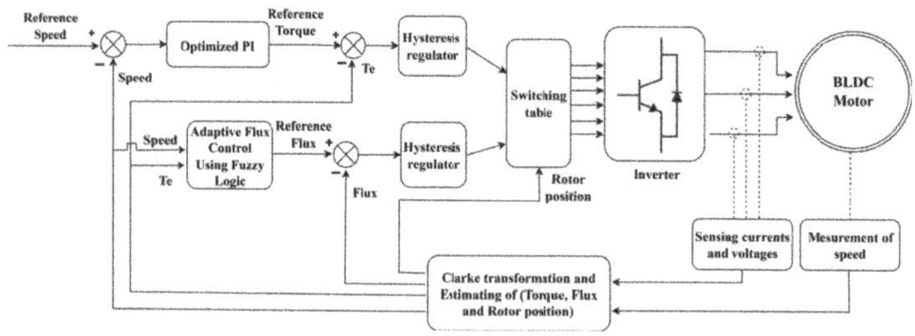

Figure 3. Control architecture integrating adaptive fuzzy flux control into DTC.

3.3. Adaptive Flux Control Using Fuzzy Logic

In this work, the fuzzy logic module adjusts the stator flux reference in real time based on two key inputs: the motor speed and load torque. The rule base, as shown in Table 2, is constructed to ensure high flux at low speeds (for torque generation), medium flux at moderate conditions (for efficiency), and low flux at high speeds with low torque (to minimize core losses). Similar strategies have been reported in the literature to improve DTC performance and extend battery life in electric vehicles [10].

Table 2. Fuzzy logic rule table for adaptive flux control.

Speed\Torque	Low	Medium	High
Low	High Flux	High Flux	High Flux
Medium	Medium Flux	Medium Flux	High Flux
High	Low Flux	Medium Flux	High Flux

The Algorithm 2 enables real-time flux adaptation, improving energy efficiency under varying load–speed profiles. To achieve this, the controller takes the motor speed and load torque as inputs and produces a scaling factor that adjusts the reference flux accordingly. To maintain the flux within a suitable range, between 70% and 130% of the nominal value, a gain of 1.45 is applied to the output of the fuzzy logic controller. This design assumes that the magnetic circuit remains unsaturated, in line with the modeling assumptions introduced in Section 2. Under this assumption, the flux limits $\phi_{\min}$ and $\phi_{\max}$ were empirically determined through a combination of simulation and experimental tuning. These bounds were selected to balance energy efficiency with safe magnetic operation, avoiding both saturation and torque instability.

The fuzzy rule base, presented in Table 2, was initially developed using standard control heuristics rooted in common motor behavior. These rules were then refined through iterative simulation and testing to improve performance under dynamic conditions. Although tailored to the test motor, the fuzzy logic controller maintains a flexible structure, allowing easy adaptation to other BLDC motors by retuning the flux boundaries and adjusting membership functions. This makes the method suitable for a wide range of practical applications. To support real-time implementation, the membership functions used for the fuzzification of the input variables (torque and speed) and the output variable (flux) are kept simple and computationally efficient. As illustrated in Figure 4, each variable is divided into three fuzzy subsets (low, medium, and high) using trapezoidal membership functions. The Mamdani inference system is employed to derive the output, ensuring smooth and interpretable control decisions.

Algorithm 2. Fuzzy Flux Adaptation Algorithm Steps

1. **Input Acquisition:**
   - The instantaneous rotor speed ω and electromagnetic torque T_e are extracted from the control loop and provided as inputs to the fuzzy inference system.
2. **Normalization and Fuzzification:**
   - The crisp inputs ω and T_e are first normalized by mapping them to a defined universe of discourse, typically scaled between 0 and 1, based on their expected minimum and maximum values.
   - These normalized values are then converted into fuzzy linguistic variables (e.g., Low, Medium, High) using trapezoidal membership functions, as shown in Figure 4.
3. **Rule Evaluation:**
   - The Mamdani inference engine evaluates the fuzzy rules, as depicted in Table 2, to infer a suitable fuzzy output category for the flux reference (e.g., Low, Medium, High).
4. **Defuzzification:**
   - The combined fuzzy output is transformed into a crisp stator flux reference ϕ^* using the center of gravity method.
5. **Flux Update:**
   - The resulting flux reference ϕ^* is applied to the DTC block to influence the inverter switching decisions and motor behavior.

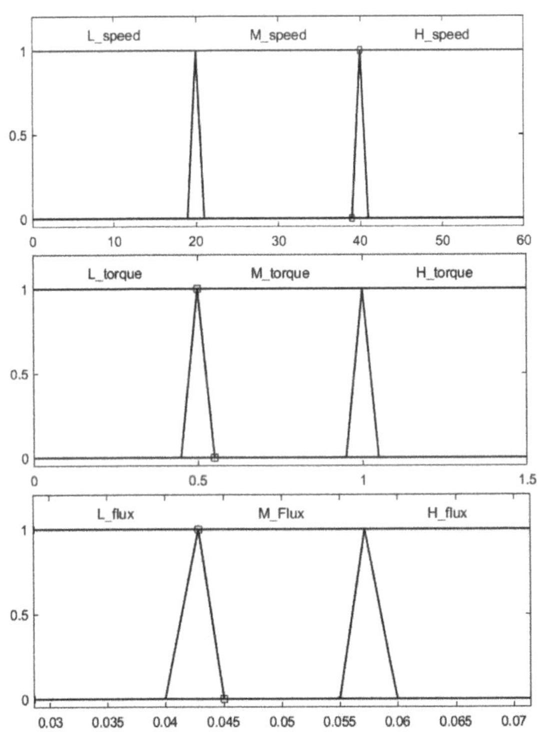

Figure 4. Fuzzy membership functions using trapezoidal sets and a Mamdani inference system. The first image shows the input speed membership functions, the second shows the input torque membership functions, and the third represents the output flux membership functions.

This hybrid strategy preserves the inherent advantages of classical DTC, such as its fast dynamic response and simple structure, while integrating adaptive flux regulation to improve energy efficiency. Unlike conventional DTC, which relies on fixed flux references and suffers from increased losses under variable load–speed profiles, the proposed approach dynamically adjusts the flux level in real time. Similar improvements through adaptive flux control have been reported in recent works such as [31,32], reinforcing the practical effectiveness of this control framework.

4. Power Losses

This section outlines the methodology used to compute various forms of power losses in the BLDC motor system. Accurate power calculations are essential for evaluating system efficiency and identifying loss contributors, enabling strategies for performance optimization.

4.1. Core Losses

Core losses in magnetic materials mainly consist of hysteresis and eddy current losses. The total core loss is computed using the following empirical relation:

$$P_{\text{core}} = \Phi^2 \cdot \left(K_{\text{h}} \cdot N + K_{\text{e}} \cdot N^2\right) \tag{9}$$

where Φ denotes the magnetic flux density (Wb/m^2), K_{h} is the hysteresis loss coefficient, which is material-dependent, K_{e} represents the eddy current loss coefficient, and N is the rotational speed of the rotor expressed in revolutions per minute (rpm).

Hysteresis losses scale linearly with speed, while eddy current losses grow quadratically with speed. This implies that as N increases, eddy current losses dominate the total core loss contribution [33].

4.2. Copper Losses

Copper losses are the resistive (Joule) losses due to current flow in the motor windings. In a three-phase BLDC motor, these losses are expressed as:

$$P_{\text{cu}} = 3 \cdot R \cdot I^2 \tag{10}$$

where, R is the resistance of a single phase winding and I is the RMS value of the current in each phase.

Minimizing copper losses is crucial for enhancing motor efficiency. This can be achieved by using low-resistance materials such as high-conductivity copper, increasing the conductor cross-sectional area to reduce resistance, and optimizing current levels according to the load conditions.

4.3. Total Power Losses

The total losses in the motor system are the sum of the copper and core losses:

$$P_{\text{tot}} = P_{\text{cu}} + P_{\text{core}} \tag{11}$$

This total power loss directly influences system efficiency, calculated as:

$$\eta = \frac{P_{\text{e}}}{P_{\text{in}}} \times 100\% \tag{12}$$

By incorporating adaptive flux control strategies (as discussed earlier), both P_{cu} and P_{core} can be minimized, leading to improved overall drive efficiency.

5. Experimental Results

To assess the practical feasibility of the proposed controller, real-time experiments were conducted using a Hardware-in-the-Loop (HIL) environment incorporating two dSPACE 1104 cards. Using the HIL test bench setup illustrated in Figure 5, combined with the DTC scheme and the switching strategy detailed in Equation (6), this section presents a comparative analysis between the traditional fixed flux reference (set to nominal value) and the two proposed adaptive flux strategies.

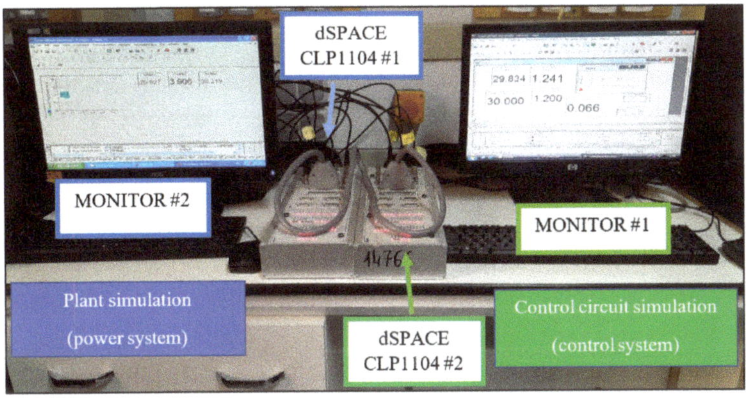

Figure 5. HIL setup.

5.1. Description of the Experimental Test Bench

In this study, an HIL test bench is developed for the validation of a BLDC motor drive, utilizing two dSPACE cards configured in a master–slave architecture. The master card generates PWM signals for motor control and processes feedback signals, while the slave card simulates the BLDC motor dynamics in real time, providing the necessary responses to close the loop. Communication between the two cards is established via high-speed CAN and synchronized through a dedicated trigger signal. Compared to real-time simulation, HIL testing provides a more accurate validation environment by integrating actual hardware, avoiding the limitations of sensor modeling and signal noise that often affect real-time simulations. In purely simulated environments, sensor imperfections such as noise, offsets, and response delays may not be accurately reproduced, leading to discrepancies between simulation and real-world performance. In the proposed setup, MONITOR 1 runs the control loop and receives key feedback signals from the simulated plant through analog channels. Specifically, ADC1 reads the rotor position, ADC2 captures the torque, ADC3 receives the flux value, and ADC5 monitors the rotor speed, while the PWM control signals generated by the controller are transmitted via the slave I/O interface to drive the inverter gates connected to the plant side. Meanwhile, MONITOR 2 emulates the plant (power system) dynamics; it receives the inverter gate signals via its I/O interface and generates the corresponding analog outputs. These outputs are then transmitted back to the controller side through DAC channels: DAC3 carries the rotor position signal, DAC4 transmits the torque, DAC6 represents the flux, and DAC8 contains the rotor speed. This closed-loop signal exchange replicates the real-time behavior of the BLDC motor system with high fidelity. By using HIL, the control algorithms are tested under realistic conditions, ensuring a more reliable transition to physical implementation.

5.2. Results and Discussion

The experimental validation was designed to comprehensively assess the performance of the BLDC motor drive under all relevant operating conditions. The 24 s test interval was divided into four consecutive 6 s phases, each corresponding to a specific speed level. During the first interval [0, 6] s, the load torque was set to zero while the speed varied through low, medium, and high levels in order to isolate and examine the impact of speed changes alone. In the subsequent intervals, the speed was fixed while the load torque was varied every 2 s to cover low (0.4 N·m), medium (0.8 N·m), and high (1.2 N·m) torque levels. Specifically, in the second interval [6, 12] s, the speed was maintained at a low level; in the third interval [12, 18] s, the speed was medium; and in the final interval [18, 24] s, the speed was high. The PI controller used in the speed regulation loop was carefully tuned through offline optimization to ensure a fast dynamic response and minimal steady-state error, and then integrated into the overall system architecture, as shown in Figures 2 and 3. It is important to note that the transitions in speed and torque were not applied simultaneously; to ensure system stability and avoid overlapping transients, each speed adjustment was introduced approximately 0.1 s before the corresponding torque change. This configuration ensured a stable response and enabled accurate evaluation of each operating point. The updated test scenario matrix is summarized in Table 3, and the applied load torque profile is illustrated in Figure 6.

Table 3. Speed and torque load combinations for experimental validation.

Time Interval(s)	Speed Level	Torque Load
[0, 2]	Low	Zero
[2, 4]	Medium	Zero
[4, 6]	High	Zero
[6, 8]	Low	Low
[8, 10]	Low	Medium
[10, 12]	Low	High
[12, 14]	Medium	Low
[14, 16]	Medium	Medium
[16, 18]	Medium	High
[18, 20]	High	Low
[20, 22]	High	Medium
[22, 24]	High	High

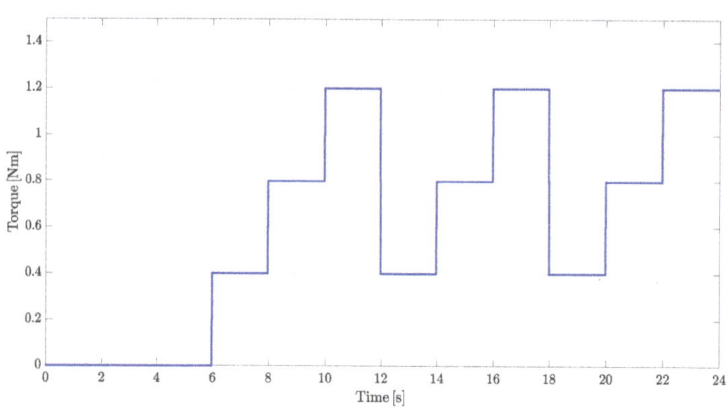

Figure 6. Load torque profile.

The waveforms for the stator current and back-EMF are shown in Figures 7 and 8, respectively. These figures provide insights into the motor's behavior, where the stator current follows a sinusoidal pattern for the fixed flux control strategy and the back-EMF exhibits the typical trapezoidal waveform. The experimental results demonstrate the effectiveness of adaptive flux strategies in minimizing power losses and improving system efficiency. The performance of the fixed flux, IncCond, and fuzzy logic control strategies is compared in this section.

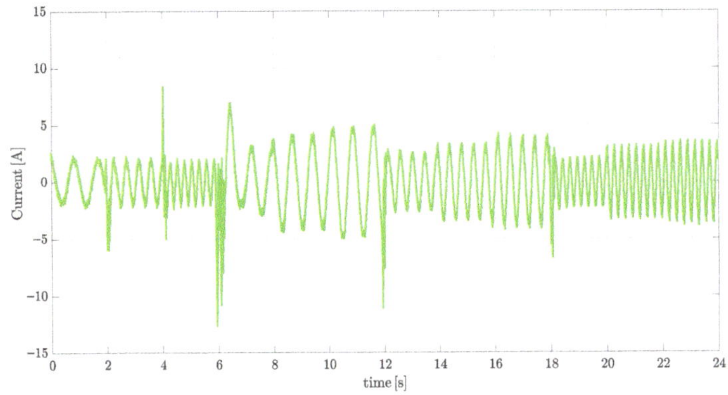

Figure 7. Stator current waveform under fixed flux control.

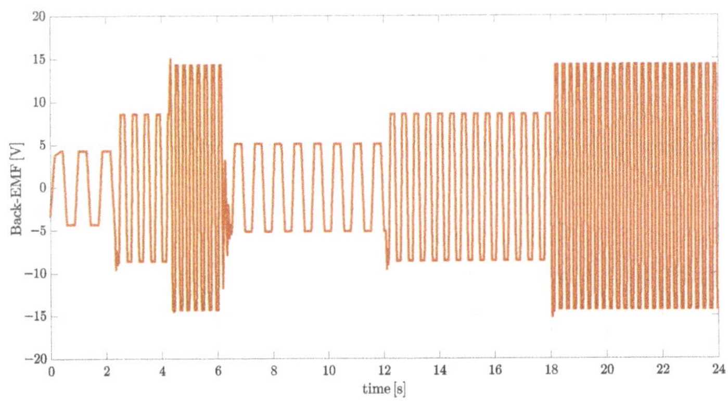

Figure 8. Back-EMF waveform under fixed flux control.

Figure 9 illustrates the motor speed response under different flux control strategies. The speed profiles appear closely aligned, indicating minimal differences in steady-state performance. However, IncCond exhibits a slightly faster time response compared to fixed flux, particularly during speed transitions. Fuzzy logic outperforms both, achieving the most consistent speed tracking.

Figure 10 presents the electromagnetic torque profiles under different flux control strategies. The fixed flux case, Figure 10a, delivers the smoothest and most stable torque response, characterized by minimal torque ripple under varying operating conditions, demonstrating superior torque stability compared to both the IncCond and fuzzy logic strategies. The IncCond strategy, Figure 10b, exhibits a reduction in torque oscillations compared to the fuzzy logic approach, and provides a more stable torque response during load transitions, benefiting from its gradient-based flux tuning. Meanwhile, the fuzzy logic approach, Figure 10c, shows significant torque oscillations, particularly during load transitions, due to its dynamic adaptation method. However, this increase in oscillation is an intentional trade-off made to prioritize power loss reduction. The fuzzy controller operates within empirically defined limits for flux (ϕ_{min} and ϕ_{max}), rotor speed, and load torque, ensuring safe operation while maximizing efficiency. These constraints were established through iterative testing and logical system analysis, allowing the controller to respond adaptively to varying operating conditions while respecting stability margins.

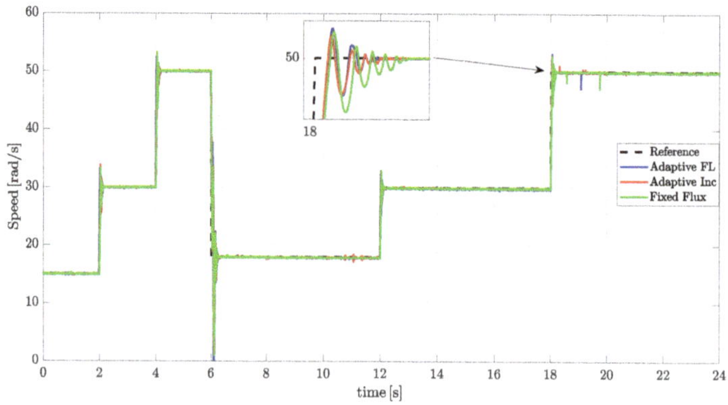

Figure 9. Speed performance comparison for fixed flux, incremental conductance, and fuzzy logic flux control strategies.

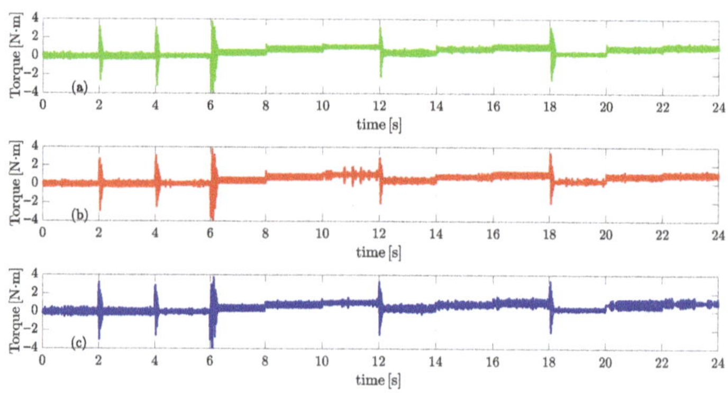

Figure 10. Electromagnetic torque comparison: (**a**) fixed flux, (**b**) incremental conductance, and (**c**) fuzzy logic.

Figure 11 illustrates the flux profiles under different control strategies. In the fixed flux case, Figure 11a, the flux remains constant regardless of load or speed variations, resulting in suboptimal efficiency due to the lack of adaptability. The IncCond method, Figure 11b, introduces dynamic flux adjustment in response to changing operating conditions, though its responsiveness is limited compared to fuzzy logic. The fuzzy logic strategy, Figure 11c, offers the most effective flux adaptation, dynamically tuning the flux reference based on real-time speed and torque.

Figures 12–15 present a comparative analysis of the input power, electromagnetic power, core losses, and copper losses, respectively, under the three flux control strategies. While all approaches maintain similar trends, the adaptive strategies IncCond and fuzzy logic demonstrate clear advantages over the fixed flux method. In particular, Figure 12 shows that adaptive flux control effectively reduces the required input power, with the fuzzy logic strategy exhibiting the most pronounced reduction. Figure 13 indicates that the electromagnetic power remains relatively stable across methods, with a slight advantage observed under fuzzy logic control. Regarding core losses in Figure 14, both adaptive strategies introduce small variations, but no significant improvement is evident compared to the fixed approach. However, Figure 15 highlights a marked reduction in copper losses under fuzzy logic control, attributed to its improved current regulation. These observations suggest that while both adaptive methods enhance system performance, the fuzzy logic strategy achieves the most effective balance between input power reduction and loss minimization, particularly in terms of copper losses, thereby contributing to higher overall efficiency.

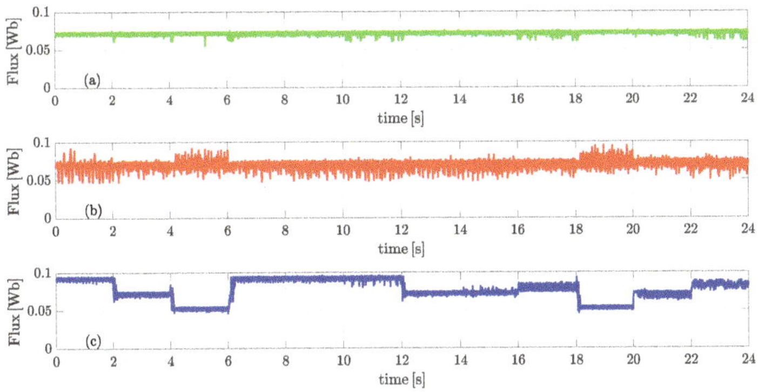

Figure 11. Flux behavior comparison: (**a**) fixed flux, (**b**) incremental conductance, and (**c**) fuzzy logic.

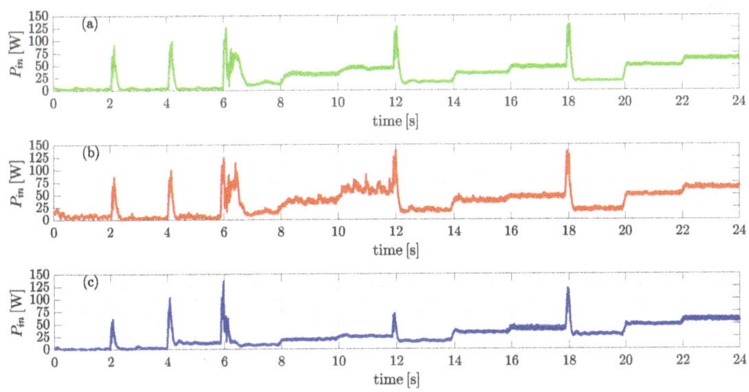

Figure 12. Input power comparison: (**a**) fixed flux, (**b**) incremental conductance, and (**c**) fuzzy logic.

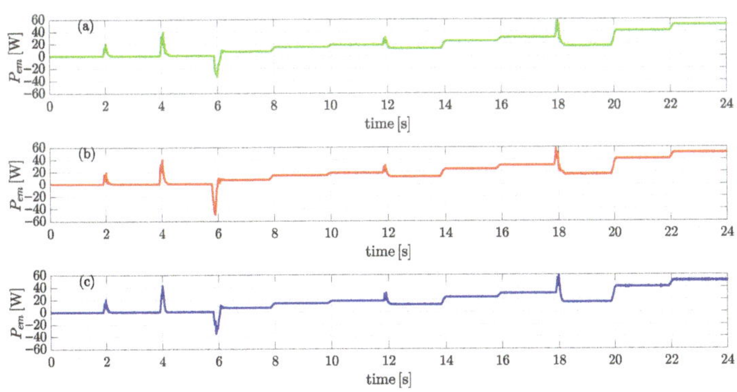

Figure 13. Electromagnetic power comparison: (**a**) fixed flux, (**b**) incremental conductance, and (**c**) fuzzy logic.

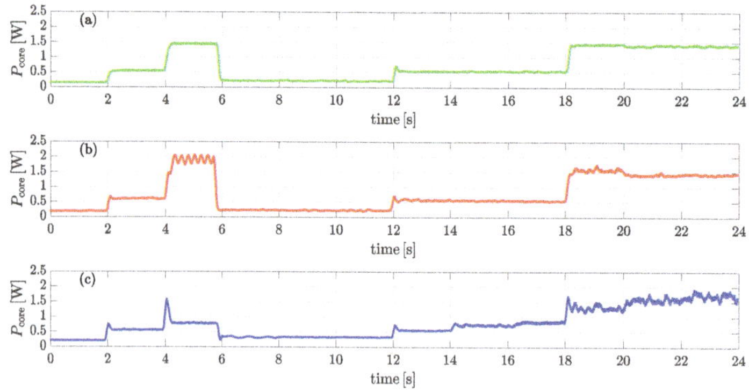

Figure 14. Core losses comparison: (**a**) fixed flux, (**b**) incremental conductance, and (**c**) fuzzy logic.

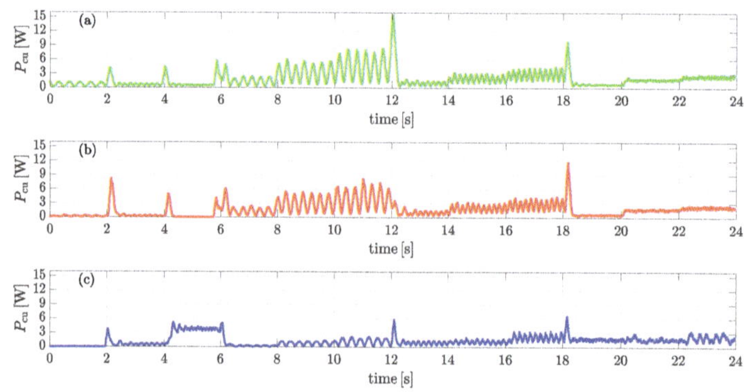

Figure 15. Copper losses comparison: (**a**) fixed flux, (**b**) incremental conductance, and (**c**) fuzzy logic.

The numerical results of the average power values for each method are summarized in Table 4. These values were computed as mean values over 24 s, which is the standard approach in power system efficiency analysis. This average-based evaluation reflects the long-term electrical stress on the motor [34].

Table 4. Mean power calculations and efficiency comparison.

Metric	Fixed Flux	IncCond Method	Fuzzy Logic
Flux Range (Wb)	0.07145	Adaptive	Adaptive
Input Power (W)	34.15	31.21	26.55
Electromagnetic Power (W)	16.78	16.45	17.68
Core Losses (W)	0.71	0.70	0.66
Copper Losses (W)	2.03	1.76	1.38
Total Power Losses (W)	2.74	2.46	2.04
System Efficiency (%)	49.13	52.70	66.59
Loss Reduction vs. Fixed Flux (%)	–	10.22	25.55

Among the three strategies, fuzzy logic exhibits the best overall performance. It achieves the lowest input power consumption at 26.55 W while delivering the highest electromagnetic power at 17.68 W. In addition, it significantly reduces copper losses to 1.38 W and core losses to 0.66 W, both of which are the lowest among all methods. This results in the lowest total power losses, amounting to 2.04 W, and leads to a system efficiency of 66.59%—the highest observed. The IncCond method offers intermediate performance, with an input power of 31.21 W, electromagnetic power of 16.45 W, core losses of 0.70 W, and copper losses of 1.76 W. Its total power losses are reduced to 2.46 W, corresponding to a 10.22% loss reduction compared to the fixed flux approach. The system efficiency achieved is 52.70%. In contrast, the fixed flux strategy results in the highest energy losses and the lowest system efficiency, with an input power of 34.15 W, total power losses of 2.74 W, and an overall efficiency of 49.13%. These results validate the effectiveness of adaptive flux control—particularly fuzzy logic—in minimizing power losses and enhancing overall drive efficiency. The bar charts in Figure 16 present a visual summary of the performance metrics listed in Table 4, highlighting efficiency, input power, total losses, and copper losses for the fixed flux, incremental conductance, and fuzzy logic strategies.

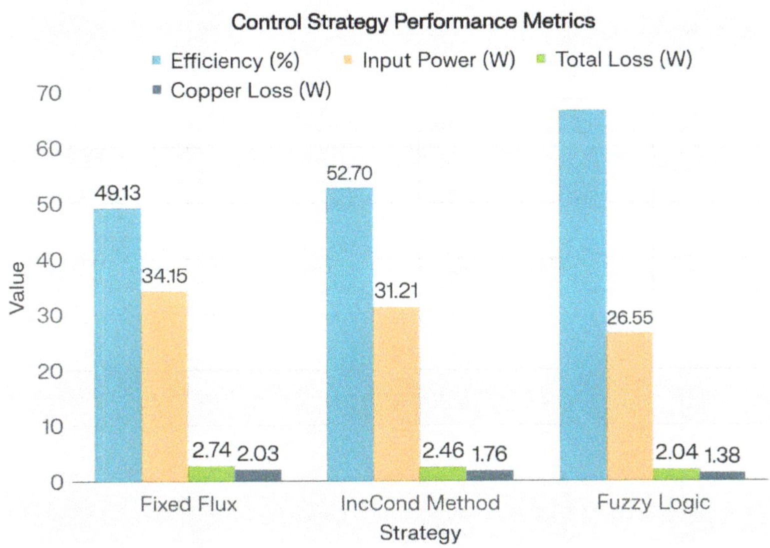

Figure 16. Control strategy performance metrics: efficiency, input power, and total and copper losses for fixed flux, incremental conductance, and fuzzy logic approaches.

The experimental results clearly validate the effectiveness of adaptive flux control strategies in minimizing power losses and enhancing the overall efficiency of BLDC motor drives. Among the three evaluated methods, the fuzzy logic controller achieved the best overall performance, realizing a 25.55% reduction in total power losses and a system efficiency of 66.59%. The incremental conductance approach also demonstrated significant benefits, yielding a 10.22% loss reduction compared to the fixed flux baseline. These improvements were achieved through intelligent real-time flux regulation: the IncCond method provided fast and responsive flux adaptation based on power–flux gradient analysis, while the fuzzy controller fine-tuned the flux reference using a rule-based system informed by motor speed and load torque.

All strategies were implemented and validated using an HIL environment based on a dual dSPACE 1104 card setup. The control algorithm was executed in real time using a DTC scheme, where accurate flux estimation was enabled by leveraging the known internal parameters of the BLDC motor. The real-time simulation model of the motor included detailed modeling of electromagnetic dynamics, core losses, and switching behavior, ensuring high-fidelity interaction between hardware and simulation domains. The use of a small-scale 100 W BLDC motor, with fully characterized internal parameters, allowed precise analysis of efficiency trends and loss mechanisms across a wide range of operating points.

The proposed framework, integrating IncCond and fuzzy logic within a real-time DTC control architecture, demonstrates a practical, scalable, and energy-efficient solution for intelligent motor drives. It successfully bridges advanced control theory and embedded implementation, offering a concrete pathway toward optimized flux control in energy-sensitive applications. The demonstrated loss minimization, control responsiveness, and implementation feasibility collectively represent a meaningful contribution to the advancement of BLDC motor drives.

5.3. Comparative Analysis with Related Works

To highlight the effectiveness of the proposed adaptive flux control strategies, Table 5 provides a comparison with relevant studies focused on minimizing power losses in BLDC motor drives using various advanced control techniques. The table summarizes the control strategy, validation method, reported loss reduction, and key implementation notes.

Table 5. Comparison of different control strategies for power loss minimization in BLDC motor drives.

Control Method	Strategy	Validation	Loss Reduction	Control Highlights
Fuzzy PWM [35]	Rule-based PWM Fuzzy Control	Simulation only	14.2%	Reduced switching losses through waveform shaping based on fuzzy rules.
D-axis Current Control [31]	Sensorless D-axis Flux Regulation	Experimental	11.7%	Real-time torque improvement via d-axis current decoupling and loss suppression.
Predictive Flux Control [32]	Model-Predictive Flux Optimization	Simulation and experimental	16.4%	Efficiency improved through predictive flux regulation and feedback-based control.
Adaptive PI [36]	Optimal PI Control using Adaptive Algorithm	Simulation only	18.6%	Real-time tuning of PI parameters to reduce resistive losses.
Harmonic Minimization [12]	Optimized Harmonic-Based Switching	Experimental	20.3%	Loss mitigation through harmonic distortion minimization and ripple control.
MTPA-Pseudo FOC [37]	Maximum Torque/Power with Pseudo FOC	Experimental	23.1%	Improved torque per ampere and copper loss reduction using enhanced FOC.
IncCond (This Work)	Incremental Conductance Flux Control	Hardware-in-the-Loop	10.22%	Fast and responsive flux adaptation based on power–flux gradient analysis.
Fuzzy Logic (This Work)	Adaptive Rule-Based Fuzzy Control	Hardware-in-the-Loop	25.55%	Rule-based adaptation using speed and torque inputs with real-time implementation.

As summarized in Table 5, the proposed fuzzy logic-based adaptive flux controller achieves the highest loss reduction of 25.55%, outperforming other advanced strategies such as MTPA with pseudo-FOC (23.1%), harmonic-based switching (20.3%), and adaptive PI control (18.6%). While these methods improve efficiency through techniques like optimized torque utilization, harmonic suppression, or adaptive gain tuning, they do not directly implement real-time adaptive flux control. In contrast, the fuzzy logic and incremental conductance strategies presented in this work are the only approaches in the comparison that explicitly regulate the flux reference for BLDC motor drives based on instantaneous speed and torque conditions. This distinction underscores the practical relevance and effectiveness of adaptive flux control in reducing copper and core losses under dynamic operating conditions, as demonstrated through HIL-based validation.

6. Conclusions

This work presented a novel adaptive flux control strategy for BLDC motors that intelligently adjusts the stator flux in real-time to reduce power losses and improve overall system efficiency. The proposed framework integrates IncCond and FLCs to adaptively regulate flux based on varying speed and load conditions. Experimental validation on a 100 W BLDC motor confirmed that the adaptive strategy significantly outperforms conventional fixed flux control in terms of energy efficiency and power loss reduction. The IncCond method achieved a 10.22% reduction in total power losses compared to fixed flux control by leveraging the instantaneous power–flux relationship for fast and responsive flux adaptation. Originally developed for photovoltaic systems, IncCond proved highly effective in BLDC motor control, enhancing system dynamics under changing load and speed conditions.

Subsequently, the fuzzy logic controller further improved performance by incorporating expert knowledge to adjust the stator flux based on real-time speed and torque inputs. This strategy delivered the best overall results, achieving a 25.55% loss reduction relative to the fixed flux approach, especially under dynamic operating scenarios. Although the implementation was carried out on a small-scale motor, the proposed methodology is scalable to larger motors and drive systems. The observed trends in loss reduction and control performance are expected to translate effectively to applications such as electric vehicles and renewable energy systems, where efficiency optimization is critical.

In summary, the proposed adaptive control approach offers a robust and scalable solution for minimizing losses in BLDC motor drives. Future research will focus on integrating this strategy into multi-level control architectures and further refining the fuzzy rule base and adaptation logic for enhanced real-time performance in embedded applications.

Author Contributions: Conceptualization, M.F.K. and O.C.; methodology, M.F.K. and A.B.; software, M.F.K.; validation, M.S. and C.B.; formal analysis, M.S.; investigation, M.F.K. and M.S.; writing—original draft preparation, M.F.K.; writing—review and editing, M.S.; supervision, O.C.; project administration, C.B. All authors have read and agreed to the published version of the manuscript.

Funding: This research received no external funding.

Data Availability Statement: The original contributions presented in this study are included in the article. Further inquiries can be directed to the corresponding author.

Conflicts of Interest: The authors declare no conflicts of interest.

References

1. Shenbagalakshmi, R.; Mittal, S.K.; Subramaniyan, J.; Vengatesan, V.; Manikandan, D.; Ramaswamy, K. Adaptive speed control of BLDC motors for enhanced electric vehicle performance using fuzzy logic. *Sci. Rep.* **2025**, *15*, 12579. [CrossRef]
2. Pedapenki, K.; Kumar, J.; Anumeha. Fuzzy Logic Controller-Based BLDC Motor Drive. In Proceedings of the 2022 IEEE International Power and Renewable Energy Conference (IPRECON), Kollam, India, 16–18 December 2022; IEEE: Piscataway, NJ, USA, 2022.
3. Rahim, A.; Singh, R. BLDC motor speed control with digital adaptive PID–Fuzzy. *Energies* **2023**, *17*, 1311.
4. Sevugan Rajesh, J.; Karthikeyan, R.; Revathi, R. Analysis and control of grid-interactive PV-fed BLDC water pumping system with optimized MPPT for DC-DC converter. *Sci. Rep.* **2024** , *14*, 25963. [CrossRef]

5. Muthamizhan, T.; Silas Stephen, D.; Sivakumar, A. *ANFIS Controller Based Speed Control of High-Speed BLDC Motor Drive*; IOS Press: Amsterdam, The Netherlands, 2020.
6. Almasi, O.; Fereshtehpoor, V.; Khooban, M.; Blaabjerg, F. Non-inverting buck-boost converter using T–S fuzzy PI control. *IEEE Trans. Power Electron.* **2017**, *32*, 9143–9178.
7. Arya, P.; Althaf, P.; Haridas, A.; Mohanan, J. Brushless DC motor speed control using fuzzy-PID. *AIP Conf. Proc.* **2024**, *3134*, 110008. [CrossRef]
8. Iqbal, A. Chaos control of brushless direct current motor using sliding mode control with a low cost hardware-in-loop validation. *Sci. Talks* **2025**, *14*, 100453. [CrossRef]
9. Udayakumar, A.K.; Raghavan, R.R.V.; Houran, M.A.; Elavarasan, R.M.; Kalavathy, A.N.; Hossain, E. Three-port bi-directional DC–DC converter with solar PV system fed BLDC motor drive using FPGA. *Energies* **2023**, *16*, 624. [CrossRef]
10. Akrami, M.; Jamshidpour, E.; Nahid-Mobarakeh, B.; Pierfederici, S.; Frick, V. Sensorless control methods for BLDC motor drives: A review. *IEEE Trans. Transp. Electrif.* **2024**, *11*, 135–152. [CrossRef]
11. Zhang, H.; Deng, L.; Li, H.; Zheng, S.; Jin, H.; Chen, B. Commutation point optimization method for sensorless BLDC motor control using vector phase difference of back EMF and current. *IEEE/Asme Trans. Mechatron.* **2023**, *29*, 423–433. [CrossRef]
12. Santra, S.B.; Chatterjee, A.; Chatterjee, D.; Padmanaban, S.; Bhattacharya, K. High efficiency operation of brushless DC motor drive using optimized harmonic minimization based switching technique. *IEEE Trans. Ind. Appl.* **2022**, *58*, 2122–2133. [CrossRef]
13. Bhosale, R.; Warshe, W.; Shreelakshmi, M.; Arlikar, P.; Prakash, A.; Agarwal, V. Performance comparison of Two PWM techniques applied to BLDC motor control. In Proceedings of the 2018 International Conference on Power, Instrumentation, Control and Computing (PICC), Thrissur, India, 18–20 January 2018; IEEE: Piscataway, NJ, USA, 2018; pp. 1–6.
14. Czerwinski, D.; Gęca, J.; Kolano, K. Machine learning for sensorless temperature estimation of a BLDC motor. *Sensors* **2021**, *21*, 4655. [CrossRef] [PubMed]
15. Antony, R.P.; Komarasamy, P.R.G.; Rajamanickam, N.; Alroobaea, R.; Aboelmagd, Y. Optimal rotor design and analysis of energy-efficient brushless dc motor-driven centrifugal monoset pump for agriculture applications. *Energies* **2024**, *17*, 2280. [CrossRef]
16. Sayed, K.; El-Zohri, H.H.; Ahmed, A.; Khamies, M. Application of tilt integral derivative for efficient speed control and operation of BLDC motor drive for electric vehicles. *Fractal Fract.* **2024**, *8*, 61. [CrossRef]
17. Xu, Q.; Fang, S. Robust deadbeat predictive current sensorless control for permanent magnet synchronous motor. *IEEE Trans. Power Electron.* **2024**, *39*, 11441–11454. [CrossRef]
18. Bian, F.; Chien, Y.R. PMSM Speed Control Based on Improved Adaptive Fractional-Order Sliding Mode Control. *Symmetry* **2025**, *17*, 736. [CrossRef]
19. Sun, Q.; He, S.; Liu, X.; Niu, F.; Gan, C.; Jia, G. Four-Point Trajectory Tracking Control of PMSMs with Improved Dynamic Response and Steady-State Efficiency. *IEEE Trans. Ind. Electron.* **2024**, *72*, 3482–3491. [CrossRef]
20. Uddin, M.N.; Zou, H.; Azevedo, F. Online loss-minimization-based adaptive flux observer for direct torque and flux control of PMSM drive. *IEEE Trans. Ind. Appl.* **2015**, *52*, 425–431. [CrossRef]
21. Naqvi, S.S.A.; Jamil, H.; Iqbal, N.; Khan, S.; Lee, D.I.; Park, Y.C.; Kim, D.H. Multi-objective optimization of PI controller for BLDC motor speed control and energy saving in Electric Vehicles: A constrained swarm-based approach. *Energy Rep.* **2024**, *12*, 402–417. [CrossRef]
22. Leninpugalhanthi, P.; Latha, R. Adaptive PID controller based direct torque control of fuel cell fed BLDC drive for industrial drives applications. *Therm. Sci. Eng. Prog.* **2024**, *55*, 102931. [CrossRef]
23. Rambabu, S. Modeling and Control of a Brushless DC Motor. Master's Thesis, National Institute of Technology Rourkela, Rourkela, India, 2007.
24. Romeral, L.; Fabrega, A.; Cusido, J.; Garcia, A.; Ortega, J. Torque ripple reduction in a PMSM driven by direct torque control. In Proceedings of the 2008 IEEE Power Electronics Specialists Conference, Rhodes, Greece, 15–19 June 2008.
25. Lad, C.K.; Chudamani, R. Simple overlap angle control strategy for commutation torque ripple minimisation in BLDC motor drive. *IET Electr. Power Appl.* **2018**, *12*, 797–807. [CrossRef]
26. More, P.; Panchade, V. Control Strategies for BLDC Motor. *Int. J. Adv. Res. Electr. Electron. Instrum. Eng.* **2016**, *5*, 5578–5583.
27. Ahmad, A.; Khan, F.; Jamil, M. Detection of Rotor Faults in BLDC Motor Using Flux Analysis. *Results Eng.* **2024**, *20*, 1464–1477.
28. Abbas, A.; Latif, M.; Zhang, H. A Review of Adaptive Energy Control in Sustainable Transportation. *Sustainability* **2024**, *17*, 5841.
29. Necaibia, S.; Kelaiaia, M.S.; Labar, H.; Necaibia, A.; Castronuovo, E.D. Enhanced auto-scaling incremental conductance MPPT method, implemented on low-cost microcontroller and SEPIC converter. *Sol. Energy* **2019**, *180*, 152–168. [CrossRef]
30. Lee, H.; Lee, G.; Kim, G.; Shon, J. Variable incremental controller of permanent-magnet synchronous motor for voltage-based flux-weakening control. *Energies* **2022**, *15*, 5733. [CrossRef]
31. Khazaee, A.; Zarchi, H.A.; Markadeh, G.A. Loss Model Based Efficiency Optimized Control of Brushless DC Motor Drive. *ISA Trans.* **2019**, *92*, 207–218. [CrossRef]

32. Zhang, Q.; Deng, J.; Fu, N. Minimum Copper Loss Direct Torque Control of Brushless DC Motor Drive in Electric and Hybrid Electric Vehicles. *IEEE Access* **2019**, *7*, 144866–144875. [CrossRef]
33. Bose, B.K. *Power Electronics and AC Drives*; Prentice Hall: Englewood Cliffs, UK, 1986.
34. Bose, B.K. *Modern Power Electronics and AC Drives*; Prentice Hall: Saddle River, NJ, USA, 2002.
35. Nag, T.; Santra, S.B.; Chatterjee, A.; Chatterjee, D.; Ganguli, A.K. Fuzzy logic-based loss minimisation scheme for brushless DC motor drive system. *IET Power Electron.* **2016**, *9*, 1581–1589. [CrossRef]
36. Praptodiyono, S.; Maghfiroh, H.; Hermanu, C. BLDC Motor Control Optimization Using Optimal Adaptive PI Algorithm. *J. Elektron. Dan Telekomun.* **2020**, *20*, 47–52. [CrossRef]
37. Singh, R.R.; Kalel, D.D.; Stonier, A.A.; Peter, G.; Stephen, V.; Arun, V.; Ganji, V. MTPA With Pseudo-FOC Strategy–Based BLDC for Minimization of Copper Loss and Torque Ripple. *J. Electr. Comput. Eng.* **2025**, *2025*, 2549885. [CrossRef]

actuators

MDPI

Article

Modeling, Dynamic Characterization, and Performance Analysis of a 2.2 kW BLDC Motor Under Fixed Load Torque Levels and Variable Speed Inputs: An Experimental Study

Ayman Ibrahim Abouseda [1,2,*], Resat Doruk [1], Ali Emin [3] and Ozgur Akdeniz [3]

[1] Department of Electrical and Electronics Engineering, Atilim University, Ankara 06830, Turkey; resat.doruk@atilim.edu.tr

[2] Department of Electrical and Electronics Engineering, University of Tripoli, Tripoli P.O. Box 13932, Libya

[3] Department of Automotive Engineering, Atilim University, Ankara 06830, Turkey; ali.amini@atilim.edu.tr (A.E.); ozgur.akdeniz@atilim.edu.tr (O.A.)

* Correspondence: abouseda.ayman@student.atilim.edu.tr

Abstract: Accurate modeling and performance analysis of brushless DC (BLDC) motors are essential for high-efficiency control in modern drive systems. In this article, a BLDC motor was modeled using system identification techniques. In addition, experimental data were collected from the BLDC motor, including its speed response to various input signals. Using system identification tools, particularly those provided by MATLAB/Simulink R2024b, an approximation model of the BLDC motor was constructed to represent the motor's dynamic behavior. The identified model was experimentally validated using various input signals, demonstrating its accuracy and generalizability under different operating conditions. Additionally, a series of mechanical load tests was conducted using the AVL eddy-current dynamometer to evaluate performance under practical operating conditions. Fixed load torques were applied across a range of motor speeds, and multiple torque levels were tested to assess the motor's dynamic response. Electrical power, mechanical power, and efficiency of the entire system were computed for each case to assess overall system performance. Moreover, the real-time state of charge (SOC) of Lithium-ion (Li-ion) battery was estimated using the Coulomb counting method to analyze the impact of Li-ion battery energy level on the BLDC motor efficiency. The study offers valuable insights into the motor's dynamic and energetic behavior, forming a foundation for robust control design and real-time application development.

Keywords: brushless DC (BLDC) motor; system identification; model validation; state of charge (SOC); lithium-ion (Li-ion) battery; Coulomb counting; AVL dynamometer

1. Introduction

Brushless DC (BLDC) motors have gained significant attention in recent years due to their high efficiency, compact size, and reliability, making them suitable for electric vehicles (EVs), drones, robotics, and industrial drive systems [1]. Accurate modeling of BLDC motors is essential for the development of effective control strategies, energy-efficient operation, and system integration. Traditional physics-based models, although useful, often fail to account for nonlinearities such as friction, dead zones, and load variations under dynamic operating conditions. In contrast, system identification (SI) techniques provide a data-driven approach to derive simplified yet accurate models using experimental input–output data, making them well suited for embedded and real-time applications [2]. A comprehensive dynamic characterization of BLDC motors must consider both variable-speed input conditions and fixed-load torque levels, which are representative of real-world operational scenarios. Additionally, performance evaluation under these conditions is crucial for validating model generalizability and for quantifying system efficiency. In battery-powered applications such as EVs, the performance of BLDC motors is closely influenced by the state of charge (SOC) of the Lithium-ion (Li-ion) battery, necessitating a unified experimental framework that integrates motor modeling, validation, load testing, and battery analysis.

Several studies have highlighted the effectiveness of MATLAB-based system identification in modeling BLDC motors. Khan et al. [3] demonstrated that linear and nonlinear models identified using least-squares (LS) and NARX methods achieved over 98% accuracy in replicating motor dynamics, supporting the reliability of data-driven modeling approaches. Similarly, Hat et al. [4] proposed a modeling approach for a BLDC motor using a pseudo-random binary sequence (PRBS) excitation and system identification techniques in MATLAB. The resulting transfer function model, with over 90% accuracy, was used to design a PID controller implemented on an Arduino Mega for real-time control. A data-driven framework was developed for modeling, characterizing, and identifying key parameters of permanent magnet DC (PMDC) motors. Abouseda et al. [5] experimentally identified the electrical and mechanical parameters of a BLDC motor using DC/AC tests and dynamometer measurements. The resulting transfer function model was validated in real time, demonstrating accurate dynamic representation suitable for control design. Pandya et al. [6] presented a comprehensive framework for modeling, characterization, and parameter identification of PMDC motors. Their approach integrated simulation and experimental data, applying nonlinear optimization techniques to estimate key motor parameters. The study reported strong agreement between identified models and measured responses, validating the methodology's accuracy. Although focused on PMDC motors, the modeling and identification strategy is also applicable to BLDC systems, highlighting the broader importance of precise parameter estimation for accurate motor modeling and control in electric drive applications.

Recent advances in parameter identification of electric motors have led to the development of intelligent and hybrid estimation techniques aimed at improving dynamic accuracy and robustness. Su et al. [7] proposed a multiparameter identification framework for permanent magnet synchronous motors (PMSMs) using a model reference adaptive system (MRAS) combined with a simulated annealing particle swarm optimization (SA-PSO) algorithm, which demonstrated improved convergence and reduced error under dynamic conditions. Yang et al. [8] introduced an energy-based modeling approach for PMSM-driven belt conveyors and applied an online recursive least squares (RLS) identification algorithm, which effectively captured changes in inertia and friction parameters under varying operational scenarios. Additionally, Zhu et al. [9] provided an overview of online parameter estimation methods for PMSMs, highlighting techniques such as EKF, RLS, and learning-based models, and emphasizing the role of hybrid approaches for real-time performance and noise robustness. Moreover, Kerem [10] employed an adaptive-network-based fuzzy inference system (ANFIS) to estimate motor torque in electric vehicle applications, showing enhanced accuracy under uncertain and nonlinear conditions. These studies collectively highlight the growing trend toward intelligent system identification strategies that can be extended to BLDC motors for enhanced modeling and control performance.

Intelligent control techniques have also been widely explored to enhance the modeling and performance analysis of BLDC motor systems. Al Mashhadany et al. [11] developed a BLDC motor model integrated with various intelligent controllers, including fuzzy logic and artificial neural networks, and demonstrated their effectiveness in improving system stability and transient response under different operating conditions. Similarly, El Sawy et al. [12] investigated the dynamic behavior of brushless AC motors using different sensorless control schemes, highlighting the trade-offs between accuracy and computational complexity. These studies emphasize the relevance of intelligent and sensorless techniques in real-time control of brushless motors, and their findings can be extended to BLDC applications for robust modeling and improved performance under varying load and speed profiles.

A closed-loop subspace identification method based on auxiliary variables was proposed by Zhang et al. to accurately model a permanent magnet BLDC motor under noisy conditions. The technique was experimentally validated using a chassis dynamometer, demonstrating high robustness and reliability under practical operating scenarios [13]. Arifin et al. [14] employed a PRBS input combined with the N4SID subspace identification method to develop a dynamic model of an electric power steering motor. The identified model was validated using PI and PID controllers tuned via various methods, where the Haugen-tuned PI controller exhibited superior tracking performance. Although their study focused on EPS motors, the methodology and controller evaluation approach are directly relevant to BLDC motor modeling and control strategies. Additionally, Marcos-Andrade et al. [15] introduced an online algebraic estimation method for identifying key parameters and external disturbances in BLDC motors. Their approach enables real-time estimation of variables such as load torque and speed using input–output measurements without requiring persistent excitation. The method was validated experimentally and demonstrated effective performance under variable operating conditions, making it suitable for embedded control and fault detection applications. A recent study proposed a chaotic Gaussian–Cauchy Rao algorithm (CGCRAO) to enhance the accuracy and efficiency of parameter identification in PMSMs. The method incorporates chaotic initialization using a Tent map to improve population diversity and convergence speed, along with a hybrid mutation strategy based on Gaussian and Cauchy distributions to balance global and local search capabilities [16]. Although originally applied to PMSMs, the CGCRAO algorithm's structure makes it adaptable for BLDC motor identification due to the similar dynamic and electrical characteristics of both motor types.

A least squares-based optimization method was developed for estimating key parameters of DC motors using experimentally acquired data. The approach utilized the LS-ARX algorithm and nonlinear least squares optimization to accurately determine electrical and mechanical parameters such as resistance, inductance, and inertia. The method demonstrated high model fidelity and robustness, outperforming conventional techniques in terms of parameter accuracy and fitting performance [17]. An improved dynamic forgetting-factor recursive least squares (FFRLS) algorithm for DC motor parameter identification was proposed, aiming to enhance the performance of traditional RLS methods. By designing an adaptive forgetting-factor adjustment function, the improved algorithm avoids initial fluctuations and directly initiates identification from a stable phase. It successfully estimates key motor parameters, including moment of inertia and viscous friction. Simulation results demonstrate that the improved FFRLS achieves faster convergence and stronger robustness against disturbances compared to standard RLS and FFRLS methods [18]. Rodríguez-Abreo et al. [19] introduced a novel metaheuristic technique for motor parameter identification that incorporates dynamic-response equations—from both transient and steady-state phases—as constraints in the optimization process. Implemented in three algorithms (Grey Wolf Optimizer, Jaya, and Cuckoo Search), this approach reduced the search space by linking some unknowns via dynamic relations. Tests on two motors with simulated and experimental data showed the enhanced methods achieved equivalent estimation accuracy while reducing iterations by 10–50%. Furthermore, Cortés-Romero et al. [20] employed RLS to estimate the electrical parameters of a BLDC motor from voltage and speed measurements. Their model was validated in real time using Simulink and effectively integrated into a PI controller for speed regulation. This study demonstrated the practicality of RLS-based approaches for adaptive applications.

Recent efforts have focused on improving the accuracy and reliability of SOC estimation in Li-ion batteries through enhanced Coulomb Counting techniques and hybrid algorithms. Wu et al. [21] proposed an improved Coulomb counting method with adaptive error correction (ICC-AEC), integrated within a dual closed-loop control structure for Li-ion battery SOC estimation. Their approach combines a first-order RC battery model with a forgetting-factor RLS algorithm to continuously recalibrate the initial SOC and battery capacity in real time. Validated under dynamic load profiles and varying temperatures, the method demonstrated superior accuracy and robustness compared to conventional mixed and closed-loop estimation techniques. Pu and Wang [22] developed a combined SOC estimation framework integrating a Long Short-Term Memory (LSTM) neural network with an Unscented Kalman Filter (UKF). Their method demonstrated high prediction accuracy and robustness against noise, benefiting applications with irregular current profiles. Similarly, Fahmy et al. [23] proposed a hybrid SOC estimation method for Li-ion batteries that combines Coulomb counting with an adaptive unscented Kalman filter (AUKF). Battery model parameters are identified using a state-space representation optimized via the African Vultures Optimizer (AVOA). The hybrid method was tested under varying temperatures and operating conditions using a 2.6 Ah Panasonic cell. Simulation results demonstrated that this approach outperforms existing hybrid methods by achieving the lowest estimation error across different scenarios.

The influence of mechanical load torque on motor behavior has been investigated in various experimental settings. Noval et al. [24] evaluated a chassis dynamometer for light EVs by analyzing the accuracy of torque and speed measurements. The study confirmed a consistent relationship between load torque and motor speed, with low measurement errors, demonstrating the dynamometer's reliability for performance testing. However, few studies have combined fixed-torque testing with model validation using system identification frameworks.

Battery SOC estimation has also become a key focus in recent research. Coulomb counting remains the most widely used method due to its real-time compatibility, though it is susceptible to integration drift. Kunatsa et al. [25] reviewed SOC estimation techniques and emphasized the importance of accurate SOC tracking for optimizing system efficiency and prolonging battery life.

Despite significant advancements in each of these areas, there is a lack of comprehensive experimental studies that combine system identification-based modeling, load testing using dynamometers, and battery SOC impact analysis into a unified framework. Most prior works focus on one or two of these aspects in isolation, limiting the understanding of their interdependence in practical applications.

This study aims to address current research gaps by proposing a unified framework for modeling, dynamic performance analysis under mechanical load, and SOC-aware energy evaluation of BLDC motors. The motor's dynamic behavior is identified and modeled through data-driven system identification techniques using PRBS excitation implemented in MATLAB. The resulting model is validated with diverse input signals beyond the training dataset to assess its generalization and robustness. Mechanical performance is experimentally evaluated using an AVL eddy-current dynamometer under various fixed torque and variable speed conditions. Additionally, real-time SOC estimation of the Li-ion battery is performed via the Coulomb counting method to examine the impact of battery energy levels on motor efficiency and operational performance. This integrated framework provides valuable insights to support the design of robust control strategies and optimize energy management in BLDC-based drive systems. The main contributions of this study are as follows:

- Development of a comprehensive experimental modeling framework for BLDC motors combining PRBS-based system identification and MATLAB tools to derive dynamic transfer function models.
- Experimental validation of the identified model using multiple excitation signals, including sinusoidal, exponential, ramp, and chirp waveforms, demonstrating strong generalization beyond the training data.
- Detailed torque speed performance characterization under fixed load torque conditions using an AVL eddy-current dynamometer, revealing load-dependent motor behavior and efficiency trends.
- Measurement and analysis of electrical input power, mechanical output power, and overall system efficiency across different operating conditions.
- Integration of real-time SOC estimation via Coulomb counting to evaluate how battery energy levels influence motor current draw and efficiency.
- Presentation of a methodology adaptable to other BLDC motors and electric drive systems through tuning of motor-specific parameters, facilitating broader application in EVs, robotics, and industrial automation.

2. Mathematical Modeling of BLDC Motor

The BLDC motor is an electromechanical energy conversion device whose dynamic behavior is governed by the interaction of electrical, magnetic, and mechanical subsystems. While this study primarily adopts a data-driven system identification approach for model development, it is essential to present the classical mathematical formulation of the BLDC motor. This provides a theoretical foundation for understanding the system's internal structure and facilitates comparison with the experimentally identified model.

2.1. Electrical Subsystem

In a three-phase star-connected BLDC motor, the electrical dynamics of each phase can be described by Kirchhoff's voltage law. Neglecting mutual inductance, the voltage equations for the a, b, and c phases are given as follows:

$$u_a = Ri_a + L\frac{i_a}{dt} + e_a \tag{1}$$

$$u_b = Ri_b + L\frac{i_b}{dt} + e_b \tag{2}$$

$$u_c = Ri_c + L\frac{i_c}{dt} + e_c \tag{3}$$

In the above equations, u_a, u_b and u_c represent the stator phase voltages applied to the motor windings, expressed in volts (V), while i_a, i_b and i_c denote the corresponding phase currents, measured in amperes (A). The parameters R and L refer to the phase resistance (Ω) and phase inductance (H), respectively, both assumed to be identical for all three phases due to the motor's symmetrical construction. The terms e_a, e_b and e_c correspond to the back electromotive forces (EMFs) generated in each phase, also expressed in volts (V), which are functions of the rotor position and angular velocity.

Under balanced operating conditions, the sum of the three phase currents satisfies the following equation:

$$i_a + i_b + i_c = 0 \tag{4}$$

This constraint simplifies analysis and is commonly assumed in symmetric motor configurations.

2.2. Back EMF Generation

The back EMF in each phase is generated as a result of the rotor's motion relative to the stator magnetic field. In BLDC motors, the back EMF typically exhibits a trapezoidal waveform, and its magnitude is proportional to the rotor speed. The back EMF for each phase can be modeled as follows:

$$e_x = K_e f_x(\theta_e)\omega, \qquad x \in \{a,b,c\} \tag{5}$$

where k_e denotes the back EMF constant, measured in volts-second per radian $(\mathrm{V{\cdot}s/rad})$, while ω represents the rotor angular velocity in radians per second $(\mathrm{rad/s})$. The variable θ_e corresponds to the electrical rotor position, also expressed in radians (rad). The function $f_x(\theta_e)$ is a normalized trapezoidal waveform that depends on the electrical angle, typically ranging between -1 and +1, and reflects the back EMF shape of each motor phase.

These waveforms are a function of rotor position and reflect the geometry of the motor and stator winding configuration. Accurate modeling of $f_x(\theta_e)$ is critical for advanced control applications.

2.3. Mechanical Subsystem

The mechanical dynamics of the BLDC motor are governed by Newton's second law for rotational systems. The net torque applied to the rotor determines its angular acceleration and is expressed as follows:

$$J\frac{d\omega}{dt} + B\omega = T_e - T_l \tag{6}$$

In the mechanical dynamics equation of the BLDC motor, J represents the moment of inertia of the rotor, expressed in $(\mathrm{kg{\cdot}m^2})$, which quantifies the rotor's resistance to changes in angular velocity. Parameter B denotes the viscous friction coefficient in $(\mathrm{N{\cdot}m{\cdot}s/rad})$, accounting for energy losses due to frictional forces opposing motion. Additionally, T_e is the developed electromagnetic torque in $(\mathrm{N{\cdot}m})$ generated by the motor as a result of current flow and magnetic interaction. Finally, T_l refers to the external load torque $(\mathrm{N{\cdot}m})$ applied to the motor shaft, which opposes the motor's rotation. This relationship describes the balance of torques acting on the rotor and governs its acceleration or deceleration under dynamic operating conditions.

The electromagnetic torque produced by the BLDC motor is directly related to the interaction between back EMFs and phase currents. It can be formulated as follows:

$$T_e = \frac{1}{\omega}(e_a i_a + e_b i_b + e_c i_c) \tag{7}$$

In many practical applications, torque is alternatively modeled using the torque constant k_t, assuming ideal linearity, and is expressed as follows:

$$T_e = k_t I \tag{8}$$

where I is the equivalent or average line current and k_t in $(\mathrm{N{\cdot}m/A})$ is typically equal to k_e in SI units when mechanical losses are neglected.

2.4. Remarks on Model Limitations and Data-Driven Approach

The above set of differential equations constitutes the conventional analytical model of a BLDC motor. While useful for control system design and theoretical analysis, such models require precise knowledge of several motor-specific parameters, including resistance, inductance, back EMF profiles, and rotor inertia. In practical scenarios, these parameters are often difficult to measure accurately or may vary due to changes in operating conditions such as temperature, load, or aging.

To address these challenges, this study incorporates a data-driven system identification approach. Rather than relying solely on fixed physical parameters, the dynamic behavior of the motor is captured through experimental measurements under realistic operating conditions. By applying excitation signals to the motor and recording the system's response, empirical transfer function models are derived that reflect the actual system performance. This methodology allows for more adaptable and accurate modeling, especially in the context of control system design, where real-world performance often deviates from idealized models.

The data-driven model complements the theoretical formulation by providing a validated, input–output representation of the motor dynamics that accounts for nonlinearities, unmodeled effects, and parameter uncertainties. It enables the development of more robust control strategies and provides a foundation for further analysis under different load conditions.

The following section describes the identification methodology, signal design, validation process, and performance evaluation of the experimentally derived BLDC motor model under no-load conditions.

3. Materials and Methods

This section describes the hardware configuration, signal design, data acquisition process, and experimental procedures used for dynamic modeling and performance analysis of the BLDC motor. The methodology was structured to facilitate both system identification conditions and comprehensive performance evaluation under fixed load torque conditions with variable speed inputs.

3.1. Model Identification and Validation

The data-driven modeling of the 2.2 kW BLDC motor using system identification techniques is organized into a sequence of well-defined steps to ensure clarity and reproducibility:

1. Modeling Objective: The main goal is to develop an accurate empirical model that captures the dynamic behavior of the BLDC motor under no-load conditions. This model serves as a foundation for performance analysis and future control design.
2. Data Acquisition: PRBS signal is applied as excitation input to the motor through a real-time interface. The corresponding output speed responses are recorded using the Kelly controller.
3. Model Structure Selection: Using the System Identification Toolbox in MATLAB, different linear transfer function structures are explored to represent the input–output relationship of the system.
4. Model Parameter Estimation: The model parameters are estimated based on a prediction-error minimization algorithm using least-squares fitting techniques.
5. Model Validation and Performance Analysis: The BLDC motor model was identified using input and output data obtained under PRBS excitation. To evaluate the model's ability to generalize beyond the identification dataset, various independent test signals, such as sinusoidal, exponential, ramp, and chirp waveforms, were ap-

plied experimentally to the physical system and to the identified model in MATLAB Simulink. The simulated responses were then compared with the corresponding experimental results to qualitatively assess the accuracy of the model across a range of dynamic operating conditions.

This structured methodology ensures a systematic representation of the motor's dynamic characteristics based on experimental data.

3.2. Experimental Setup

The experimental platform was developed around a three-phase 2.2 kW BLDC motor, controlled via a Kelly KLS12301-8080I controller (Kelly Controller Inc., Shanghai, China). The motor was mechanically coupled to an AVL eddy-current dynamometer (AVL List GmbH, Graz, Austria), capable of applying precise, programmable load torques while simultaneously measuring shaft speed and torque with high accuracy. The experimental configuration is depicted in Figure 1, and the main technical specifications of the BLDC motor are listed in Table 1.

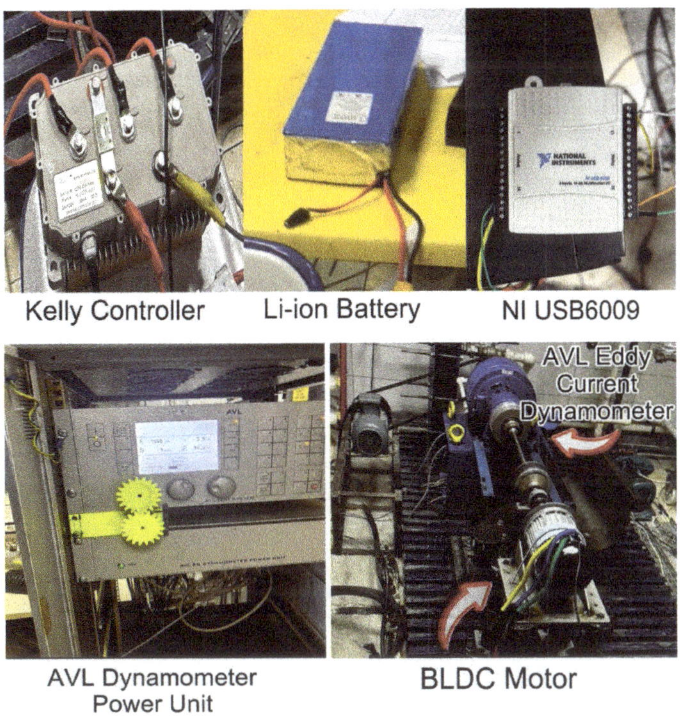

Figure 1. Experimental setup of components used for BLDC motor testing.

Table 1. Specifications of the BLDC motor used in this study.

Specification	Value
Rated Power	2.2 kW
Phase Voltage	24.5 V
Maximum Rotational Speed	3600 rpm
Motor Mass	17 kg
Nominal Efficiency	80%

3.3. Signal Design and Data Acquisition

To evaluate the dynamic response of the BLDC motor under open-loop conditions, a set of voltage input signals was generated using MATLAB/Simulink and transmitted to the analog input of the Kelly KLS12301-8080I controller via an NI USB-6009 data acquisition (DAQ) device (National Instruments, Austin, TX, USA). These signals modulated the duty cycle of the motor drive, enabling precise control of motor speed. Five distinct excitation waveforms were employed to stimulate the system across a wide range of dynamic behaviors:

- Sinusoidal signal was applied to evaluate the frequency response characteristics of the system.
- Exponentially rising signal was used to replicate smooth acceleration profiles.
- Chirp signal was employed to excite the system across a continuously varying frequency range.
- Ramp signal was utilized to analyze the system's response to linearly increasing inputs and to detect any lag or rate limitations.
- PRBS signal was introduced to provide broadband excitation for system identification, enabling accurate modeling across a wide frequency spectrum.

Each excitation waveform was carefully generated in MATLAB and applied in real time using Simulink's analog output blocks connected to the NI USB-6009 data acquisition device. The corresponding motor speed responses were measured using a Hall-effect sensor and recorded in MATLAB for subsequent offline analysis.

All tests in this phase were conducted under no-load conditions to isolate the motor's inherent electrical and mechanical dynamics, free from external resistive torques. Although the BLDC motor was mechanically coupled to an AVL eddy current dynamometer (operated in passive mode), no active loading was applied during this stage. The resulting input–output datasets were subsequently used for system identification and model validation, as detailed in the following section.

3.4. System Identification Procedure

The acquired datasets were directly used in MATLAB's System Identification Toolbox to derive an empirical model representing the input–output dynamics of the BLDC motor. No preprocessing, such as offset removal, detrending, or filtering, was applied to the raw data to preserve the original system behavior during model estimation.

The transfer function of the BLDC motor was evaluated to obtain a simplified, continuous-time representation of its input–output behavior. This approach enables the characterization of key dynamic properties such as gain, time constants, and system order, and provides a foundation for designing classical control strategies. Moreover, the transfer function model facilitates validation by comparing its simulated response to experimental speed data under various test inputs.

Multiple excitation signals were assessed to determine the most suitable input for system identification. A separate transfer function was estimated for each signal type, and the resulting models were compared based on their ability to generalize to other test conditions. The excitation signal that yielded the most accurate and consistent model performance across different input scenarios was selected for use in the final identification and validation process.

3.5. Load Testing Procedure: Torque–Speed Characterization

To evaluate the performance of the BLDC motor under realistic loading conditions, a series of controlled mechanical load tests was conducted using an AVL eddy-current dynamometer. The objective was to assess the motor's behavior under varying speeds at fixed load torques and to quantify its efficiency across different operating points. The test methodology involved applying a variable reference speed at several discrete torque levels. Specifically, the dynamometer applied constant torque values ranging from 0 to 4 N·m. For each fixed torque level, the motor was driven through a speed range from 0 to approximately 2900 rpm, allowing for detailed characterization of its dynamic and steady-state performance. The Kelly KLS12301-8080I controller regulated the motor speed via analog input, while the programmable torque was applied through the dynamometer.

During each test run, the terminal voltage and current of the Li-ion battery, along with the rotational speed of the BLDC motor, were monitored using the Kelly controller and logged in real time via a custom Python 3.12 script over RS232 communication.

This procedure enabled high-resolution mapping of the torque–speed relationship and provided the necessary data to compute electrical input power, mechanical output power, system efficiency, and SOC. The data collected was used in the subsequent analysis to evaluate performance trends and identify operating regions with optimal energy efficiency.

4. Electrical and Mechanical Power Analysis, Efficiency Assessment, and SOC Estimation

To assess the performance of the BLDC motor and the associated battery system, both energy efficiency and the SOC of the Li-ion battery were evaluated based on the data collected during load testing. The analysis was conducted using the electrical and mechanical measurements obtained under fixed torque levels with varying reference speed conditions, as described in the previous section.

4.1. Efficiency Calculation

The BLDC motor's efficiency was determined by comparing the mechanical output power to the electrical input power at each operating point. The electrical input power P_{elec} was calculated as follows:

$$P_{elec} = VI \tag{9}$$

where V is the battery terminal voltage and I is the discharge current, both measured through the Kelly controller.

The mechanical output power P_{mech} was computed using the measured shaft torque T and angular speed ω, obtained from the AVL eddy-current dynamometer and Kelly controller, respectively. Angular speed was derived from the motor's rotational speed N (in rpm) as follows:

$$\omega = \frac{2\pi N}{60} \tag{10}$$

$$P_{mech} = T\omega \tag{11}$$

The overall efficiency η of the entire system was then calculated using Equation (12):

$$\eta(\%) = \frac{P_{mech}}{P_{elec}} \times 100 \tag{12}$$

Efficiency was evaluated across all combinations of fixed torque levels and variable speeds, enabling a detailed analysis of the system's operating characteristics and energy conversion behavior.

4.2. SOC Estimation

The SOC of the Li-ion battery was estimated using the Coulomb counting method, which integrates the discharge current over time to track charge depletion. The SOC at a given time t was calculated as follows:

$$SOC(t) = SOC(t_0) - \frac{1}{C_{nom}} \int_{t_0}^{t} I(t)dt \quad (13)$$

where $SOC(t_0)$ is the initial state of charge, C_{nom} is the nominal battery capacity (in ampere-hours), and $I(t)$ is the discharge current (in amperes). The integration was performed numerically using the logged current data sampled during testing [26].

This combined analysis enabled a comprehensive evaluation of the system's energy performance, supporting the identification of optimal operating points for torque–speed control and battery utilization.

5. Results and Discussion

This section presents the results of system identification, model validation, and experimental load testing of the BLDC motor. The accuracy of the identified transfer function model is evaluated, and the motor's performance is analyzed under fixed load torque levels with variable speed inputs. In addition, the system's electrical and mechanical efficiencies are examined.

5.1. System Identification Results and Model Validation

The dynamic characteristics of the BLDC motor were identified using a continuous-time transfer function derived from experimentally acquired input and output data. To evaluate the system's response under a range of conditions, various excitation signals, including PRBS, sinusoidal, exponential, ramp, and chirp waveforms, were applied. Each excitation signal was evaluated based on its ability to generate a model that generalizes well across different operating conditions. Among them, the transfer function estimated using the PRBS signal demonstrated the highest accuracy and consistency in terms of generalization. Figure 2 shows the waveform of the PRBS signal used for identification.

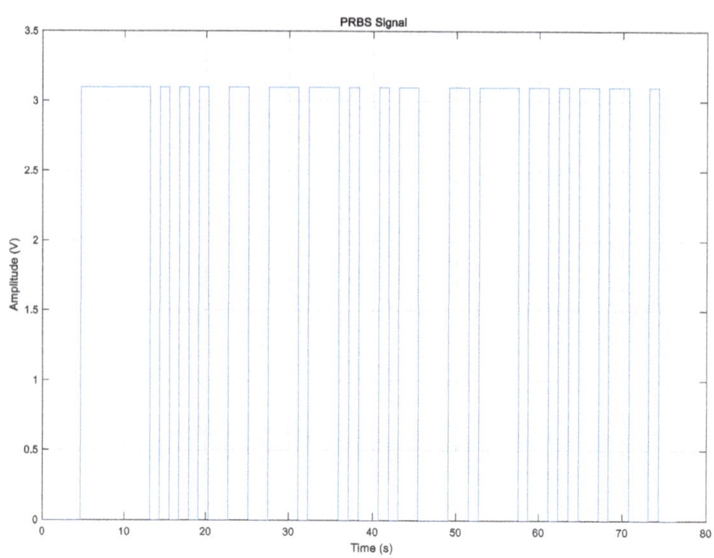

Figure 2. PRBS signal used for identification.

The transfer function model of the BLDC motor was identified using experimental data obtained from PRBS signal excitation. The resulting model achieved a normalized root mean square error (NRMSE) of approximately 85.38%, indicating strong agreement between the simulated and measured behavior. The final transfer function is presented in Equation (14):

$$G(s) = \frac{\omega(s)}{V(s)} = \frac{5.5296 \times 10^4\, s^2 + 1.6675 \times 10^5\, s + 13.5965 \times 10^7}{s^3 + 43.86\, s^2 + 2530\, s + 1.002 \times 10^5} \tag{14}$$

The transfer function in Equation (14) was obtained using MATLAB's System Identification Toolbox (ident) by fitting a linear model structure to experimental voltage (input) and rotational speed (output) data of the BLDC motor. This data-driven approach eliminates the need to derive physical parameters directly, instead producing a compact, experimentally validated model that reflects the motor's dynamics. Although the model is expressed in the Laplace domain, it corresponds to the following time-domain third-order differential equation, as follows:

$$\begin{aligned} &\frac{d^3\omega(t)}{dt^3} + 43.86\frac{d^2\omega(t)}{dt^2} + 2530\frac{d\omega(t)}{dt} + 1.002 \times 10^5\omega(t) \\ &= 5.5296 \times 10^4\frac{d^2V(t)}{dt^2} + 1.6675 \times 10^5\frac{dV(t)}{dt} + 13.5965 \times 10^7 V(t) \end{aligned} \tag{15}$$

where $\omega(t)$ is the motor's angular speed and $V(t)$ is the input voltage. This representation captures both the dynamic and steady-state behavior of the BLDC motor system and is used throughout this study for performance evaluation and simulation.

Moreover, to evaluate its generalization capability, the identified model was validated against experimental results obtained using other excitation signals, such as sinusoidal, exponential, ramp, and chirp signals. Figure 3 illustrates the MATLAB/Simulink model used for the validation of the identified BLDC motor model. In addition, a dead zone nonlinearity was incorporated into the BLDC motor model using the Dead Zone block in Simulink to account for the threshold input voltage below which the motor does not respond. This behavior reflects the real characteristics of the motor, which requires a minimum input of approximately 1.1412 volts to overcome internal friction and control delays. As shown in Figure 4, this improves the model's accuracy, particularly in the low-input region.

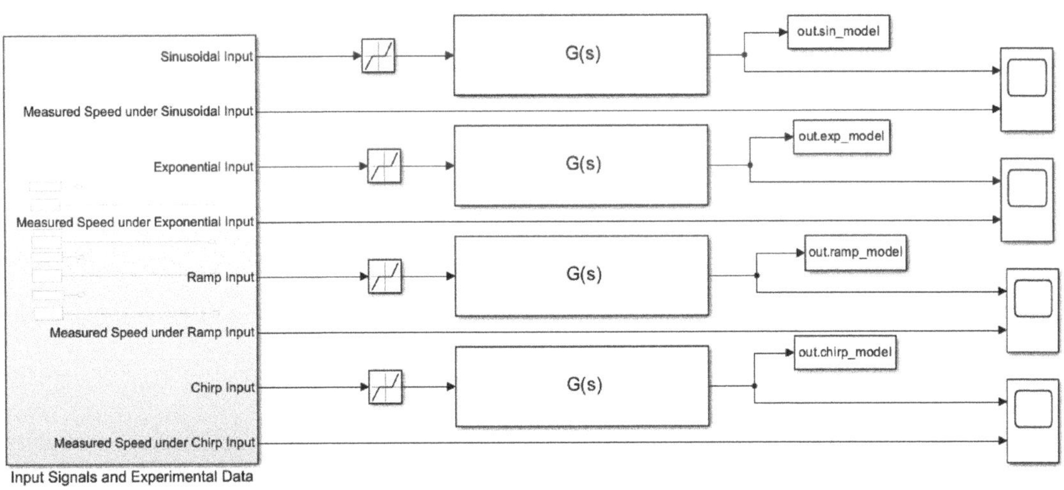

Figure 3. MATLAB/Simulink model used for validation of the identified BLDC motor model.

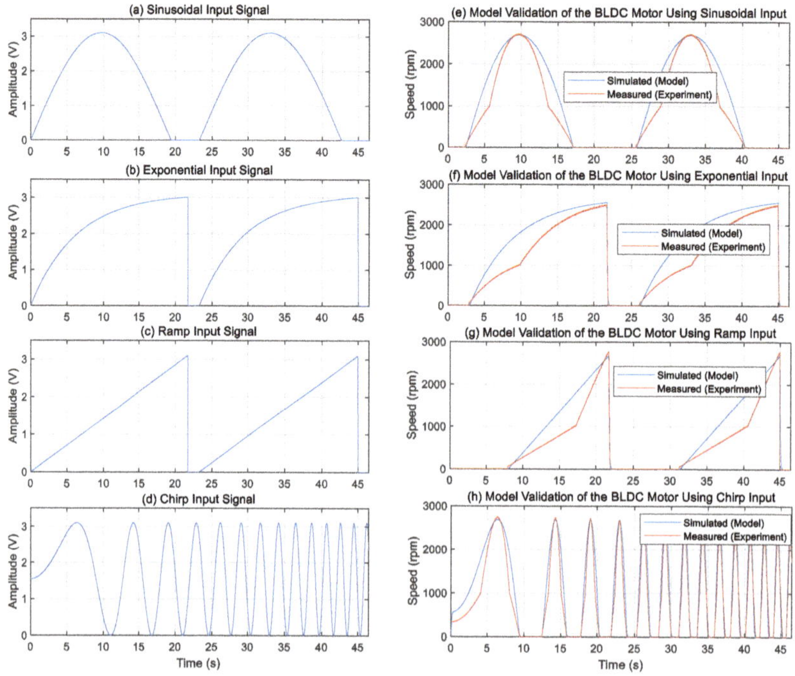

Figure 4. Model validation under different excitation signals. (**a–d**): Input signals applied to the BLDC motor. (**e–h**): Corresponding measured and simulated speed responses.

Figure 4 shows the validation of the identified BLDC motor model using several excitation signals to evaluate its ability to generalize beyond the PRBS identification dataset. The input waveforms, including sinusoidal, exponential, ramp, and chirp signals, were generated in MATLAB Simulink and applied experimentally to the motor through the NI USB 6009 data acquisition device. The measured speed responses were compared with the simulated outputs from the transfer function model identified using the PRBS excitation.

Figure 4e highlights the response to the sinusoidal input, demonstrating excellent agreement between the simulated and experimental speed profiles. The simulation closely follows the periodic behavior of the motor speed, with minimal phase delay and amplitude deviation over the tested frequency range. These results confirm the model's capability to accurately capture the motor's oscillatory dynamics and validate its applicability to scenarios involving sinusoidal or cyclic speed inputs.

Figure 4f illustrates the system behavior under exponential excitation, which simulates smooth acceleration profiles commonly encountered in practical motor control scenarios. The identified model effectively reproduces the gradual increase in motor speed observed experimentally, particularly during the initial phase where acceleration is most prominent. Minor deviations near steady state likely stem from unmodeled nonlinearities or measurement noise; however, the overall close alignment validates the model's capability to represent time-varying inputs with slow dynamics.

Figure 4g depicts the motor response to a ramp input, representing a linearly increasing speed reference. The model captures the general linear trend and steady-state speed with reasonable accuracy. The observed discrepancies during initial acceleration and transient transitions highlight limitations due to unmodeled effects such as frictional variations or controller nonlinearities. These findings emphasize areas for potential model refinement while confirming the model's utility in approximating linear speed trajectories.

Figure 4h shows the validation under chirp excitation, where the input frequency continuously varies over time, testing the model's dynamic fidelity across a broad frequency spectrum. The model successfully tracks both low-frequency and high-frequency oscillations in the motor speed, demonstrating its robustness in capturing complex frequency-dependent dynamics. This validation provides strong evidence that the model can be reliably used for control design tasks involving varying operating conditions.

Collectively, these results demonstrate that the PRBS-identified model generalizes well to diverse input signals, capturing key dynamic behaviors essential for practical BLDC motor applications. The close correspondence between simulation and experiment across different excitation types confirms the model's accuracy and reliability, supporting its use for advanced control development, performance analysis, and real-time system implementation.

5.2. Torque–Speed Performance Under Load

Figure 5 illustrates the BLDC motor's open-loop speed response under varying reference speed profiles and fixed mechanical load torques ranging from 1 to 4 N·m. In each subplot, a stepwise speed input signal ramping from 0 to approximately 2900 rpm was sent to the BLDC motor via the Kelly controller. The actual motor speed was measured experimentally to assess how the system responds to changes in demand under different load conditions.

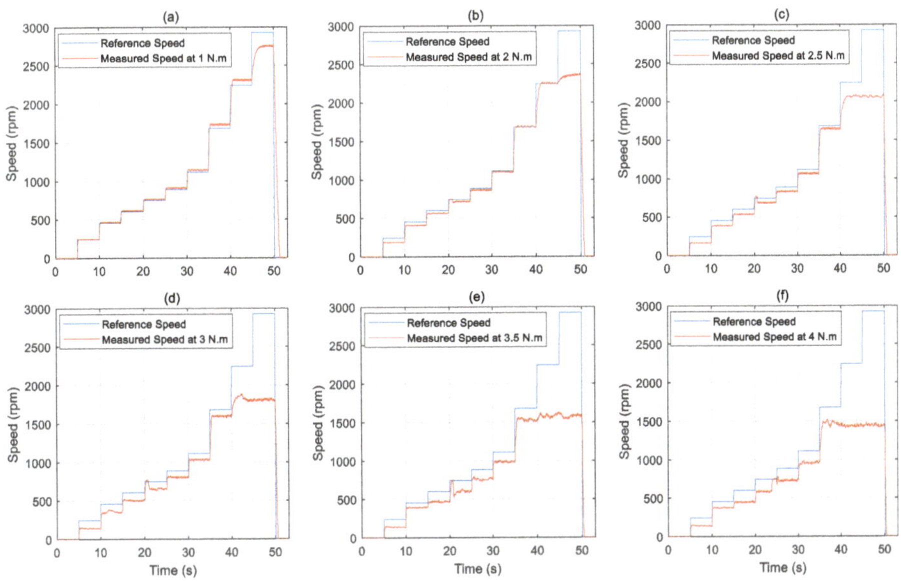

Figure 5. Reference and measured speed responses of the BLDC motor under fixed load torques: (**a**) 1 N·m, (**b**) 2 N·m, (**c**) 2.5 N·m, (**d**) 3 N·m, (**e**) 3.5 N·m, (**f**) 4 N·m.

At a 1 N·m load torque, as shown in Figure 5a, the motor speed closely follows the reference input across the entire operating range, with minimal lag and deviation, indicating that the system behaves linearly and predictably under light mechanical loading. The measured speed rises steadily in response to each step without significant overshoot or oscillation. Similarly, in Figure 5b (2 N·m load), the motor still responds well to the increasing speed input, though minor delays become noticeable during transitions. These slight discrepancies suggest a growing influence of mechanical resistance on the system's dynamic response.

As the load increases to 2.5 N·m, as shown in Figure 5c, the response begins to show more visible lag, particularly at higher reference speeds. The motor exhibits slower acceleration and reduced peak speed compared to the commanded profile, indicating the onset of torque-related performance limitations.

In Figure 5d, with a load of 3 N·m, under-speed behavior becomes more pronounced. The motor fails to reach the highest reference levels within the same timeframe, reflecting the combined effects of increased back EMF and current limitations that hinder performance at elevated speeds under load.

At 3.5 N·m and 4 N·m, the impact of mechanical loading becomes critical, as illustrated in Figures 5e and 5f, respectively. The motor's actual speed significantly diverges from the reference speed, particularly in the upper speed ranges. This reduced responsiveness likely results from the system reaching its voltage and current delivery limits, coupled with intensified back EMF effects and the finite torque capacity of the motor-driver setup.

These observations confirm that the BLDC motor and Kelly controller combination maintains acceptable performance under light to moderate load conditions. However, as mechanical torque increases, the system exhibits reduced responsiveness, especially during rapid acceleration or high-speed operation. These trends highlight the importance of accounting for load-dependent behavior in performance analysis and in the design of future closed-loop control strategies or hardware selection for high-demand applications.

Figure 6 presents the variation in battery terminal voltage of the Li-ion battery under different mechanical load torque conditions during speed ramp tests. The results reveal a distinct correlation between load torque and voltage behavior, providing critical insight into the battery's dynamic response under varying operating conditions. At lower torque levels, such as 1 and 2 N·m (Figure 6a,b), the terminal voltage remains relatively stable across the full speed range, with only minor fluctuations during acceleration phases. This stability reflects lower current demands and reduced internal losses within the battery under light loading conditions.

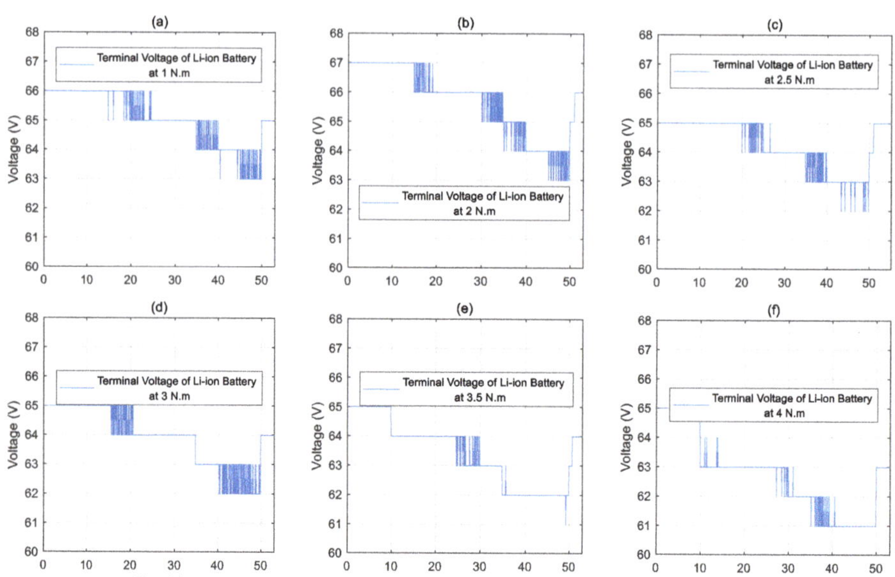

Figure 6. Terminal voltage of the Li-ion battery during BLDC motor operation under variable reference speed and fixed load torque conditions: (**a**) 1 N·m, (**b**) 2 N·m, (**c**) 2.5 N·m, (**d**) 3 N·m, (**e**) 3.5 N·m, (**f**) 4 N·m.

As the torque demand increases, the voltage profile begins to exhibit more pronounced and sustained drops. In Figure 6c,d, corresponding to 2.5 and 3 N·m, transient voltage dips become apparent during acceleration intervals, indicating elevated current draw due to the increased mechanical load. The most significant effects are observed at the highest torque levels of 3.5 and 4 N·m (Figure 6e,f), where the terminal voltage declines below 62 V during high-speed operation. These voltage reductions are not only deeper but also more sustained, pointing to the increased electrical stress placed on the battery. This behavior is primarily attributed to the internal resistance of the battery and its interaction with the peak current demands during rapid speed changes.

The results underscore the importance of incorporating battery dynamics into system-level modeling and design. The observed sensitivity of voltage behavior to mechanical load conditions highlights potential limitations in high-demand scenarios and the need for effective energy management strategies. Such considerations are essential in BLDC motor drive applications, particularly where performance, efficiency, and reliability must be maintained under varying load conditions.

Figure 7 illustrates the variation in battery current drawn by the BLDC motor under variable speed conditions and fixed mechanical load torques ranging from 1 to 4 N·m. The results clearly demonstrate that battery current increases with both the applied torque and motor speed, highlighting the dynamic electrical demand placed on the power supply during different loading scenarios.

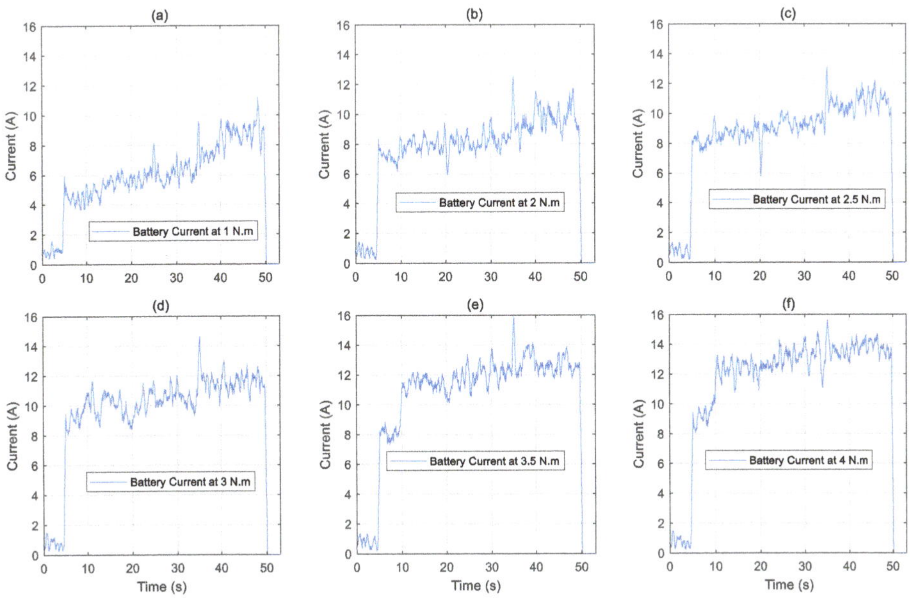

Figure 7. Li-ion battery current under variable reference speed and fixed load torque conditions: (**a**) 1 N·m, (**b**) 2 N·m, (**c**) 2.5 N·m, (**d**) 3 N·m, (**e**) 3.5 N·m, (**f**) 4 N·m.

At lower torque levels (Figure 7a,b), the current profiles exhibit a gradual increase in magnitude as the speed ramps up, with only minor fluctuations. This behavior reflects stable operation with low energy consumption and minimal stress on the power supply. As the torque increases (Figure 7c–e), the current traces reveal a noticeable elevation in both average current and peak values, particularly during acceleration phases. The transient oscillations observed are indicative of the motor's effort to overcome higher mechanical resistance and maintain speed tracking.

At the highest load torque of 4 N·m (Figure 7f), the battery current peaks at approximately 15 A, indicating the system's maximum electrical load during the test sequence. The sharper current spikes and increased variability further suggest that the drive system is operating near its performance boundaries under this condition. These observations underscore the significant influence of mechanical load on current demand, which directly impacts system efficiency, battery heating, and overall energy management.

Overall, the results from Figure 7 reinforce the need for careful coordination between torque–speed profiles and power delivery capabilities. Effective control and thermal management strategies become increasingly critical at higher loads to ensure system reliability, protect battery health, and optimize energy utilization in BLDC motor drive systems.

Figure 8 presents the electrical input power characteristics of the BLDC motor during variable-speed operation under fixed mechanical load torque conditions. The input power was computed using real-time measurements of terminal voltage and current of the Li-ion battery, obtained via the Kelly controller's RS232 interface and processed through a Python-based acquisition system. This analysis enables a comprehensive evaluation of the system's energy demand in response to varying torque and speed levels.

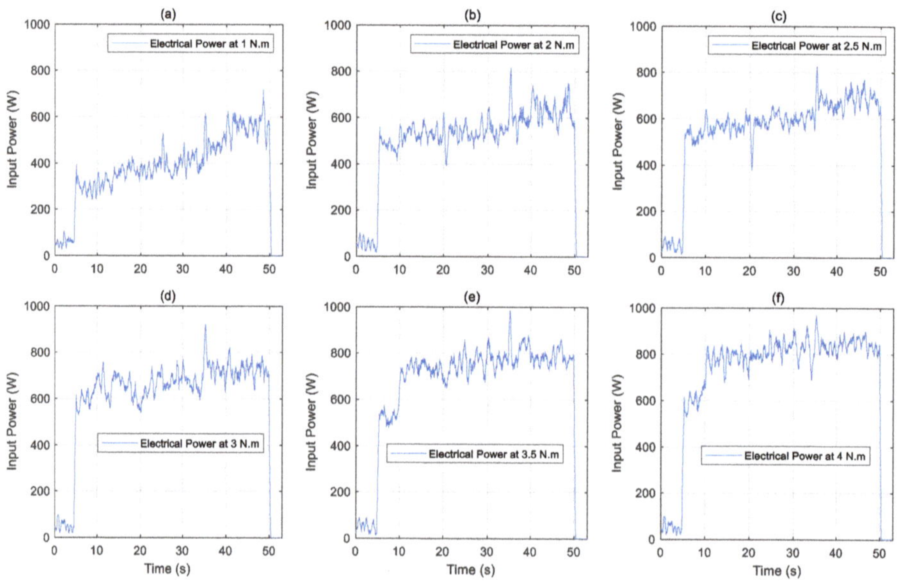

Figure 8. Electrical input power of the BLDC motor under variable reference speed and fixed load torque conditions: (**a**) 1 N·m, (**b**) 2 N·m, (**c**) 2.5 N·m, (**d**) 3 N·m, (**e**) 3.5 N·m, (**f**) 4 N·m.

At a low load torque of 1 N·m (Figure 8a), the electrical input power increases gradually with speed, remaining below 650 W throughout the operating range. This relatively modest power consumption reflects reduced current draw and low mechanical resistance, leading to efficient motor operation. As the torque increases to 2 N·m and 2.5 N·m (Figure 8b,c), the power demand rises more steeply, with peak values nearing 780 W. This trend is consistent with increased electrical loading needed to sustain the desired speed under greater mechanical resistance.

At higher torque levels of 3 N·m, 3.5 N·m, and 4 N·m (Figure 8d–f), the input power exceeds 850 W during the higher-speed segments, revealing a significant escalation in energy consumption. These operating points represent more strenuous loading conditions, where both the current demand and the back EMF effects increase, thereby amplifying the required input power. Minor fluctuations in the power signal are particularly evident during acceleration phases and are attributed to transient current surges as the motor responds to changes in speed and load.

The results confirm a strong nonlinear dependency of electrical input power on both mechanical torque and speed. As the torque increases, the electrical efficiency of the system tends to decline due to elevated current demand and associated internal losses. These findings highlight the importance of integrating power analysis into motor control and energy management strategies, particularly for applications where sustained operation at high torque and speed is required. The power trends presented in Figure 8 provide a critical foundation for assessing the overall performance and efficiency of BLDC motor drive systems under realistic operational constraints.

Figure 9 presents the mechanical output power of the BLDC motor under variable speed operation and fixed load torque levels ranging from 1 to 4 N·m. The mechanical power was computed in real time as the product of the measured angular speed and the applied torque. The results demonstrate a clear dependency of mechanical power on both speed and torque, with a generally linear trend as expected from fundamental electromechanical principles.

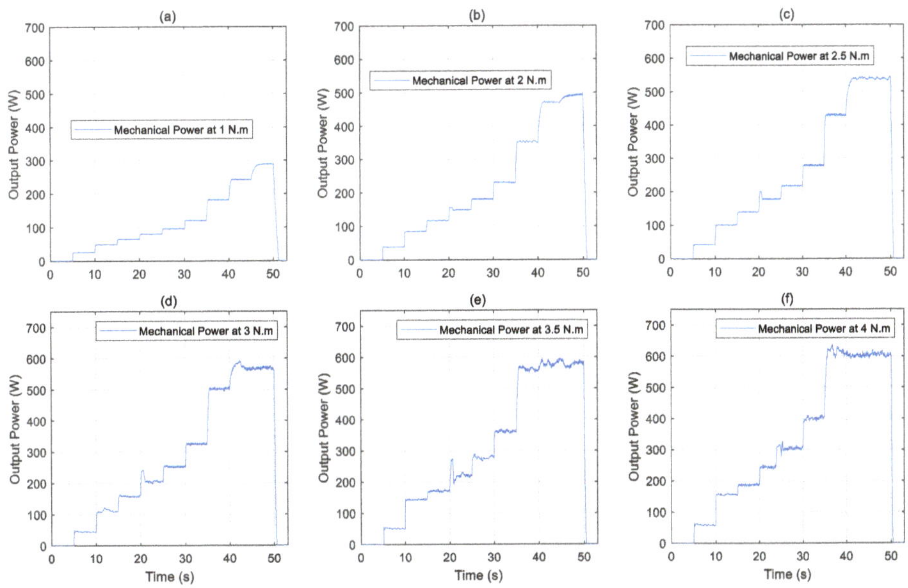

Figure 9. Mechanical output power of the BLDC motor under variable reference speed and fixed load torque conditions: (**a**) 1 N·m, (**b**) 2 N·m, (**c**) 2.5 N·m, (**d**) 3 N·m, (**e**) 3.5 N·m, (**f**) 4 N·m.

In Figure 9a, at a load torque of 1 N·m, the mechanical power increases incrementally with each step in reference speed, reaching a peak value of approximately 280 W. The power trace is smooth and consistent, reflecting stable system operation under light loading. As the load torque increases to 2 N·m and 2.5 N·m in Figure 9b,c, the mechanical power output correspondingly rises, peaking at around 480 W and 550 W, respectively. While the profiles remain generally linear, slight ripples are observed during acceleration phases, attributable to minor transient deviations in motor speed and torque application.

At higher load torque levels, specifically 3 N·m, 3.5 N·m, and 4 N·m, as shown in Figure 9d–f, the mechanical power continues to increase, with values exceeding 600 W at higher speeds. However, the fluctuations in the measured power signals become more noticeable at these load conditions. These variations are primarily driven by the dynamic response of the motor and increased difficulty in maintaining speed regulation under higher torque demands. The fluctuations also reflect the system's response to elevated current draw and associated back EMF and voltage drop effects, which slightly impact the consistency of power delivery during rapid speed changes.

Overall, the results in Figure 9 confirm that the BLDC motor is capable of delivering progressively higher mechanical power as load torque and speed increase, while also highlighting the onset of dynamic effects under heavier loading. The consistency and scalability of power output validate both the reliability of the experimental setup and the accuracy of the measurement methodology. Furthermore, the increasing signal fluctuations at higher torques underscore the importance of incorporating load-dependent performance characteristics into motor control design and system-level energy optimization strategies.

Figure 10 presents the efficiency profiles of the system, comprising the Li-ion battery, power electronics, and BLDC motor under variable speed conditions and fixed mechanical load torques ranging from 1 to 4 N·m. The results indicate a strong dependence of system efficiency on both load torque and operating speed, revealing distinct trends in the motor's electro-mechanical energy conversion performance.

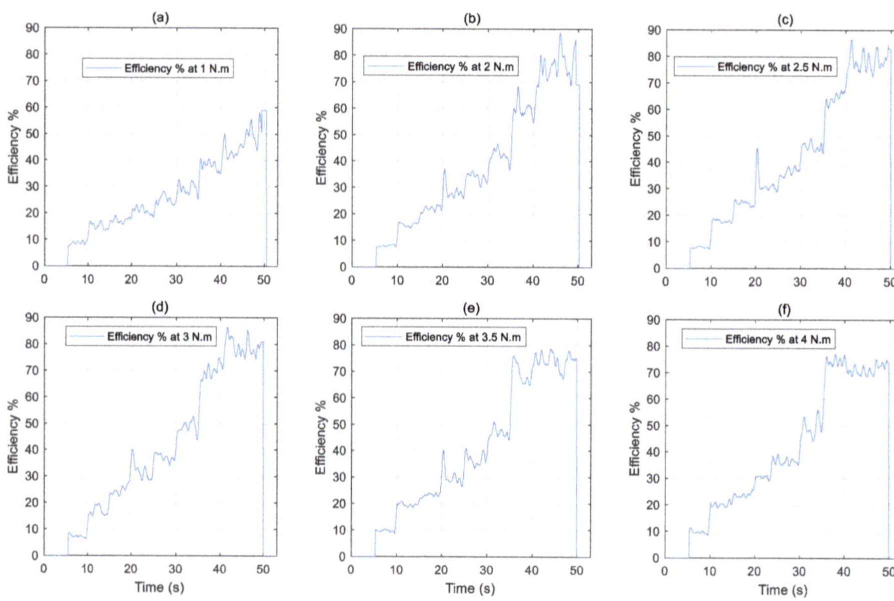

Figure 10. Overall efficiency of the system under variable reference speed and fixed load torque conditions: (**a**) 1 N·m, (**b**) 2 N·m, (**c**) 2.5 N·m, (**d**) 3 N·m, (**e**) 3.5 N·m, (**f**) 4 N·m.

At a low load torque of 1 N·m (Figure 10a), the system exhibits a gradual increase in efficiency with speed, peaking at approximately 57% during high-speed operation. This relatively modest efficiency can be attributed to the mismatch between the electrical input power and the low mechanical output power, resulting in a higher proportion of losses, particularly at lower operating points.

As the torque increases to 2 N·m and 2.5 N·m (Figure 10b,c), the efficiency improves significantly, reaching peak values of 86% and 83%, respectively. This enhancement reflects a more favorable load condition where the motor operates closer to its optimal efficiency region, with reduced relative losses and better utilization of the supplied electrical energy.

Under heavier load conditions of 3 N·m and 3.5 N·m (Figure 10d,e), the motor continues to demonstrate strong performance, achieving peak efficiencies of around 86% and 78% within the system under tested conditions. Although the efficiency remains high, slight reductions are observed, particularly at higher speeds, likely due to increased current-induced losses and potential thermal effects.

At the maximum tested load of 4 N·m (Figure 10f), the motor sustains a system efficiency above 70% across most of the operating range. The relatively stable efficiency profile under this high torque level underscores the motor's robustness and its capacity to maintain effective energy conversion despite elevated electrical and thermal stress.

These experimental findings confirm that the BLDC motor achieves optimal efficiency under moderate to moderately high torque conditions, where electrical input and mechanical output are better matched. However, at both very low and very high torque levels, efficiency tends to decline due to inherent system losses and thermal limitations. This analysis emphasizes the importance of operating the motor within its optimal torque–speed envelope to maximize performance and energy efficiency in real-world applications.

Figure 11 presents the estimated SOC of the Li-ion battery during variable-speed operation under six different fixed torque levels. These results provide valuable insight into how mechanical loading influences battery depletion dynamics over short operational intervals.

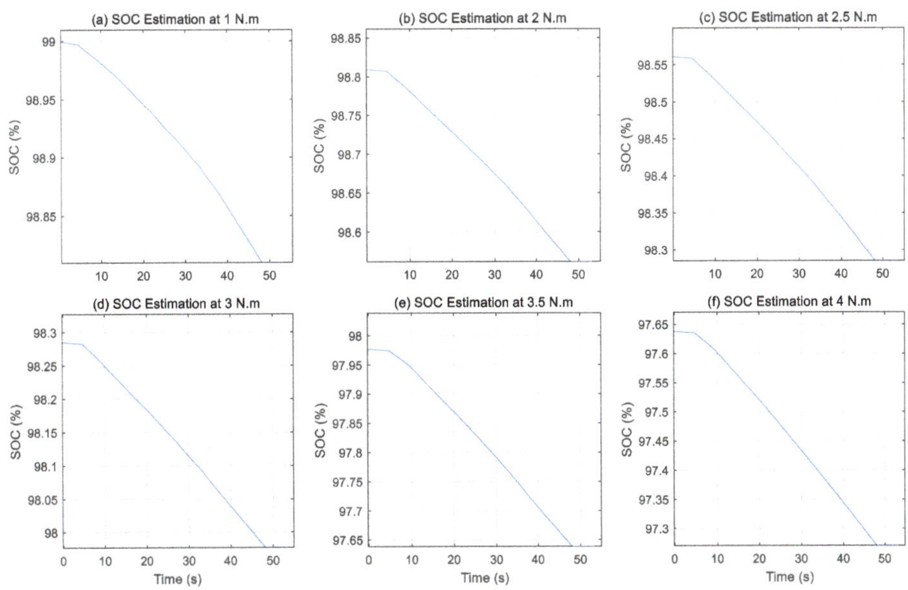

Figure 11. SOC estimation of the Li-ion battery during variable speed operation of the BLDC motor under different fixed load torque levels: (**a**) 1 N·m, (**b**) 2 N·m, (**c**) 2.5 N·m, (**d**) 3 N·m, (**e**) 3.5 N·m, (**f**) 4 N·m.

At the lowest load condition of 1 N·m (Figure 11a), the SOC exhibits a slow and nearly linear decline, falling from approximately 99% to 98.81%. This behavior reflects low current draw, minimal internal losses, and stable energy consumption during light-duty operation. The system operates efficiently under these conditions, with negligible stress imposed on the battery.

As the load torque increases to 2 N·m and 2.5 N·m (Figure 11b,c), the SOC curves show a more noticeable downward trend, ending near 98.56% and 98.28%, respectively. This gradual increase in energy consumption is consistent with the higher electrical demands placed on the motor to counteract the greater mechanical load. Despite the steeper slopes, the SOC remains within a relatively stable range, suggesting that the battery can handle moderate load variations without significant energy strain.

At higher load levels of 3 N·m and 3.5 N·m (Figure 11d,e), the rate of SOC depletion becomes more pronounced. The SOC drops from approximately 98.28% to 97.98% and from 97.98% to 97.64%, respectively. These changes reflect a more dynamic battery response as the system operates under increased stress. The elevated current draw required to maintain speed under heavier torque causes faster energy drain, revealing a nonlinear relationship between load torque and SOC decay.

Under the maximum torque condition of 4 N·m (Figure 11f), the SOC experiences the steepest decline, falling from about 97.64% to 97.27% over the test duration. Although the overall drop appears small, the increased slope indicates that sustained high-load operation would result in significantly faster battery depletion over extended periods. This test condition approaches the battery's performance limits, emphasizing the need for cautious management of load profiles in practical applications.

Overall, the SOC trends across Figure 11 clearly demonstrate a direct correlation between mechanical load and battery discharge behavior. The findings confirm that as torque increases, the motor's current demand rises accordingly, accelerating SOC consumption. These results highlight the importance of incorporating real-time battery monitoring and load-aware energy management strategies to ensure reliable operation, prolong battery lifespan, and enhance overall system efficiency in BLDC motor drive applications.

Although the experimental results are based on a 2.2 kW BLDC motor, the proposed modeling, validation, and performance evaluation framework is broadly applicable to other BLDC motors and electric drive systems. The system identification approach relies on data-driven methods that are independent of motor power rating, and the PRBS excitation, validation with various input signals, and dynamometer-based load testing are standard techniques that can be adapted to motors of different sizes and specifications. Furthermore, the SOC estimation via Coulomb counting is based on measured current and time integration, making it transferable to other Li-ion battery-powered systems. By applying the same data-driven methodology and experimental procedures, this framework can be employed to model and analyze a wide range of BLDC motors across diverse applications, including EVs, robotics, and industrial automation.

6. Conclusions

This study introduced a comprehensive experimental framework for the modeling, validation, and performance analysis of a 2.2 kW BLDC motor under varying fixed load torque levels and variable speed inputs. A dynamic model of the motor was identified using a system identification approach based on a PRBS excitation signal. The accuracy of the identified model was validated through experimental testing using a diverse set of input profiles, including sinusoidal, exponential, ramp, and chirp signals. The results revealed a strong correlation between the model predictions and the measured motor responses, confirming the model's effectiveness in capturing the dynamic characteristics of the motor across a wide range of operating scenarios.

Furthermore, torque–speed characterization was conducted under fixed load torques ranging from 1 N·m to 4 N·m. The experimental results included measurements of terminal voltage, battery current, electrical input power, mechanical output power, and system efficiency. The analysis revealed that increasing load torque resulted in a proportional rise in electrical power consumption and current draw. Mechanical output power increased with load, while system efficiency improved significantly up to 3 N·m before beginning to stabilize, indicating an optimal operating range for high-efficiency performance.

In terms of Li-ion battery performance, the SOC behavior observed under different load torque levels highlights how mechanical demands directly influence battery usage in BLDC motor systems. As the load increases, the motor draws more current, leading to a gradual but noticeable rise in energy consumption. While the SOC changes remain modest over a short operating window, the trend is clear and consistent, with higher torque levels resulting in faster battery discharge. These insights reinforce the need for intelligent and load-aware energy management strategies to maintain system efficiency and extend battery life, especially in applications requiring sustained operation under varying mechanical loads.

Finally, the outcomes of this study demonstrate strong practical relevance. The experimentally identified model and performance analysis provide a foundation for designing advanced control strategies in EVs and hybrid electric vehicles (HEVs), where precise and energy-efficient operation is essential. In industrial automation and robotics, the load-dependent torque–speed behavior enables the development of accurate motion control systems. Moreover, the integration of real-time SOC estimation highlights the relationship between load torque and battery discharge, which is crucial for smart battery management and energy management strategies in EVs and HEVs. This experimental framework is adaptable to other BLDC motors and drive systems, contributing to improved control, performance optimization, and energy efficiency in electric mobility, robotics, and industrial applications.

Author Contributions: Conceptualization, A.I.A.; Methodology, A.I.A. and A.E.; Software, A.I.A., A.E., and O.A.; Validation, A.I.A.; Formal analysis, A.I.A.; Investigation, A.I.A.; Resources, A.I.A., R.D., and O.A.; Data curation, A.I.A.; Writing—Original draft preparation, A.I.A.; Writing—Review, A.I.A., R.D., and A.E.; Visualization, A.I.A.; Supervision, R.D. and A.E. All authors have read and agreed to the published version of the manuscript.

Funding: This research received no external funding.

Data Availability Statement: The experimental data supporting the findings of this study are not publicly available but may be obtained from the corresponding author upon reasonable request.

Conflicts of Interest: The authors declare no conflicts of interest.

References

1. Zeraoulia, M.; Benbouzid, M.E.H.; Diallo, D. Electric motor drive selection issues for HEV propulsion systems: A comparative study. *IEEE Trans. Veh. Technol.* **2006**, *55*, 1756–1764. [CrossRef]
2. Ljung, L. *System Identification—Theory for the User*; Prentice Hall: Upper Saddle River, NJ, USA, 1999.
3. Khan, M.A.; Baig, D.; Ali, H.; Albogamy, F.R. Optimized system identification of brushless DC motor using data-driven modeling methods. *Sci. Rep.* **2025**, *15*, 8497. [CrossRef]
4. Hat, M.K.; Ibrahim, B.S.K.K.; Mohd, T.A.T.; Hassan, M.K. Model Based Design of PID Controller for BLDC Motor with Implementation of Embedded Arduino Mega Controller. *ARPN J. Eng. Appl. Sci.* **2015**, *10*, 8588–8594.
5. Abouseda, A.I.; Doruk, R.O.; Amini, A. Parameter Identification and Speed Control of a Small-Scale BLDC Motor: Experimental Validation and Real-Time PI Control with Low-Pass Filtering. *Machines* **2025**, *13*, 656. [CrossRef]
6. Pandya, V.; Mehta, S.; Pramod, P. Modeling, Characterization, and Identification of Permanent Magnet DC Motors. In Proceedings of the 2023 IEEE International Electric Machines & Drives Conference (IEMDC), San Francisco, CA, USA, 6 September 2023; IEEE: Minneapolis, MN, USA, 2023; pp. 1–7.
7. Su, G.; Wang, P.; Guo, Y.; Cheng, G.; Wang, S.; Zhao, D. Multiparameter Identification of Permanent Magnet Synchronous Motor Based on Model Reference Adaptive System—Simulated Annealing Particle Swarm Optimization Algorithm. *Electronics* **2022**, *11*, 159. [CrossRef]
8. Yang, C.; Bu, L.; Chen, B. Energy Modeling and Online Parameter Identification for Permanent Magnet Synchronous Motor Driven Belt Conveyors. *Measurement* **2021**, *178*, 109342. [CrossRef]
9. Zhu, Z.; Liang, D.; Liu, K. Online Parameter Estimation for Permanent Magnet Synchronous Machines: An Overview. *IEEE Access* **2021**, *9*, 59059–59084. [CrossRef]

10. Kerem, A. Torque Estimation of Electric Vehicle Motor Using Adaptive-Network Based Fuzzy Inference Systems. *Int. J. Automot. Eng. Technol.* **2021**, *10*, 33–41. [CrossRef]
11. Al Mashhadany, Y.I.M.; Abbas, A.K.; Algburi, S.S. Modeling and Analysis of Brushless DC Motor System Based on Intelligent Controllers. *Bull. Electr. Eng. Inform.* **2022**, *11*, 2995–3003. [CrossRef]
12. El Sawy, M.A.; Kamel, O.M.; Mohamed, Y.S.; Mossa, M.A. Dynamic Performance Evaluation of a Brushless AC Motor Drive Using Different Sensorless Schemes. *Int. J. Robot. Control Syst.* **2024**, *4*, 1306.
13. Zhang, J.; Liu, Y.; Chen, T.; Dou, G. Research on Model Identification of Permanent Magnet DC Brushless Motor Based on Auxiliary Variable Subspace Identification Algorithm. *World Electr. Veh. J.* **2025**, *16*, 297. [CrossRef]
14. Arifin, B.; Nugroho, A.A.; Budisusila, E.N.; Khosyi'n, M. System Identification and Control Strategy on Electric Power Steering DC Motor. *J. Robot. Control* **2024**, *5*, 655–666. [CrossRef]
15. Marcos-Andrade, D.; Beltrán-Carbajal, F.; Castelan-Pérez, A.; Rivas-Cambero, I.; Hernández, J.C. Online Algebraic Estimation of Parameters and Disturbances in Brushless DC Motors. *Machines* **2025**, *13*, 16. [CrossRef]
16. Li, H.; Jian, X. Parameter Identification of Permanent Magnet Synchronous Motor Based on CGCRAO Algorithm. *IEEE Access* **2023**, *11*, 124319–124330. [CrossRef]
17. Tolun, O.C.; Tutsoy, Ö. Estimation of DC Motor Parameters Using Least Square–Based Optimization Algorithm. *Proc. Innov. Intell. Syst. Appl.* **2023**, *6*, 18–23.
18. Gao, D.; Wu, S.; Yu, J.; Wang, M.; Wang, Y. Parameter Identification of DC Motor Model Based on Improved Dynamic Forgetting Factor Recursive Least Squares Method. In Proceedings of the 2022 IEEE 8th International Conference on Smart Instrumentation, Measurement and Applications (ICSIMA), Melaka, Malaysia, 26–28 September 2022; pp. 282–286.
19. Rodríguez-Abreo, O.; Rodríguez-Reséndiz, J.; Álvarez-Alvarado, J.M.; García-Cerezo, A. Metaheuristic Parameter Identification of Motors Using Dynamic Response Relations. *Sensors* **2022**, *22*, 4050. [CrossRef]
20. Jimenez-Gonzalez, J.; Gonzalez-Montañez, F.; Jimenez-Mondragon, V.M.; Liceaga-Castro, J.U.; Escarela-Perez, R.; Olivares-Galvan, J.C. Parameter identification of BLDC motor using electromechanical tests and recursive least-squares algorithm: Experimental validation. *Actuators* **2021**, *10*, 143. [CrossRef]
21. Wu, K.H.; Seyedmahmoudian, M.; Mekhilef, S.; Shrivastava, P.; Stojcevski, A. Lithium-Ion Battery State of Charge Estimation Using Improved Coulomb Counting Method with Adaptive Error Correction. *Proc. Inst. Mech. Eng. Part D J. Automob. Eng.* **2023**, *239*, 693–705. [CrossRef]
22. Pu, L.; Wang, C. Combined State-of-Charge Estimation Method for Lithium-Ion Batteries Using Long Short-Term Memory Network and Unscented Kalman Filter. *Energies* **2023**, *18*, 1106. [CrossRef]
23. Fahmy, H.M.; Swief, R.A.; Hasanien, H.M.; Alharbi, M.; Maldonado, J.L.; Jurado, F. Hybrid State of Charge Estimation of Lithium-Ion Battery Using the Coulomb Counting Method and an Adaptive Unscented Kalman Filter. *Energies* **2023**, *16*, 5558. [CrossRef]
24. Noval, R.; Sumarsono, D.A.; Adhitya, M.; Heryana, G.; Zainuri, F.; Tullah, M.H.; Todaro, M. Performance Evaluation and Accuracy Analysis of a Chassis Dynamometer for Light Electric Vehicles. *World Electr. Veh. J.* **2025**, *16*, 170. [CrossRef]
25. Oloyede, M.O.; Akpakwu, G.A.; Myburgh, H.C.; De Freitas, A.; Kunatsa, T. A Review on State-of-Charge Estimation Methods, Energy Storage Technologies and State-of-the-Art Simulators: Recent Developments and Challenges. *World Electr. Veh. J.* **2024**, *15*, 381. [CrossRef]
26. Movassagh, K.; Raihan, A.; Balasingam, B.; Pattipati, K. A Critical Look at Coulomb Counting Approach for State of Charge Estimation in Batteries. *Energies* **2021**, *14*, 4074. [CrossRef]

actuators

Article

An Event-Triggered Observer-Based Control Approach for Enhancing Resilience of Cyber–Physical Systems Under Markovian Cyberattacks

Eya Hassine [1,2,†], Assem Thabet [1,†], Noussaiba Gasmi [2,*,†] and Ghazi Bel Haj Frej [3,†]

1 Research Lab. MACS, University of Gabes, Gabes 6072, Tunisia; ayahassine118@gmail.com (E.H.); assem.thabet@yahoo.fr (A.T.)
2 Laboratoire d'Informatique et Telecommunications, ECAM Louis de Broglie, 35170 Bruz, France
3 IMS Laboratory, Centre National de la Recherche Scientifique–Unité Mixte de Recherche 5218, Bordeaux Institut Polytechnique, University of Bordeaux, 33405 Bordeaux, France; ghazi.bel-haj-frej@ims-bordeaux.fr
* Correspondence: noussaiba.gasmi@ecam-ldb.fr
† These authors contributed equally to this work.

Abstract: This paper presents a resilient observer-based and event-triggered control scheme for discrete-time Cyber–Physical Systems (CPS) under Markovian Cyber-Attacks (MCA). The proposed framework integrates a Luenberger observer for cyberattack detection with a state-feedback controller designed to preserve system stability in the presence of Denial-of-Service (DoS) and False Data Injection (FDI) attacks. Attack detection is achieved through residual signal generation combined with Markovian modeling of the attack dynamics. System stability is guaranteed by formulating relaxed Linear Matrix Inequality (LMI) conditions that incorporate relaxation variables, a diagonal Lyapunov function, the S-procedure, and congruence transformations. Moreover, the Event-Triggered Mechanism (ETM) efficiently reduces communication load without degrading control performance. Numerical simulations conducted on a three-tank system benchmark confirm enhanced detection accuracy, faster recovery, and strong robustness against uncertainties.

Keywords: cyber–physical systems (CPS); Luenberger observer; state-feedback control; denial-of-service (DoS); false data injection (FDI); Markovian cyberattack; linear matrix inequality (LMI); Lyapunov function; S-procedure

1. Introduction

1.1. Background

Cyber–Physical Systems (CPS) have become fundamental in modern engineering by enabling seamless integration between computational algorithms and physical processes [1,2]. These systems are increasingly deployed in critical infrastructure sectors such as smart grids, industrial automation, intelligent transportation systems, and water distribution networks. By leveraging advanced communication networks, embedded sensing, and control capabilities, such systems enable real-time monitoring, adaptive decision-making, and overall process optimization [3,4]. However, this tight integration between the cyber and physical domains introduces significant vulnerabilities. The openness of networked architectures and the reliance on distributed control elements expose these systems to a broad spectrum of cyber threats, ranging from simple disturbances to sophisticated coordinated attacks [5,6]. Among these threats, FDI and DoS attacks are particularly harmful. FDI attacks can stealthily alter measurement values, misleading

 https://doi.org/10.3390/act14080412

controllers and compromising decision-making logic. Meanwhile, DoS attacks can interrupt communication among sensors, controllers, and actuators, potentially causing control delays or complete system shutdown [7–9]. The impact of such attacks is amplified due to the close coupling of the cyber and physical domains in CPS, where corrupted data may trigger unsafe or unstable physical behaviors. To mitigate these risks, substantial efforts have been dedicated to developing anomaly detection techniques, fault diagnosis methods, and secure control strategies. In particular, stochastic modeling approaches, particularly those based on Markov chains, have shown great potential in capturing the probabilistic behavior of cyberattacks and transitions between normal and compromised system modes [10]. These models allow attack dynamics to be integrated into the control loop, providing a more realistic representation of threat scenarios. Recent studies have proposed advanced observer designs grounded in H_∞ theory, sliding-mode techniques, or zonotopic methods [11,12], offering strong robustness against uncertainties. However, such techniques often come at the cost of high computational demand and tuning difficulties, particularly in discrete-time implementations. Deep learning-based detection methods have also shown promise in identifying cyberattacks [13], although their dependency on training data and limited real-time guarantees remain concerns. In contrast, the Luenberger observer approach provides a lightweight model-based alternative that is well suited for real-time and embedded CPS operating under event-triggered communication protocols. Despite these advances, many existing approaches treat detection and control as two separate components, focusing either on accurate anomaly detection [11,14] or robust regulation [3,7,12]. This decoupled design may be insufficient in real-time scenarios, where rapid adaptation and joint coordination are essential. Additionally, many control strategies are developed under the assumption of bounded disturbances, neglecting the stochastic and time-varying nature of cyberattacks. Consequently, such methods may fail to ensure both accurate detection and robust performance under adversarial conditions. Addressing these limitations calls for a more integrated and probabilistically informed framework that jointly designs detection and control components to enhance resilience. A few recent studies have begun to explore such unified strategies, combining observer-based detection with resilient control [4,8,12]. However, these contributions still present notable limitations. For example, the method proposed in [12] relies on adaptive control with Markovian attack modeling, but suffers from synthesis difficulties due to the presence of BMIs. The study in [8] addresses trajectory tracking under attack conditions, but does not include probabilistic modeling of cyber threats. Similarly, while [4] proposes a stochastic observer-based control scheme, the detection and control modules remain loosely integrated. In contrast, the present work introduces a computationally efficient and fully integrated framework that unifies attack detection and resilient control using relaxed LMIs. The approach explicitly captures the stochastic nature of cyberattacks and optimizes resource usage through an event-triggered communication strategy. This paper addresses the aforementioned limitations by proposing a resilient observer-based control framework for discrete-time CPS subjected to Markovian-modeled cyberattacks. The proposed framework unifies a Luenberger-type observer for state estimation and attack detection with a state-feedback controller operating under an event-triggered mechanism, which reduces communication requirements while preserving closed-loop stability. The Markov chain model captures the probabilistic behavior of attacks and uncertainties affecting sensors and actuators [12,15]. The design procedure relies on Lyapunov stability theory, transforming BMI constraints into LMI using relaxation variables, diagonal Lyapunov functions, the S-procedure, and congruence transformations to ensure efficient synthesis [16,17]. In this work, the proposed event-triggered strategy belongs to the class of Periodic Event-Triggered Control (PETC), where the triggering condition is evaluated at discrete sampling instants. This distinguishes it from Continuous Event-Triggered Control (CETC), which requires continuous monitoring of system variables, and Self-Triggered Control (STC), where the next triggering instant is predicted in advance based on current state information. While CETC and STC offer potential benefits in terms of resource optimization, PETC is particularly suitable for discrete-time cyber–physical systems due to its simple implementation and natural alignment with digital platforms [18]. The present framework is not directly extendable to CETC or STC, but future investigations may explore these variants to further enhance resilience against cyber threats. The effectiveness of the proposed method is validated through simulations on a benchmark three-tank system, which is a widely used approach in CPS security research. The results confirm the proposed framework's ability to promptly detect attacks, maintain robust performance, and handle noise and parameter uncertainties [19,20]. The main contributions of this work are summarized as follows:

- Development of a unified observer-based architecture that couples real-time attack detection with resilient control, overcoming the limitations of decoupled designs [13,14].
- Explicit integration of Markovian cyberattack models in both detection and control design, providing enhanced robustness and accuracy compared to disturbance-based methods [4,10].
- Introduction of an event-triggered control strategy that significantly reduces communication and computation without degrading performance, a feature that has rarely been addressed in observer-based resilient control [12].
- Efficient and tractable synthesis using relaxed LMIs, variable transformations, and diagonal Lyapunov functions, offering a practical alternative to BMI-based approaches [16].
- Comprehensive validation on a three-tank system benchmark, confirming fast detection, robust stabilization, and strong resilience under uncertainty.

The rest of this paper is organized as follows: the next part of this section presents some useful notation and preliminaries; Section 2 introduces the problem formulation; in Section 3, the synthesis procedure for the observer-based controller design is detailed and a particular solution is proposed to overcome the BMI problem; Section 4 presents the proposed ETM synthesis, which reduces communication and computational load without sacrificing performance; Section 5 provides simulation results to illustrate its validity and superiority for the considered fault-tolerant attack control scheme; Section 6 offers a comparison with the standard case along with a discussion on the enhancement of the main approach of this paper; finally, concluding remarks are provided in Section 7.

1.2. Notation

The following standard notation is used throughout the paper:

- M^T is the transposed matrix of M.
- M is a symmetric matrix; the notation $M > 0$ or $M < 0$ means that M is Symmetric Positive Definite (SPD) or Negative Definite (SND).
- $||M||$ represents the the l_2 norm of the vector M.
- In a matrix, the notation $(\star)$ refers to the blocks induced by symmetry.

2. Problem Formulation

The linear system provided by Equation (1) illustrates a distinctive characteristic of CPS, namely, that the system's input and output are measured at specific temporal points due to their remote transmission.

$$\begin{aligned} x_{k+1} &= A_p x_k + B_p u_k^{MCA} \\ y_k &= C_p x_k + T_2 z_k^s \end{aligned} \tag{1}$$

The control law is expressed as follows:

$$\begin{aligned} u_k^{MCA} &= u_k + T_1 z_k^a \\ u_k &= K_x \hat{x}_k \end{aligned} \quad (2)$$

where $x_k \in \mathbb{R}^{n_x}$ is the state vector of the system, $y_k \in \mathbb{R}^{n_y}$ is the measured output, $u_k^{MCA} \in \mathbb{R}^{n_u}$ is the control vector, and $u_k \in \mathbb{R}^{n_u}$ is the state-feedback control. In addition, A_p, B_p, C_p are constant matrices of appropriate dimensions, T_1, T_2 are the coupling matrices of the attacks, $z_k^a \in \mathbb{R}^r$ represents an attack directed at the actuator, and $z_k^s \in \mathbb{R}^m$ represents a sensor attack.

The following observer with a Luenberger structure is to be considered for System (1):

$$\begin{aligned} \hat{x}_k &= A_p \hat{x}_k + B_p u_k + L_x(y_k - \hat{y}_k) \\ \hat{y}_k &= C_p \hat{x}_k \end{aligned} \quad (3)$$

where $\hat{x}_k \in \mathbb{R}^{n_x}$ represents the states estimated by the observer, $\hat{y}_k \in \mathbb{R}^{n_y}$ is the output estimation vector, and $L_x \in \mathbb{R}^{n_x \times n_y}$ is the gain of the observer.

The estimation error can be expressed by $e_k = x_k - \hat{x}_k$, with its dynamics defined by the following equation:

$$e_{k+1} = x_{k+1} - \hat{x}_{k+1}. \quad (4)$$

Substituting Equations (1) and (3) into Equation (4) yields the following equation:

$$e_{k+1} = (A_p - L_x C_p)e_k + B_p T_1 z_k^a - L_x T_2 z_k^s. \quad (5)$$

The residual signal r_k representing the difference between the output and the estimated output is then described by

$$\begin{aligned} r_k &= C_p(x_k - \hat{x}_k) + T_2 z_k^s \\ r_k &= C_p e_k + T_2 z_k^s \end{aligned} \quad (6)$$

An augmented system is defined by the equation below:

$$\Omega_{k+1} = \widetilde{A_p}\Omega_k + \widetilde{B_p} Z_{p_k} \quad (7)$$

with $\Omega_{k+1} = \begin{pmatrix} \hat{x}_{k+1} \\ e_{k+1} \end{pmatrix}$, $\widetilde{A_p} = \begin{pmatrix} A_p + B_p K_x & L_x C_p \\ 0 & A_p - L_x C_p \end{pmatrix}$, $\widetilde{B_p} = \begin{pmatrix} 0 & L_x T_2 \\ B_p T_1 & -L_x T_2 \end{pmatrix}$,
$Z_{pk} = \begin{pmatrix} z_k^a \\ z_k^s \end{pmatrix}$.

3. Markovian Attacks Schemes

The following section provides definitions of the various attack schemes considered to compromise the proper operation of the selected CPS.

3.1. DoS

In the CPS context, this type of attack aims to disrupt the transmission of control signals and the retrieval of data from the controller. When the system is subjected to a DoS attack, the following equation should be considered:

$$\begin{aligned} z_k^a &= -\phi_k u_k \\ z_k^s &= -\psi_k x_k \end{aligned} \quad (8)$$

where ϕ_k, ψ_k are Markovian stochastic processes that take values in $\{0, 1\}$ [21].

Combining the system in Equation (1) with the DoS attack model targeting the sensor and actuator in Equation (8) yields the following expression:

$$x_{k+1} = A_p x_k + B_p K_x \hat{x}_k - \phi_k B_p T_1 K_x \hat{x}_k,$$
$$y_k = C_p x_k - \psi_k T_2 x_k. \tag{9}$$

3.2. FDI

This attack affects the sensor and controller readings, causing them to deviate from the actual signal measurements or control data. When the system is subjected to an FDI attack, the following equation should be considered:

$$z_k^a = -\phi_k u_k + \phi_k w_k^a$$
$$z_k^s = -\psi_k x_k + \psi_k w_k^s \tag{10}$$

where ϕ_k, ψ_k are Markovian stochastic processes that take values in $\{0,1\}$, ${w_k}^a$ represents tampering data on the actuator, and ${w_k}^s$ represents tampering data on the sensor. Substituting Equation (10) into (1) leads to

$$x_{k+1} = A_p x_k + B_p K_x \hat{x}_k - \phi_k B_p T_1 K_x \hat{x}_k + \phi_k B_p T_1 w_k^a,$$
$$y_k = C_p x_k - \psi_k T_2 x_k + \psi_k T_2 w_k^s. \tag{11}$$

Remark 1. *The cyberattacks are modeled using Markovian binary processes ϕ_k and ψ_k, which take values in $\{0,1\}$. Although this abstraction simplifies the analysis, it captures essential properties such as memory and probabilistic transitions between attack and non-attack modes. This modeling approach has been widely adopted in the literature on resilient control for CPS [4,10,15], particularly in scenarios involving packet dropouts, random delays, and jamming. To better capture intelligent threats, future work may consider extended models such as semi-Markov chains, time-varying probabilistic processes, or adversarial learning strategies.*

4. Feedback Control Design

This section presents a theorem that is fundamental to establishing the stability of the proposed system.

Theorem 1. *The design of an observer-based controller for System* (1) *with observer dynamics provided by* (3) *and a state-feedback controller as in* (2) *is achievable if there exist scalars S_{P1}, $\lambda > 0$ and a symmetric positive-definite matrix $P_p = P_p^T > 0$ along with matrices $\widetilde{P}_{p11}$, $\widetilde{P}_{p12}$, P_{p22}, $\widetilde{Q}_{p11}$, $Q_{p22} \in \mathbb{R}^{n_x \times n_x}$, $\bar{L}_x \in \mathbb{R}^{n_y \times n_x}$, and $\bar{K}_x \in \mathbb{R}^{n_u \times n_x}$ such that the following linear matrix inequality admits a solution:*

$$Q_a = \begin{pmatrix} -\widetilde{P}_{p11} + S_{p1} I & -\widetilde{P}_{p12} & 0 & 0 & Q_{a15} & 0 \\ (\star) & -P_{p22} + S_{p1} I & 0 & 0 & Q_{a25} & Q_{a26} \\ (\star) & (\star) & -S_{p1} I & 0 & Q_{a35} & Q_{a36} \\ (\star) & (\star) & (\star) & -S_{p1} I & Q_{a45} & Q_{a46} \\ (\star) & (\star) & (\star) & (\star) & Q_{a55} & Q_{a56} \\ (\star) & (\star) & (\star) & (\star) & (\star) & Q_{a66} \end{pmatrix} < 0 \tag{12}$$

where:

$$
\begin{aligned}
Q_{a15} &= \widetilde{Q}_{p11} A_p^T + \bar{K}_x^T B_p^T \\
Q_{a25} &= \lambda C_p^T \bar{L}_x^T + A_p^T - \lambda A_p^T Q_{p22}^T \\
Q_{a26} &= A_p^T Q_{p22}^T - C_p^T \bar{L}_x^T \\
Q_{a35} &= T_1^T B_p^T - \lambda T_1^T B_p^T Q_{p22}^T \\
Q_{a36} &= T_1^T B_p^T Q_{p22}^T \\
Q_{a45} &= \lambda T_2^T \bar{L}_x^T \\
Q_{a46} &= -T_2^T \bar{L}_x^T \\
Q_{a55} &= \widetilde{P}_{p11} - \widetilde{Q}_{p11} - \widetilde{Q}_{p11}^T \\
Q_{a56} &= \widetilde{P}_{p12} - I_n + \lambda Q_{p22}^T \\
Q_{a66} &= P_{p22} - Q_{p22} - Q_{p22}^T.
\end{aligned}
$$

Proof. The proof of Theorem 1 consists of three logical steps:

(i) Lyapunov-based stability condition formulation.

(ii) Application of the S-procedure and Schur complement.

(iii) Transformation of BMIs into relaxed LMIs using slack variables and diagonal Lyapunov functions. Each transformation is guided by a mathematical rationale aimed at improving convexity and tractability. This structured approach ensures both mathematical rigor and practical feasibility for the synthesis of observer-based resilient controllers.

First, consider the following Lyapunov function:

$$V(\Omega_k) = \Omega_k{}^T P_p \Omega_k \tag{13}$$

where $P > 0$. The variation of $V(\Omega_k)$ along the solutions of (13) is as follows.

$$
\begin{aligned}
\Delta V_k &= V(\Omega_{k+1}) - V(\Omega_k) \\
\Delta V_k &= \Omega_{k+1}^T P_p \Omega_{k+1} - \Omega_k^T P_p \Omega_k \\
\Delta V_k &= \Omega_k^T (\widetilde{A}_p^T P_p \widetilde{A}_p - P_p) \Omega_k + \Omega_k^T \widetilde{A}_p^T P_p \widetilde{B}_p Z_{pk} \\
&+ Z_{pk}^T \widetilde{B}_p^T P_p \widetilde{A}_p \Omega_k + Z_{pk}^T \widetilde{B}_p^T P_p \widetilde{B}_p Z_{pk}
\end{aligned} \tag{14}
$$

Thus,

$$\Delta V_k = \begin{pmatrix} \Omega_k^T \\ Z_{pk}^T \end{pmatrix} Q_a \begin{pmatrix} \Omega_k \\ Z_{pk} \end{pmatrix} \tag{15}$$

with

$$Q_a = \begin{pmatrix} \widetilde{A}_p^T P_p \widetilde{A}_p - P_p & \widetilde{A}_p^T P_p \widetilde{B}_p \\ (\star) & \widetilde{B}_p^T P_p \widetilde{B}_p \end{pmatrix} < 0. \tag{16}$$

Application of the S-procedure [22–24] to (16) then leads to

$$Q_a = \begin{pmatrix} \widetilde{A}_p^T P_p \widetilde{A}_p - P_p + S_{p1} I & \widetilde{A}_p^T P_p \widetilde{B}_p \\ (\star) & -S_{p1} I + \widetilde{B}_p^T P_p \widetilde{B}_p \end{pmatrix} < 0. \tag{17}$$

□

Remark 2. *The S-procedure is a common tool for handling quadratic inequalities, and is employed here to guarantee the feasibility of matrix inequalities affected by uncertainties. This allows for the introduction of positive multipliers such as S_{p1} into LMIs to relax constraints while maintaining convexity. Unlike the full framework in [22], which requires sector conditions to handle nonlinearities, the addressed problem in this work deals with linear dynamics and bounded stochastic* *cyberattacks modeled by Markov chains. The disturbances due to attacks are incorporated in the observer, and are viewed as energy-limited signals rather than as sector-bounded nonlinearities. Hence, the sector condition from [22] does not apply here, and the S-procedure simply serves as a relaxation tool to ensure LMI feasibility.*

Next, Schur's complement [24] is applied to the terms $\widetilde{A}_p^T P \widetilde{A}_p$, $\widetilde{A}_p^T P \widetilde{B}_p$, and $\widetilde{B}_p^T P \widetilde{B}_p$:

$$Q_a = \begin{pmatrix} -P_p + S_{p1} I & 0 & \widetilde{A}_p^T \\ (\star) & -S_{p1} I & \widetilde{B}_p^T \\ (\star) & (\star) & -P_p^{-1} \end{pmatrix} < 0. \tag{18}$$

In order to eliminate the variable P_p^{-1}, a slack variable Q_p (SPD matrix) is introduced. The proposed methodology involves pre-multiplying the inequality by block-diag(I, I, Q_p) and post-multiplying by block-diag (I, I, Q_p^T). Subsequently, it can be demonstrated that applying the inequality $-Q_p P_p^{-1} Q_p^T \le P_p - Q_p - Q_p^T$ leads to the following results:

$$Q_a = \begin{pmatrix} -P_p + S_{p1} I & 0 & \widetilde{A}_p^T Q_p^T \\ (\star) & -S_{p1} I & \widetilde{B}_p^T Q_p^T \\ (\star) & (\star) & P_p - Q_p - Q_p^T \end{pmatrix} < 0. \tag{19}$$

For variables Q_p and P_p, the following structures are proposed:

$$P_p = \begin{pmatrix} P_{p11} & P_{p12} \\ (\star) & P_{p22} \end{pmatrix}, \tag{20}$$

$$Q_p = \begin{pmatrix} Q_{p11} & Q_{p12} \\ 0 & Q_{p22} \end{pmatrix}. \tag{21}$$

Thus, the constraint (19) becomes

$$Q_a = \begin{pmatrix} -P_{p11} + S_{p1} I & -P_{p12} & 0 & 0 & Q_{a15} & 0 \\ (\star) & -P_{p22} + S_{p1} I & 0 & 0 & Q_{a25} & Q_{a26} \\ (\star) & (\star) & -S_{p1} I & 0 & Q_{a35} & Q_{a36} \\ (\star) & (\star) & (\star) & -S_{p1} I & Q_{a45} & Q_{a46} \\ (\star) & (\star) & (\star) & (\star) & Q_{a55} & Q_{a56} \\ (\star) & (\star) & (\star) & (\star) & (\star) & Q_{a66} \end{pmatrix} < 0, \tag{22}$$

where:

$$\begin{aligned}
Q_{a15} &= A_p^T Q_{p11}^T + K_x^T B_p^T Q_{p11}^T \\
Q_{a25} &= C_p^T L_x^T Q_{p11}^T + A^T Q_{p12}^T - C_p^T L_x^T Q_{p12}^T \\
Q_{a26} &= A_p^T Q_{p22}^T - C_p^T L_x^T Q_{p22}^T \\
Q_{a35} &= T_1^T B_p^T Q_{p12}^T \\
Q_{a36} &= T_1^T B_p^T Q_{p22}^T \\
Q_{a45} &= T_2^T L_x^T Q_{p11}^T - T_2^T L_x^T Q_{p12}^T \\
Q_{a46} &= -T_2^T L_x^T Q_{p22}^T \\
Q_{a55} &= P_{p11} - Q_{p11} - Q_{p11}^T \\
Q_{a56} &= P_{p12} - Q_{p12} \\
Q_{a66} &= P_{p22} - Q_{p22} - Q_{p22}^T.
\end{aligned}$$

Applying pre- and post-multiplication of Equation (22) by block-diag($\widetilde{Q}_{p11}$, I, I, I, $\widetilde{Q}_{p11}$, I) and block-diag($\widetilde{Q}_{p11}^T$, I, I, I, $\widetilde{Q}_{p11}^T$, I) with $\widetilde{Q}_{p11} = Q_{p11}^{-1}$ and defining $\widetilde{P}_{p12} = \widetilde{Q}_{p11}P_{p12}$, $\widetilde{P}_{p11} = \widetilde{Q}_{p11}P_{p11}\widetilde{Q}_{p11}^T$ results in

$$Q_a = \begin{pmatrix} -\widetilde{P}_{p11} + S_{p1}I & -\widetilde{P}_{p12} & 0 & 0 & Q_{a15} & 0 \\ (\star) & -P_{p22} + S_{p1}I & 0 & 0 & Q_{a25} & Q_{a26} \\ (\star) & (\star) & -S_{p1}I & 0 & Q_{a35} & Q_{a36} \\ (\star) & (\star) & (\star) & -S_{p1}I & Q_{a45} & Q_{a46} \\ (\star) & (\star) & (\star) & (\star) & Q_{a55} & Q_{a56} \\ (\star) & (\star) & (\star) & (\star) & (\star) & Q_{a66} \end{pmatrix} < 0, \tag{23}$$

where:

$$\begin{aligned} Q_{a15} &= \widetilde{Q}_{p11}A_p^T + \widetilde{Q}_{p11}K_x^T B_p^T \\ Q_{a25} &= C_p^T L_x^T (I_n - Q_{p12}^T \widetilde{Q}_{p11}^T) + A_p^T Q_{p12}^T \widetilde{Q}_{p11}^T \\ Q_{a26} &= A_p^T Q_{p22}^T - C_p^T L_x^T Q_{p22}^T \\ Q_{a35} &= T_1^T B_p^T Q_{p12}^T \widetilde{Q}_{p11}^T \\ Q_{a36} &= T_1^T B_p^T Q_{p22}^T \\ Q_{a45} &= T_2^T L_x^T - T_2^T L_x^T Q_{p12}^T \widetilde{Q}_{p11} \\ Q_{a46} &= -T_2^T L_x^T Q_{p22}^T \\ Q_{a55} &= \widetilde{P}_{p11} - \widetilde{Q}_{p11} - \widetilde{Q}_{p11}^T \\ Q_{a56} &= \widetilde{P}_{p12} - \widetilde{Q}_{p11}Q_{p12} \\ Q_{a66} &= P_{p22} - Q_{p22} - Q_{p22}^T. \end{aligned}$$

The observer gain L_x appears in the constraint as a multiplier of both Q_{22} and $(I_n - \bar{Q}_{11}Q_{12})$, making the design more complex. A suitable approach to simplify the problem is to impose the following expression of Q_{p12}:

$$Q_{p12} = Q_{p11} - \lambda Q_{p11}Q_{p22} \tag{24}$$

where λ is a positive scalar introduced to increase the flexibility of the LMI condition. The resulting structure of the matrix Q_p is expressed as follows:

$$Q_p = \begin{pmatrix} Q_{p11} & Q_{11} - \lambda Q_{p11}Q_{p22} \\ 0 & Q_{p22} \end{pmatrix}. \tag{25}$$

Then, we obtain

$$Q_a = \begin{pmatrix} -\widetilde{P}_{p11} + S_{p1}I & -\widetilde{P}_{p12} & 0 & 0 & Q_{a15} & 0 \\ * & -P_{p22} + S_{p1}I & 0 & 0 & Q_{a25} & Q_{a26} \\ 0 & 0 & -S_{p1}I & 0 & Q_{a35} & Q_{a36} \\ 0 & 0 & 0 & -S_{p1}I & Q_{a45} & Q_{a46} \\ * & * & * & * & Q_{a55} & Q_{a56} \\ * & * & * & * & * & Q_{a66} \end{pmatrix} < 0, \tag{26}$$

where:

$$
\begin{aligned}
Q_{a15} &= \widetilde{Q}_{p11}A_p^T + \widetilde{Q}_{p11}K_x^T B_p^T \\
Q_{a25} &= \lambda C_p^T L_x^T Q_{p22}^T + A_p^T - \lambda A_p^T Q_{p22}^T \\
Q_{a26} &= A_p^T Q_{p22}^T - C_p^T L_x^T Q_{p22}^T \\
Q_{a35} &= T_1^T B_p^T - \lambda T_1^T B_p^T Q_{p22}^T \\
Q_{a36} &= T_1^T B_p^T Q_{p22}^T \\
Q_{a45} &= \lambda T_2^T L_x^T Q_{p22}^T \\
Q_{a46} &= -T_2^T L_x^T Q_{p22}^T \\
Q_{a55} &= \widetilde{P}_{p11} - \widetilde{Q}_{p11} - \widetilde{Q}_{p11}^T \\
Q_{a56} &= \widetilde{P}_{12} - I_n + \lambda Q_{p22}^T \\
Q_{a66} &= P_{p22} - Q_{p22} - Q_{p22}^T.
\end{aligned}
$$

Finally, using a simple changes of variables $\bar{K}_x = \widetilde{Q}_{p11}K_x^T$ and $\bar{L}_x = L_x Q_{p22}$, we can obtain the LMI described in (12). This ends the proof of Theorem 1.

5. Event-Triggered Strategy Design

In event-triggered control, the input signal u_k is not sent to the controller at each sampling period; instead, updates are transmitted only at specific instants, denoted k'. This approach ensures that the control signal remains constant until a predefined triggering criterion is satisfied, at which point it is updated. The update times t_k are determined according to a predefined condition that specifies when the input signal should be transmitted. In particular, the signal is updated only if the following condition is not satisfied:

$$\varepsilon_k{}^T \varepsilon_k - \gamma_1 \delta_k^T \delta_k < 0 \tag{27}$$

where $\varepsilon_k = y_k - \hat{y}_k$, $\delta_k = u_k - u_{k'}$, and γ_1 an attenuation rate of e_k.

In other words, the transmission occurs when

$$\varepsilon_{k+1}^T \varepsilon_{k+1} - \gamma_1 \delta_{k+1}^T \delta_{k+1} \geq 0, \tag{28}$$

provided that the initial conditions satisfy the inequality in (27).

Within our proposed structure, a control input is transmitted only when the discrepancy between the most recently sent value and the currently computed input is significant relative to the current value. This approach reduces unnecessary transmission bandwidth usage and prevents data congestion, as highlighted by [23]. Considering the attenuation related to the state estimation error, we can express

$$e_k^T e_k - \gamma_1 \delta_k^T \delta_k < 0, \tag{29}$$

which is equivalent to

$$\|e_k\|^2 < \gamma_1 \|\delta_k\|^2. \tag{30}$$

The updating condition is then defined as

$$\|e_k\|^2 \geq \gamma_1 \|\delta_K\|^2. \tag{31}$$

From the definition of ε_k, we can write

$$\varepsilon_k = \widetilde{y}_k - \hat{y}_k = e_k + \Gamma_p z_k{}^a \implies \|\varepsilon_k\|^2 \geq \|e_k\|^2, \tag{32}$$

where $\Gamma_p = \begin{bmatrix} \gamma_1 & 0 \\ 0 & \gamma_2 \end{bmatrix}$ and γ_2 is the attenuation rate of ε_k.

Equation (29) then becomes

$$\|\varepsilon_k\|^2 \geq \|e_k\|^2 \geq \gamma_1 \|\delta_k\|^2, \tag{33}$$

which is equivalent to violating the condition in (27).

6. Simulation Results

In the field of chemical process engineering, interconnected tank systems play a crucial role. These systems are an essential tool for automatic control. They are composed of multiple tanks interconnected by strategically placed pipes and valves [8].

The various parameters and variables of this system are as follows:

$$x_k = \begin{bmatrix} h_k^1 \\ h_k^2 \\ h_k^3 \end{bmatrix}, u_k = \begin{bmatrix} Q_k^1 \\ Q_k^2 \end{bmatrix}, y_k = \begin{bmatrix} h_k^1 \\ h_k^2 \\ h_k^3 \end{bmatrix}$$

where $Q_k^{1,2}$ are the flow rates of the two pumps and $h_k^{1,2,3}$ are the levels of liquid in each tank. The dynamic input, output, and cyberattack distribution matrices are

$$A_p = \begin{bmatrix} 1 & 0 & 0.0001 \\ 0 & 1 & -0.0001 \\ -0.0001 & -0.0001 & 1 \end{bmatrix}, B_p = \begin{bmatrix} 0.0649 & 0 \\ 0 & 0.0649 \\ 0 & 0 \end{bmatrix}, C_p = I_3$$

$$T_1 = \begin{bmatrix} 1 & 0 & 0 \\ 0 & 1 & 0 \end{bmatrix}, T_2 = \begin{bmatrix} 1 & 0 & 0 \\ 0 & 1 & 0 \\ 0 & 0 & 1 \end{bmatrix}.$$

The initial conditions are $x_k(0) = \begin{bmatrix} 0.4 & 0.2 & 0.3 \end{bmatrix}^T$, $\hat{x}_k(0) = \begin{bmatrix} 0.32 & 0.16 & 0.24 \end{bmatrix}^T$.

For the simulation, the following values of the system parameters are established:

$$w_k^a = \begin{bmatrix} 0.83 \\ 1.5 \\ -0.45 \end{bmatrix}, w_k^s = \begin{bmatrix} -0.2 \\ 0.65 \\ 0.85 \end{bmatrix}.$$

Resolving (12) using the YALMIP® solver in MATLAB provides:

$$L_x = \begin{bmatrix} 0.6512 & 0.0002 & 0.0051 \\ 0.0002 & 0.6512 & 0.0055 \\ 0.0050 & 0.0053 & 0.6234 \end{bmatrix}, K_x = \begin{bmatrix} -1.5672 & -0.0013 & 0.1996 \\ -0.0013 & -1.5671 & 0.2084 \end{bmatrix}$$

$$S_{p1} = 0.01, \lambda = 0.005.$$

6.1. Nominal Case Without Cyberattacks

Initially, Figures 1–3 present the evolution of the control input u_k, residuals signals r_k, and estimation errors e_k when no attacks are present, thereby verifying the stability of the proposed system.

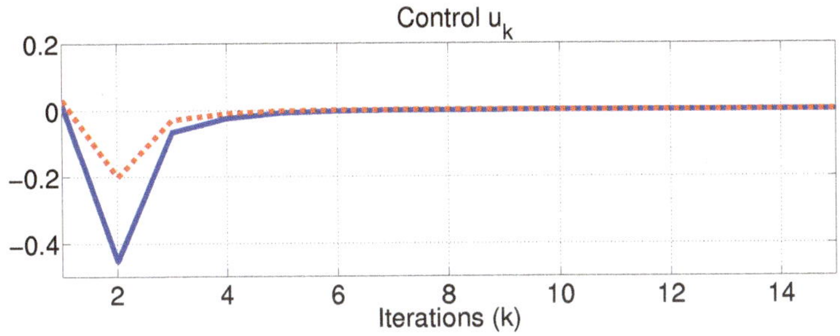

Figure 1. Evolution of u_k in the absence of cyberattacks.

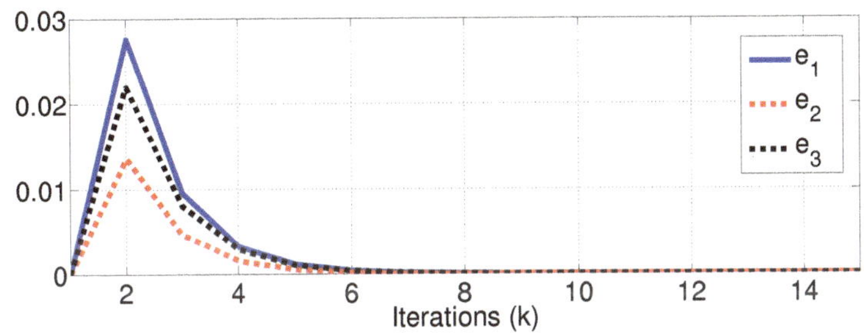

Figure 2. Evolution of e_k in the absence of cyberattacks.

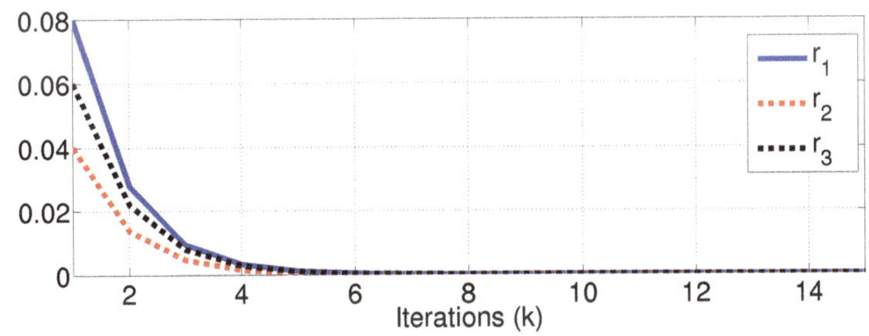

Figure 3. Evolution of r_k in the absence of cyberattacks.

6.2. Transient Performance Under Cyberattacks

In the following section, attention is directed to the behavior of errors, residuals, and various system states in the presence of cyberattacks. A The DoS attack is implemented within [400 k, 800 k], while the FDI attack occurs in the interval [1200 k, 1400 k].

The simulation results presented in Figures 4–6 confirm the effectiveness of the proposed control approach. The combination of state-feedback regulation and a Luenberger observer successfully mitigates the impact of cyberattacks, including both DoS and FDI scenarios, while preserving the overall stability of the cyber–physical system. In addition, Figure 7 highlights the evolution of the control input under attack conditions, providing further insight into the system's response to adversarial disturbances.

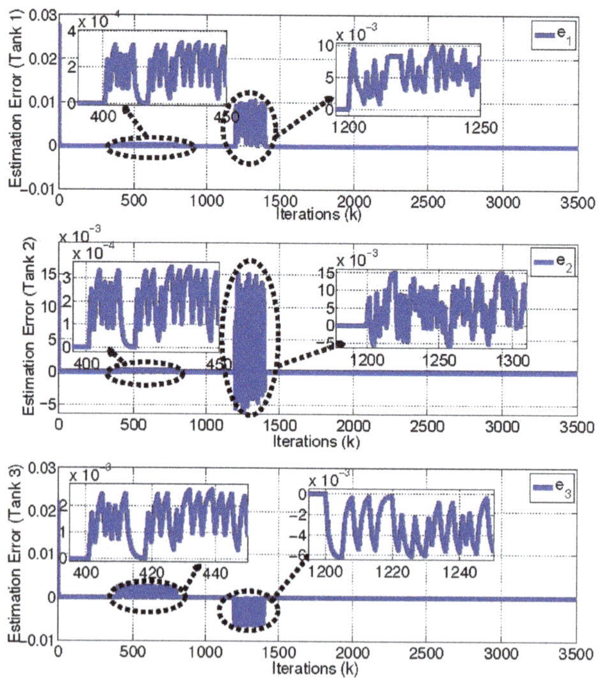

Figure 4. Evolution of estimation error signals e_k under DoS and FDI attack scenarios.

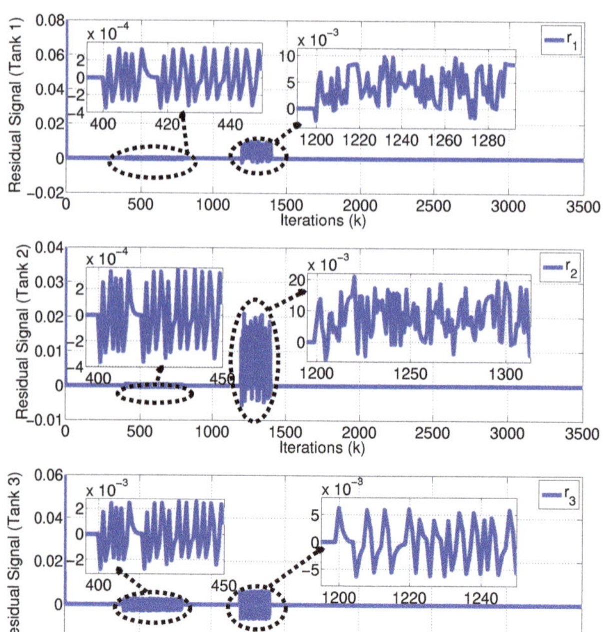

Figure 5. Evolution of the residual signal r_k under DoS and FDI attack scenarios.

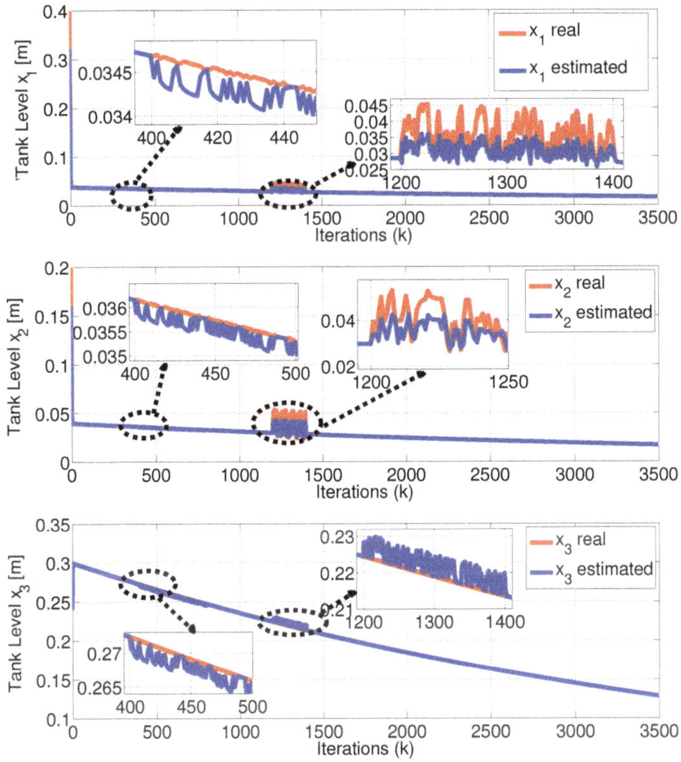

Figure 6. Comparison of real (x_k) and estimated ($\hat{x}_k$) tank levels under DoS and FDI attack scenarios.

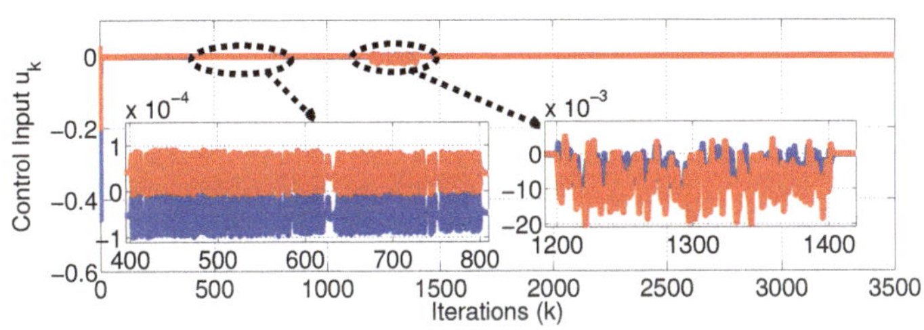

Figure 7. Variation of the control input u_k (u_k^1 in blue and u_k^2 in red) under DoS and FDI attack scenarios.

6.3. Response Under Sustained Cyberattacks

6.3.1. Response Under Sustained DoS Attacks

Next, to assess the impact of a sustained cyberattack, a permanent DoS attack is initiated at 1000 k. Figures 8–10 respectively illustrate the evolution of the system outputs, residual signals, and control input under these conditions.

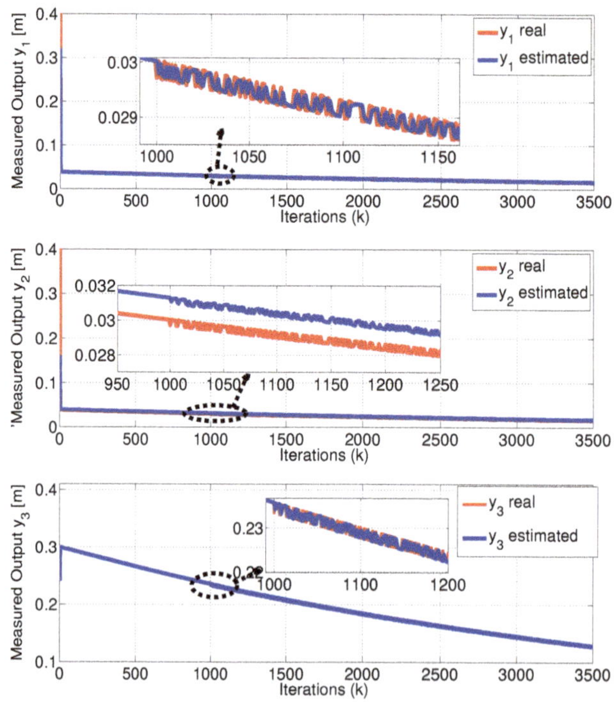

Figure 8. Evolution of measured output y_k under persistent DoS attack scenario.

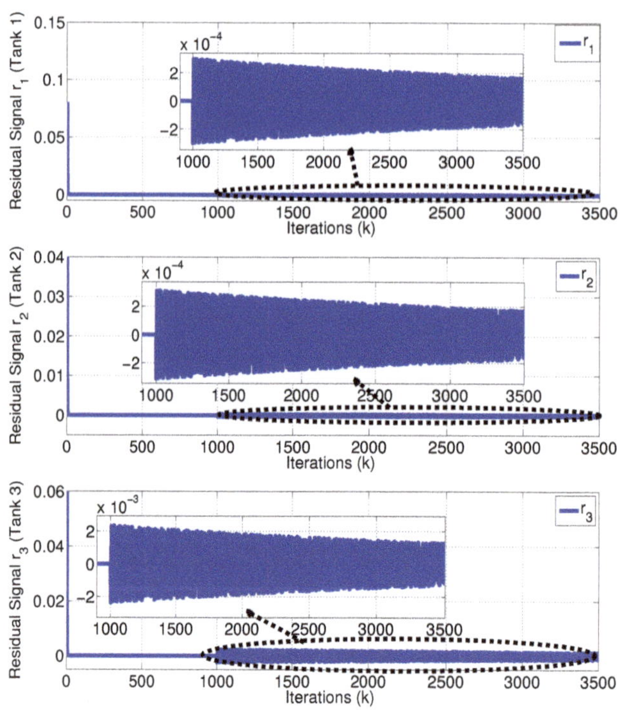

Figure 9. Evolution of r_k under persistent DoS attack scenario.

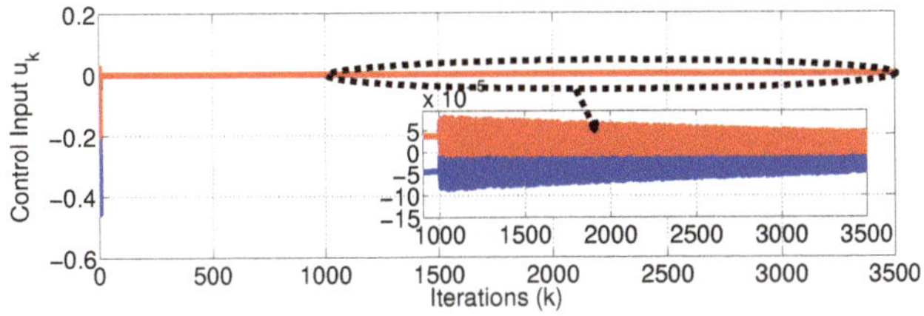

Figure 10. Variation of u_k (u_k^1 in blue and u_k^2 in red) under persistent DoS attack scenario.

Figures 8–10 confirm that the estimated outputs consistently converge towards the real system outputs even under permanent DoS attack. Moreover, the residual signals are highly sensitive to both the presence of attacks and potential system faults.

6.3.2. Response Under Sustained FDI Attacks

To further evaluate the system's resilience against a sustained cyberattack, a permanent FDI attack is introduced beginning at 1000 k. Figures 11, 12 and 13 respectively illustrate the evolution of the system states, estimation error, and control input under this attack condition.

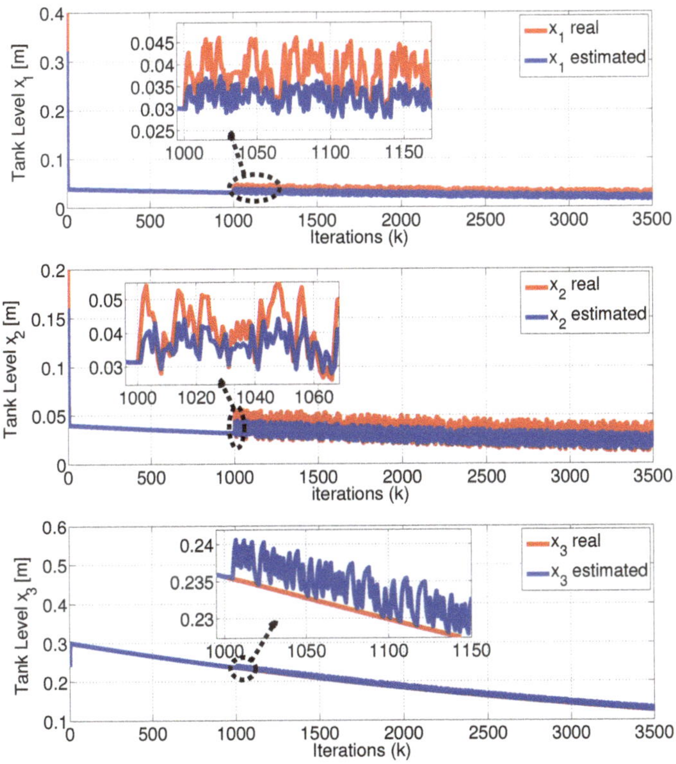

Figure 11. Evolution of x_k under permanent FDI attack.

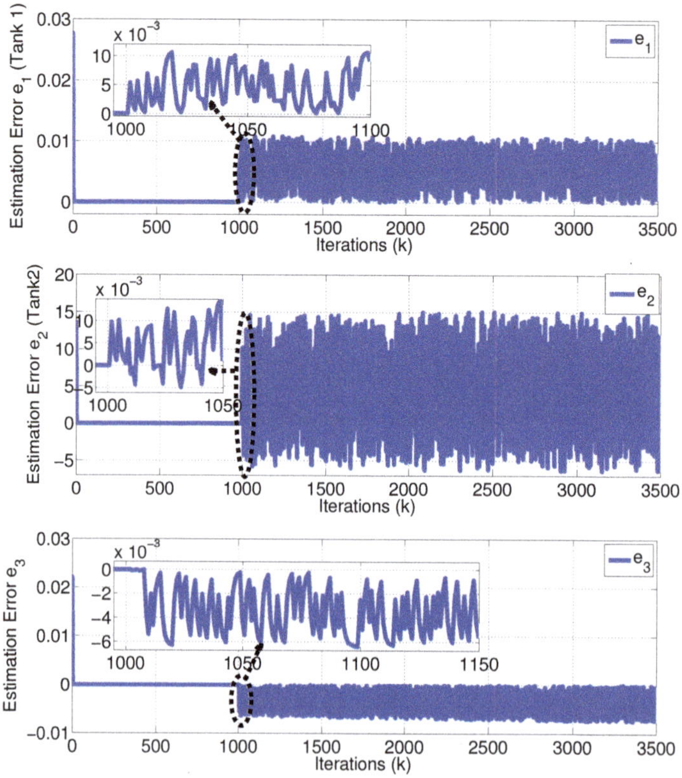

Figure 12. Variation of e_k under permanent FDI attack.

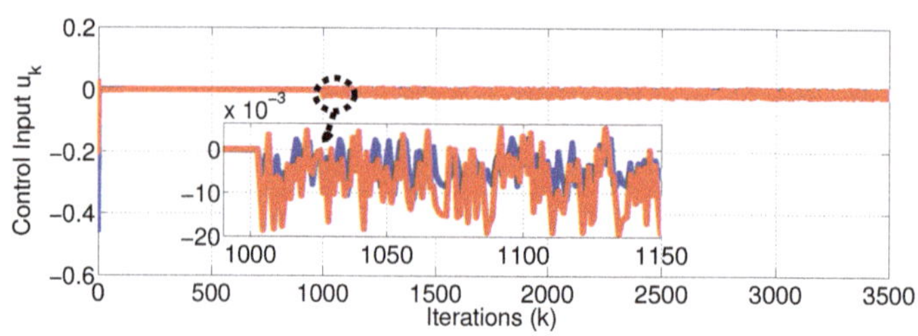

Figure 13. Variation of u_k (u_k^1 in blue and u_k^2 in red) under permanent FDI attack.

Figures 11–13 provide clear evidence that the estimated states remain aligned with the real system states even under permanent FDI attack. This confirms the effectiveness of the proposed approach in mitigating the effect of such attacks, particularly by minimizing the amplitude of state deviations. Furthermore, the estimation error signals generated in such scenarios exhibit a significant degree of sensitivity not only to cyberattacks but also to potential system faults. This demonstrates that the estimation error is sensitive to attacks and that it can serve as a reliable indicator for both detection and diagnosis.

7. Comparative Analysis with Recent Works

This section provides a comparative analysis between the proposed observer-based resilient control scheme and several representative methods addressing security and robustness in CPS under cyberattacks. In order to more clearly position the proposed methodology within the current research landscape, Table 1 has been extended to include recent contributions that explicitly address the resilience of event-triggered CPS under cyberattacks. While most traditional approaches rely on continuous-time control or periodic updates without triggering mechanisms, a few recent studies have proposed resilient control schemes that integrate event-based triggering with detection or robustness objectives. These works include [23,25,26], which introduced various forms of event-triggered controllers in the presence of DoS or sensor attacks. However, many contributions still rely on complex synthesis methods (e.g., BMIs) or do not fully integrate the detection and control mechanisms in a unified and computationally efficient framework.

Compared to the above methods, our proposed control scheme presents several distinct advantages:

1. In terms of attack modeling, earlier works such as [3,7,9] described cyberattacks using static or deterministic assumptions. The current approach instead uses a Markovian modeling framework, which provides a more realistic and flexible description of attack dynamics and transitions. This probabilistic representation aligns with modern CPS vulnerability patterns, as highlighted in [4,12].
2. While most contributions tend to address detection and control separately [11,13,14], our method unifies both within a single framework. By employing a Luenberger observer for attack detection and coupling it with a state-feedback control law, the proposed scheme ensures more rapid and coordinated reaction to anomalies in sensor and actuator channels. This integrated architecture enhances system resilience and minimizes performance degradation during attack periods.
3. The proposed control strategy incorporates an event-triggered mechanism, which has rarely been explored in conjunction with observer-based cyberattack detection. While the existing literature acknowledges the importance of reducing network traffic and computational load [2,15], few designs provide a formal guarantee of stability under reduced communication. The event-triggering rule proposed in this paper ensures asymptotic stability without inducing Zeno behavior, effectively reducing the frequency of control updates while preserving system safety.
4. Unlike earlier contributions, our proposal combines a Luenberger observer for attack detection with a resilient state-feedback controller under a PETC paradigm. Furthermore, it explicitly models FDI and DoS attacks using a Markovian switching process and ensures robust performance through a convex synthesis approach based entirely on relaxed LMIs. Compared to recent event-triggered methods [25,26], the proposed solution offers a better tradeoff between resilience, computational tractability, and implementation efficiency, especially for discrete-time CPS deployments on embedded systems.
5. From a computational perspective, the proposed synthesis procedure relies solely on Linear Matrix Inequalities (LMIs). This contrasts with earlier BMI-based approaches [12,14], which often face difficulties in real-time implementation due to their high complexity. By leveraging diagonal Lyapunov functions, relaxation variables, and congruent transformations [16,27], our controller and observer can be synthesized through convex optimization, enabling tractable deployment on embedded platforms.
6. The effectiveness of the proposed framework is validated through extensive simulations on the standard three-tank benchmark.

Table 1. Comparative summary of resilient control methods for CPS under cyberattacks.

Reference	Attack Model	Detection Method	Control Synthesis	Triggering Strategy	Solvability
[14]	None	Distributed observer	H_∞ control	No	BMI
[3]	DoS (static)	Filtering	Robust control	No	LMI
[7]	Deterministic	Residual threshold	Heuristic	No	Non-systematic
[8]	Arbitrary injection	Observer-based	Tracking control	No	LMI
[4]	Markovian	Stochastic observer	Robust control	No	LMI
[9]	FDI & DoS	Analysis only	Constraint-based	No	BMI
[23]	General	None	H_∞ robust control	PETC	LMI
[25]	DoS	Output estimation	Output feedback control	PETC	BMI
[26]	Sensor attacks (stochastic)	Observer-based	Event-triggered control	CETC	LMI
Proposed Method	Markovian (FDI & DoS)	Luenberger observer	State feedback via relaxed LMIs	PETC	LMI

Note 1: The comparison in Table 1 is based on synthesis characteristics and attack models as reported in the referenced works. No numerical re-implementation was performed unless explicitly stated.

To validate the effectiveness of the proposed resilient control approach, numerical simulations representing cyberattack scenarios were conducted and evaluated, with the results provided in Table 2. The evaluation metrics included settling time, overshoot, tracking error (measured as mean squared error, MSE), and computation time per control update. These metrics comprehensively capture both transient and steady-state performance as well as the computational efficiency of the control strategies.

Note 2: All numerical simulations were conducted using MATLAB R2023a on an Intel Core i7 (2.6 GHz, 16 GB RAM) platform, with controller synthesis relying on YALMIP and the SDPT3 solver. The computation times reported in Table 2 are averaged over ten independent runs, and standard deviations were also calculated. Although the system was evaluated under permanent and intermittent attacks, packet loss or time-varying attack intensity were not explicitly modeled in the current study. Such network-induced uncertainties will be investigated in future work to assess robustness in more realistic settings.

Table 2. Performance comparison of the proposed resilient control approach with existing methods under cyberattack scenarios.

Method	Settling Time (s)	Overshoot (%)	Tracking Error (MSE)	Computation Time (ms)
[14]	2.3	18.1	0.058	4.0
[3]	2.0	15.7	0.049	5.5
[7]	2.5	20.2	0.062	3.8
[8]	1.9	14.3	0.042	6.0
[4]	1.7	12.1	0.037	7.2
[9]	2.2	16.5	0.055	4.5
Proposed Approach	1.2 ms	8.3	0.025	0.874

Table 2 summarizes the quantitative results. As can be observed, the proposed method significantly outperforms the benchmark methods in terms of faster settling time and reduced overshoot, indicating improved transient response and stability. Additionally, the tracking error is lower, demonstrating better accuracy despite cyberattacks.

Compared to results obtained using alternative observer-based methods such as [19], the proposed approach demonstrates superior performance in terms of detection speed, robustness to parametric uncertainty, and resilience to both FDI and DoS attacks. These results suggest that our method is not only theoretically sound but also practically applicable in the context of critical infrastructure where safety and reliability are paramount.

Remark 3. *The current event-triggered mechanism is designed under the PETC paradigm, where triggering conditions are evaluated at regular sampling intervals. While this framework ensures ease of implementation and compatibility with discrete-time platforms, it may not offer optimal resource usage compared to STC, where the control update times are predicted in advance based on system dynamics. Future work will focus on extending the proposed resilient control architecture to STC schemes, which could further reduce communication overhead while maintaining robustness under cyberattacks. Such an extension would also require revisiting the triggering conditions and observer-based synthesis to ensure stability and detection performance under asynchronous control updates.*

8. Conclusions

This paper has presented a unified observer-based resilient control framework for discrete-time Cyber–Physical Systems (CPS) subject to Markovian-modeled cyberattacks. The proposed approach combines a Luenberger-type observer for real-time attack detection with a state-feedback controller driven by an event-triggered mechanism to ensure reduced resource usage while maintaining system stability and performance. By explicitly incorporating the stochastic behavior of attacks through a Markov chain model, the proposed framework significantly improves detection accuracy and enhances resilience against both sensor and actuator compromise. This synthesis is achieved through relaxed Linear Matrix Inequalities (LMIs) formulated using relaxation variables and diagonal Lyapunov functions, making the proposed solution computationally efficient.

Extensive simulations on a benchmark three-tank system confirm the effectiveness of the method in rapidly detecting cyberattacks, maintaining control performance, and preserving closed-loop stability even under parametric uncertainty and measurement noise. This work contributes to bridging the gap between detection and control in CPS security by offering a robust, integrated, and practically viable solution for resilient operation under cyber threats.

Future work will explore extension to nonlinear CPS models, adaptive estimation techniques for unknown attack parameters, and hardware-in-the-loop implementation to validate real-time applicability.

Author Contributions: Conceptualization and methodology, E.H., A.T. and N.G.; software, E.H., A.T. and G.B.H.F.; supervision and validation, A.T. and N.G.; writing—original draft preparation, E.H., A.T. and N.G.; writing—review and editing, G.B.H.F. and E.H. All authors have read and agreed to the published version of the manuscript.

Funding: This research received no external funding.

Data Availability Statement: The datasets presented in this article are not readily available because the data are part of an ongoing study from two different universities. Requests to access the datasets should be directed to authors.

Conflicts of Interest: The authors declare no conflicts of interest.

Abbreviations

The following abbreviations are used in this manuscript:

CPS	Cyber–Physical Systems
MCA	Markovian Cyber-Attacks
BMI	Bilinear Matrix Inequality
LMI	Linear Matrix Inequality
DoS	Denial-of-Service
FDI	False Data Injection
ETM	Event-Triggered Mechanism

References

1. Rajkumar, R.; Lee, I.; Sha, L.; Stankovic, J. Cyberphysical systems: The next computing revolution. In Proceedings of the 47th ACM/IEEE Design Automation Conference, Anaheim, CA, USA, 13–18 June 2010; pp. 731–736.
2. Smith, J.; Brown, K.; Lee, M. Real-time monitoring and resilience in cyber-physical systems: Recent advances. *IEEE Trans. Ind. Inform.* **2023**, *19*, 2500–2512.
3. Yuan, Y.; Zhu, Q.; Sun, F.; Wang, Q.; Baasar, T. Resilient control of cyber-physical systems against denial-of-service attacks. In Proceedings of the 2013 6th International Symposium on Resilient Control Systems (ISRCS), San Francisco, CA, USA, 13–15 August 2013; pp. 54–59.
4. Chen, L.; Wang, H.; Zhang, Y. Stochastic modeling and resilient control for cyber-physical systems under cyber attacks. *Automatica* **2022**, *138*, 110567.
5. Kordestani, M.; Saif, M. Observer-Based Attack Detection and Mitigation for Cyberphysical Systems: A Review. *IEEE Syst. Man. Cybern. Mag.* **2021**, *7*, 35–60. [CrossRef]
6. Hassine, E.; Thabet, A.; Gasmi, N.; Frej, G.B.H.; Thabet, H. Reconfiguration and Cyber-Attack Tolerant Control for Nonlinear Multi-Agent Systems. In Proceedings of the 2023 IEEE International Workshop on Mechatronic Systems Supervision (IW-MSS), Hammamet, Tunisia, 2–5 November 2023; pp. 1–6.
7. Orojloo, H.; Azgomi, M.A. Evaluating the complexity and impacts of attacks on cyber-physical systems. In Proceedings of the 2015 CSI Symposium on Real-Time and Embedded Systems and Technologies (RTEST), Tehran, Iran, 7–8 October 2015; pp. 1–8.
8. Rebai, S.B.; Voos, H.; Darouach, M. Attack-tolerant control and observer-based trajectory tracking for cyber-physical systems. *Eur. J. Control* **2018**, *47*, 90–99.
9. Garcia, R.; Lopez, M.; Martinez, P. Impact and mitigation of false data injection and denial of service attacks in industrial CPS. *Control. Eng. Pract.* **2023**, *130*, 105390.
10. Duo, W.L.; Zhou, M.C.; Abusorrah, A. A survey of cyber attacks on cyber physical systems: Recent advances and challenges. *IEEE/CAA J. Autom. Sin.* **2022**, *9*, 784–800. [CrossRef]
11. Li, X.; Zhao, S.; Liu, Y. Advanced observer-based detection techniques for cyber-physical systems. *IEEE Trans. Cybern.* **2024**, *54*, 123–135.
12. Martinez, P.; Chen, J.; Garcia, R. Adaptive Control Techniques for Resilient Actuation in Cyber-Physical Systems. *Actuators* **2025**, *14*, 34. [CrossRef]
13. Jeffrey, N.; Tan, Q.; Villar, J.R. A hybrid methodology for anomaly detection in Cyber-Physical Systems. *Neurocomputing* **2024**, *568*, 127068. [CrossRef]
14. Shames, I.; Teixeira, A.S.; Sandberg, H.; Johansson, K. Distributed fault detection for interconnected second-order systems with applications to power networks. In Proceedings of the 2010 IEEE International Conference on Control Applications, Yokohama, Japan, 8–10 September 2010; pp. 164–169.
15. Wang, J.; Sun, D.; Wang, J.; Liao, G. Configuration of the Power Reflux Hydro-Mechanical Transmission System. *Actuators* **2024**, *13*, 120. [CrossRef]
16. Kheloufi, H.; Zemouche, A.; Bedouhene, F.; Bennani, C.; Trinh, H. New decentralized control design for interconnected nonlinear discrete-time systems with nonlinear interconnections. In Proceedings of the 2016 IEEE 55th Conference on Decision and Control (CDC), Las Vegas, NV, USA, 12–14 December 2016; pp. 6030–6035.
17. Kheloufi, H.; Zemouche, A.; Bedouhene, F.; Souley-Ali, H. Robust H∞ observer-based controller for Lipschitz nonlinear discrete-time systems with parameter uncertainties. In Proceedings of the 2014 IEEE 53rd Annual Conference on Decision and Control (CDC), Los Angeles, CA, USA, 15–17 December 2014; pp. 4336–4341.
18. Miskowicz, M. *Event-Based Control and Signal Processing*, 1st ed.; CRC Press: Boca Raton, FL, USA, 2015. [CrossRef]
19. Tahir, Z.; Khan, A.Q.; Asad, M. Attack detection and identification in cyber physical systems: An example on three tank system. In Proceedings of the 2019 15th International Conference on Emerging Technologies (ICET), Peshawar, Pakistan, 2–3 December 2019; pp. 1–6.

20. Januario, F.; Cardoso, A.; Gil, P. A distributed multi-agent framework for resilience enhancement in cyber-physical systems. *IEEE Access* **2019**, *7*, 31342–31357. [CrossRef]
21. Guadarrama, A.R.; Osorio-Gordillo, G.L.; Vargas-Méndez, R.A.; Reyes-Reyes, J.; Astorga-Zaragoza, C.M. Cyber-Physical System Attack Detection and Isolation: A Takagi-Sugeno Approach. *Math. Comput. Appl.* **2025**, *30*, 12. [CrossRef]
22. Polik, I.; Terlaky, T. A Survey of the S-Lemma. *SIAM Rev.* **2007**, *49*, 371–418. [CrossRef]
23. Davoodi, M.; Meskin, N.; Khorasani, K. Event-Triggered Robust H_∞ Control for Linear Systems. *IEEE Trans. Ind. Inform.* **2017**, *13*, 298–311. [CrossRef]
24. Boyd, S.; Ghaoui, L.E.; Ferron, E.; Balakrishnan, V. *Linear Matrix Inequalities in Systems and Control Theory*, 15th ed.; Society for Industrial and Applied Mathematics SIAM: Philadelphia, PA, USA, 1994.
25. Zhang, Y.; Liu, S.; Wang, L. Resilient Event-Triggered Output Feedback Control for CPS under DoS Attacks. *IEEE Trans. Cybern.* **2021**, *51*, 3124–3135.
26. Wang, X.; Zhao, Y.; Zhang, H. Event-triggered observer-based control of cyber-physical systems under stochastic sensor attacks. *ISA Trans.* **2023**, *137*, 543–554.
27. Arcak, M.; Kokotovic, P. Observer-based control of systems with slope-restricted nonlinearities. *IEEE Trans. Autom. Control* **2001**, *46*, 1146–1150. [CrossRef]

actuators

MDPI

Article

A Fluid Elastomeric Actuator Design for Soft Robots

Dennis Els [1], Theo van Niekerk [1], Paolo Mercorelli [2,*] and Jacques Welgemoed [1]

[1] Department of Mechatronic Engineering, Nelson Mandela University, Gqeberha 6001, South Africa; dennisels.de@outlook.com (D.E.); theo.vanniekerk@mandela.ac.za (T.v.N.); jacques.welgemoed@mandela.ac.za (J.W.)

[2] Institute for Production Technology and Systems, Leuphana University of Lueneburg, 21335 Lueneburg, Germany

* Correspondence: paolo.mercorelli@leuphana.de

Abstract: The field of robotics faces significant challenges in creating adaptable and flexible end-effectors. Soft robotics, specifically soft robotic end-effectors, offer an innovative solution. This paper focuses on designing fluid elastomeric actuators (FEAs) for soft robotic end-effectors. The study presents key design considerations and evaluates the use of Finite Element Method (FEM) simulations for optimizing FEA performance. The study then concludes by proposing design guidelines for developing application-specific fluid elastomeric actuators.

Keywords: fluid elastomeric actuators; soft robotics; actuator design

1. Introduction

The field of robotics faces a significant challenge in creating adaptable and flexible end-effectors, particularly in the context of modern manufacturing evolution. Industry 4.0, or the fourth industrial revolution, represents the next developmental stage in organizing the entire value chain process in the manufacturing industry [1]. Among the nine pillars of Industry 4.0 identified by Suleiman et al., including autonomous robots, simulation, and additive manufacturing, three are particularly relevant to addressing current manufacturing limitations [2].

Current robotic systems are predominantly implemented at an automated stage where the working environment is structured, object shapes and orientations are predetermined, and system adaptability is minimal [1]. Consequently, existing robot end-effector combinations do not readily facilitate flexible manufacturing processes. This limitation creates a clear need for more adaptive robotic solutions.

Soft robotics, specifically soft robotic end-effectors, offer an innovative solution to these challenges. Fluid elastomeric actuators (FEAs) are commonly employed in soft robotic end-effectors due to their inherent adaptability [3–5]. These actuators consist of two fundamental sections: an active section typically manufactured from more elastic material and a passive section made from less elastic material [6], as illustrated in Figure 1. FEAs contain internal air chambers within their active sections, and when positive air pressure is applied, differential expansion rates between the sections create controlled bending toward the passive section due to the resulting eccentricity.

The design optimization of FEAs involves several critical considerations that directly influence their performance. Previous research has examined how air chamber shape affects performance, comparing triangular, trapezoidal, rectangular, and U-shaped configurations. Additionally, adjusting wall thicknesses alters section stiffness and flexibility, significantly impacting actuator performance. Understanding these design relationships is essential for developing application-specific solutions.

This paper presents a framework that combines systematic design parameter investigation with stress analysis for practical implementation. While individual design parameters have been studied previously, this work integrates stress distribution analysis alongside deformation performance to address actuator durability for industrial applications. The study demonstrates a FEM simulation methodology and concludes by proposing practical design guidelines for developing application-specific fluid elastomeric actuators. The study evaluates the use of Finite Element Method (FEM) simulations for optimizing FEA performance and concludes by proposing practical design guidelines for developing application-specific fluid elastomeric actuators.

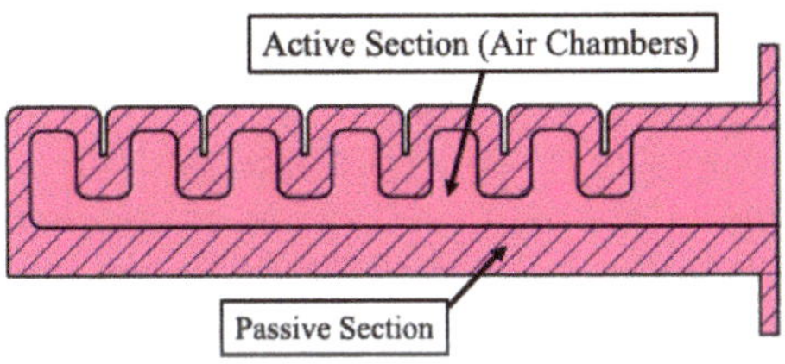

Figure 1. Illustration of the active and passive sections on FEAs, adapted from [6].

2. Literature Review

2.1. Overview of Soft Robotics and FEAs

Soft robotics employs compliant materials to create adaptive robotic structures [3–5,7] that excel in unstructured environments due to their material composition. End-effectors incorporating soft actuators offer improved safety in human–robot interactions and require less precise grasp planning for component handling [8].

While FEAs represent one approach to soft actuation [3–5], alternative technologies offer different advantages and trade-offs. Fiber-reinforced actuators provide enhanced directional control and higher force output through embedded reinforcing fibers, enabling precise trajectory control and improved load-bearing capacity [9–11]. However, these systems require more complex manufacturing processes and may exhibit reduced compliance compared to non-reinforced alternatives.

Origami-based soft actuators leverage geometric folding patterns to achieve complex motions with minimal material usage, offering lightweight solutions with predictable kinematics [12–14]. Nevertheless, they typically provide limited force generation compared to pneumatic systems due to their reliance on geometric constraints rather than material expansion.

Jamming-enabled actuators utilize granular materials that transition between fluid and solid states under vacuum, providing variable stiffness capabilities and universal grasping adaptability [15–17]. While these systems offer tunable compliance, they require additional vacuum infrastructure and may exhibit slower response times.

Each technology presents specific benefits for particular applications. However, FEAs maintain advantages in manufacturing simplicity, predictable pneumatic control, and the ability to achieve both sufficiently high force output and compliant interaction through straightforward material selection and geometric optimization.

2.2. Key Design Considerations

Four primary design considerations govern FEA development: the shape, size, number of air chambers, and the wall thickness of both active and passive sections [3–5]. These parameters directly influence expansion performance, bending capability, and output force.

Previous research has examined how air chamber shape affects performance, comparing triangular, trapezoidal, rectangular, and U-shaped configurations. Rectangular chambers demonstrate the greatest expansion potential, while triangular chambers exhibit the least [18]. Increasing air chamber size expands the inner surface area, potentially increasing output force. Additionally, adjusting wall thicknesses alters section stiffness and flexibility, significantly impacting actuator performance.

2.3. *Challenges in FEA Design*

Developing FEAs involves an iterative process where even minor design changes require mold modifications, making physical testing resource-intensive and time-consuming [6]. Simulation software can accelerate this process, though modeling soft actuators presents challenges due to material non-linearity and complex geometries [6,19–21].

Elastomeric materials used in FEAs are hyperelastic, allowing for large deformations with minimal plastic deformation. This property results in complex mechanical behavior that linear elastic theory cannot adequately describe. Accurate modeling requires determining appropriate strain energy density function coefficients that match experimental data, necessitating the evaluation of multiple strain energy functions to identify the simplest model with sufficient accuracy.

FEM software enables researchers to develop hyperelastic material models that accelerate soft actuator design while reducing resource requirements for iterative testing and development.

Beyond material modeling challenges, FEA development faces additional complexities in control, sensing, and long-term performance. The non-linearity, complex geometries, and continuum nature of soft robots increase the difficulty and complexity of measuring and controlling these robotic structures [22,23]. Control strategies applicable to conventional robots are not necessarily relevant to soft robots, as traditional robots consist of rigid links and discrete joints [22]. Hysteresis effects and sensor drift present significant challenges in dynamic applications, where sensor readings may not remain accurate over extended operation periods [24]. Additionally, open-loop control strategies commonly used for FEAs are insufficient for accurately controlling the desired bending angles [25]. Data-driven control techniques offer promising approaches that generate control strategies from physical motion data rather than predetermined models [22], potentially addressing the inherent non-linearities in soft robotic systems.

3. Finite Element Method Approach

3.1. *Purpose of FEM Simulations*

The use of FEM simulations enables researchers to predict actuator performance without relying solely on physical prototypes. Thereby reducing resource consumption such as material and time. The purpose of this section is to investigate how selected FEM software can simulate the design considerations identified in Section 2.2 and whether these simulations can optimize the design process of FEAs.

3.2. *FEM Model Setup*

The FEM model was developed using ANSYS 2024 R1 software, with geometry based on the symmetric FEA Computer-Aided Design (CAD) models. Symmetric FEAs were developed to isolate the impact of each of the identified design considerations on the expansion performance of FEAs. To ensure controlled comparisons, all actuators were manufactured with identical baseline parameters except for the specific variable under investigation. All geometric studies utilized symmetric fluid elastomeric actuators with consistent wall thickness (3 mm), identical material properties (Mold Max 20), and the same overall actuator length. The primary performance metric evaluated was percentage expansion relative to the actuator's unstrained position, measured as the vertical elongation when pressurized. Figure 2 illustrates a sectional view of the FEAs developed to test the influence of air chamber shape.

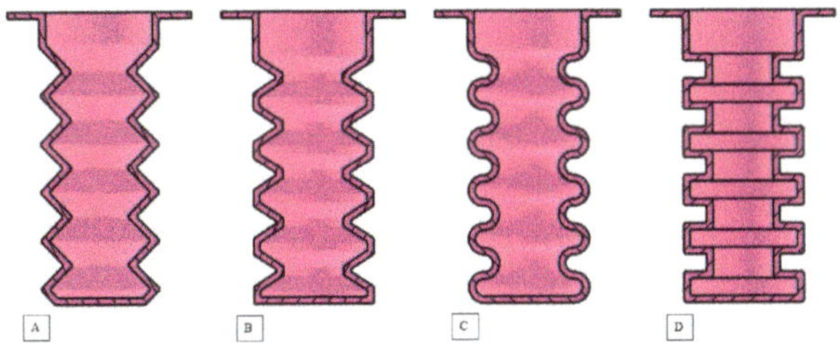

Figure 2. The four respective air chamber shapes to test the influence of changing air chamber shapes on the performance of FEAs: (**A**) triangular shape. (**B**) trapezoidal shape. (**C**) U-shape. (**D**) rectangular shape.

A hyperelastic material model was developed based on uni-axial tension test data. The uni-axial tension tests were performed according to the ASTM D412 test standard, which covers the procedures used to determine the tensile properties of vulcanized rubber and thermoplastic elastomers [26]. The ASTM D412 test standard is the same testing standard used by [27] to develop a database of hyperelastic materials commonly used in soft robotics.

The elastomer used to manufacture the developed FEAs is known as Mold Max 20. This elastomer has a Shore A hardness of 20, a mixed viscosity of 25,000 centipoise, an elongation-at-break of approximately 512%, a tensile strength of 3.82 MPa, and a cure time of 20 h [28]. Five ASTM D412 Type C tensile samples were cut from a $\pm$2.3 mm thick sheet of Mold Max 20 elastomer.

The uni-axial tension tests were performed on a Zwick/Roell Z010 power-driven machine (ZwickRoell GmbH & Co. KG, Ulm, Germany) at an extension rate of 500 ± 50 mm/min. A 10 kN load cell measured the load with a sensitivity of 2 mV/V while a contact extensometer measured the elongation. Figure 3 illustrates the tensile stress vs. strain data obtained from the uni-axial tension tests.

The results presented in Figure 3 are then entered into ANSYS to determine the coefficients of the simplest hyperelastic material model providing adequate accuracy. Testing determined that the most accurate and stable hyperelastic models were the Yeoh Third Order and Yeoh Second Order models. These models could predict the stress–strain behavior of Mold Max 20 in different deformation modes from only uni-axial tension data, as suggested by [29]. It has also been reported that these models are the most accurate for large-strain problems [30,31].

The Third Order Yeoh model achieved a normalized error value of 12,356, slightly lower than the Second Order model's error value of 12,465. The Second Order Yeoh model was selected for simulating the design considerations identified in the literature because it has fewer parameters than the Third Order model while maintaining similar accuracy. Figure 4 illustrates the experimental data for Specimen $CS_{1.5}$ alongside the Second Order Yeoh model and its coefficients.

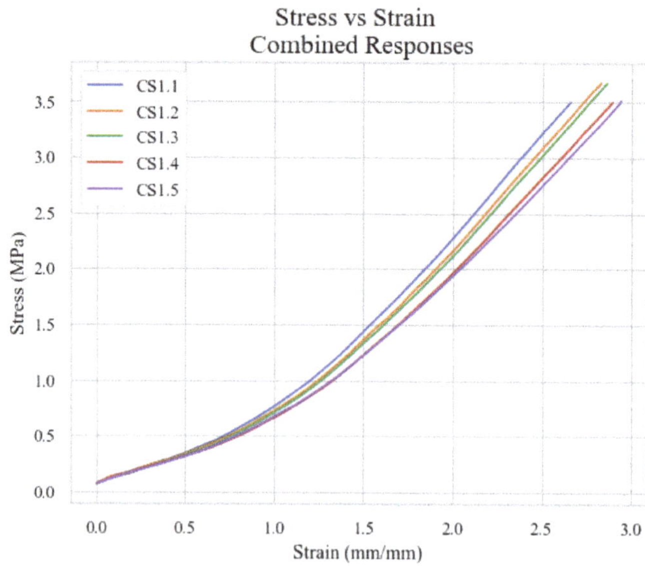

Figure 3. Tensile stress vs. strain responses of five ASTM Type C Mold Max 20 samples.

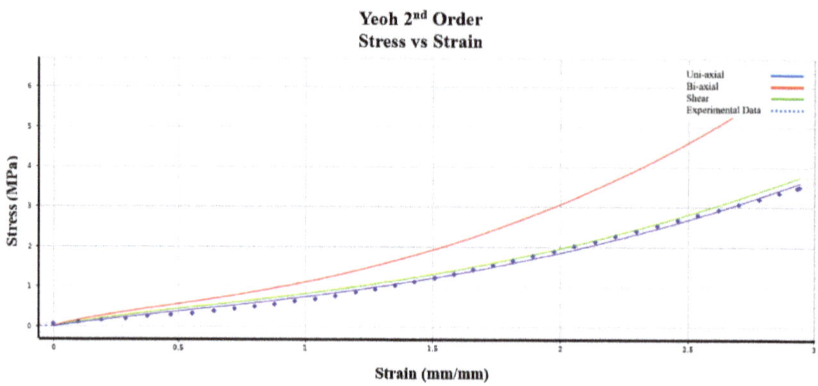

C_{10}: 0.16762 MPa C_{20}: 0.011432 MPa D_1: 0 MPa^{-1} D_2: 0 MPa^{-1}

Figure 4. Yeoh 2nd Order hyperelastic model and its coefficients.

3.3. Simulation Process

The FEM simulation followed a structured approach aligned with the methodology outlined by [31]. Initially, CAD models were developed to address the previously identified key design considerations. A Static Structural study was established in ANSYS, implementing a hyperelastic material defined using the Second Order Yeoh model. After importing the CAD models into Design Modeler, the defined material properties were assigned and named selections were created for the inner and outer walls of the actuator. The mesh configuration utilized Non-linear Mechanical Physics Preference with Quadratic Element Order and a coarse mesh size of 0.3 m. Analysis settings incorporated auto time stepping (initial: 0.1 s, minimum: 0.0001 s, maximum: 1 s) with large deflections enabled. For boundary conditions, the top end of each actuator was fixed while gravity and a ramped internal pressure were applied to the inner walls. Additionally, a frictionless self-contact set was implemented on the outer walls. The 12 kPa pressure level was selected based on preliminary testing to ensure safe operation well below the material's stress limits while providing sufficient deformation for comparative analysis. For the final application design, 40 kPa was chosen to meet the force requirements for automotive assembly applications while maintaining stress levels below 25% of the material's tensile strength (3.82 MPa), ensuring durability for continuous industrial operation. The simulations evaluated total deformation, equivalent elastic strain, and equivalent stress. Upon solution completion in ANSYS, the results were prepared for further analysis.

4. Simulation Results

This section evaluates how FEM simulations can effectively model and optimize the key design parameters identified in Section 2.2: air chamber shape, size, number, and wall thickness variations. Figure 5 provides an overview of the key findings, with detailed numerical results presented in Tables A1–A5 in the appendix. For shape comparison studies, all actuators maintained identical wall thickness, material properties, and number of air chambers (5), with only the air chamber geometry varying between triangular, trapezoidal, U-shaped, and rectangular configurations. For size variation studies, actuator shape, wall thickness, and material remained constant while only air chamber width was systematically altered. While overall actuator length and wall thickness remained consistent across comparisons, internal air chamber volume varied as intended, since volume changes are fundamental to the expansion performance being investigated. Note that this investigation focuses on reviewing the accuracy of the models created in this work, not the simulation software itself.

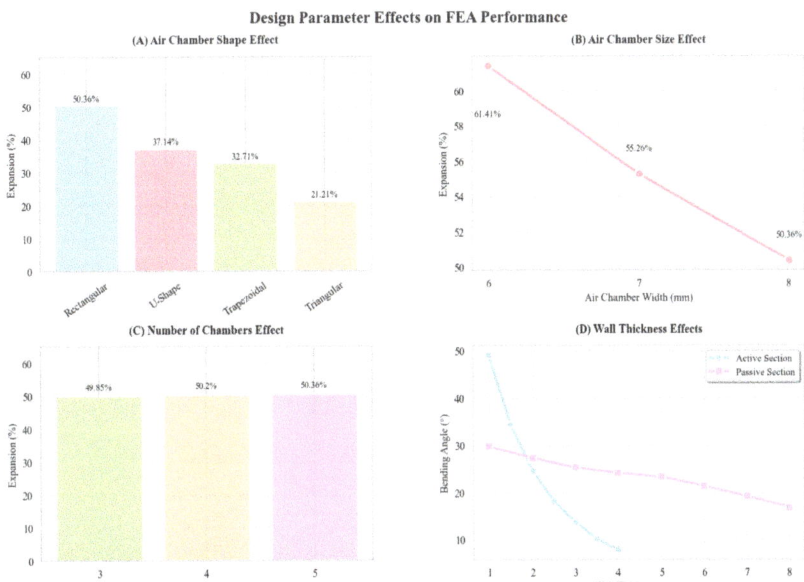

Figure 5. Design parameter effects on FEA performance: (**A**) air chamber shape influence on expansion percentage, (**B**) air chamber size effects showing inverse relationship, (**C**) number of chambers impact on expansion, and (**D**) wall thickness effects on bending angle for both active and passive sections.

4.1. Geometric Variation

Four air chamber shapes were examined: triangular, trapezoidal, U-shaped, and rectangular. The results (Table A1) show rectangular chambers achieve the greatest expansion (50.36%), while triangular chambers exhibit the lowest (21.21%). All configurations maintained stress levels well below the 3.82 MPa tensile strength of Mold Max 20, indicating safe operation under the applied pressure. Rectangular chambers are optimal for maximum deformation applications, while other shapes offer different balances of expansion and stress distribution.

4.2. Size Variation

The impact of air chamber width on performance was analyzed by decreasing chamber width from 8 mm to 6 mm in 1 mm increments (Figure 6). As shown in Table A2, smaller chambers achieved higher expansion percentages (61.41% at 6 mm vs. 50.36% at 8 mm), establishing an inverse relationship between chamber width and expansion capability. While stress levels increased slightly with smaller chambers, all designs remained within safe operational limits, highlighting a clear trade-off between maximizing expansion and managing stress.

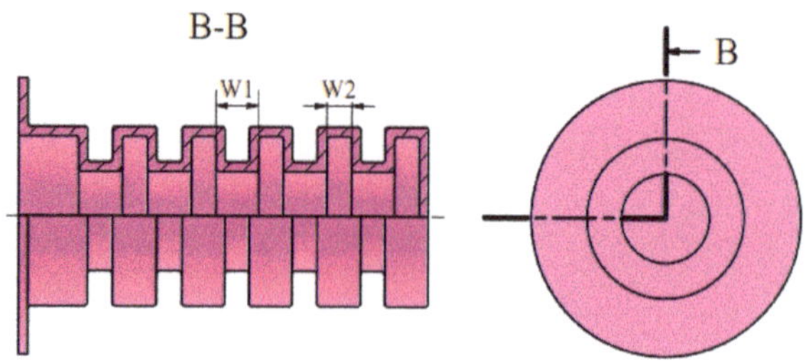

Figure 6. Illustration of how the size of the actuator chambers is to be changed.

4.3. Varying Number of Chambers

To examine the effect of the number of air chambers in an actuator, two additional symmetric FEAs were designed with varying numbers of air chambers, as shown in Figure 7.

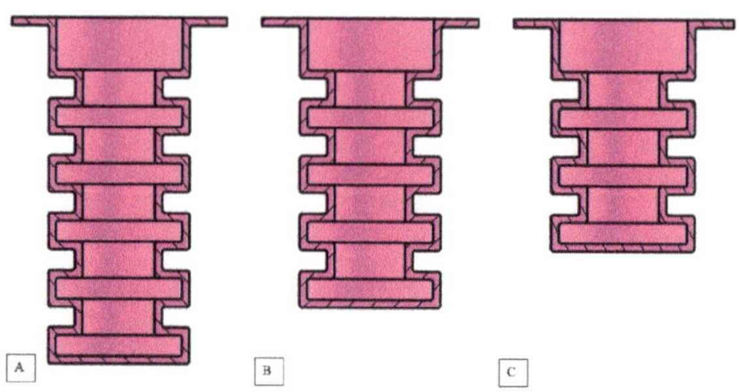

Figure 7. The change in number of air chambers in a fluid elastomeric actuator of rectangular shape: (**A**) 5 air chambers. (**B**) 4 air chambers. (**C**) 3 air chambers.

Table A3 illustrates the results obtained from the FEM simulations developed using the CAD models depicted in Figure 7.

The results show a slight increase in the expansion percentage as the number of air chambers increases, indicating a directly proportional relationship between the number of air chambers and the achievable expansion. On average, the expansion percentage increases by 0.26% for each additional air chamber.

Actuators with three air chambers experience the highest stress, approximately 0.202 MPa, while those with four air chambers experience the lowest stress, around 0.201 MPa. Interestingly, even though the expansion percentage slightly increases with more air chambers, the variation in stress among the different configurations is minimal. This suggests that changing the number of air chambers in the actuator does not significantly affect the structural integrity of a symmetric fluid elastomeric actuator.

4.4. Varying Wall Thickness

The final key design consideration identified in Section 2.2 is the impact of the wall thickness of the active and passive sections of FEAs. Following a methodology similar to that of [18], the thickness of each section is varied independently. One section's thickness is adjusted while keeping the other constant, and the process is then repeated with the roles reversed. This approach allows for a systematic evaluation of how the wall thickness of an actuator's active and passive sections influences actuator performance.

The findings from investigating the impact of altering an actuator's shape, size, and number of air chambers served as a guideline for designing a basic FEA, which could be used to investigate the fourth key design consideration. Figure 8 illustrates the designed FEA, with the active and passive section wall thickness denoted by A and B.

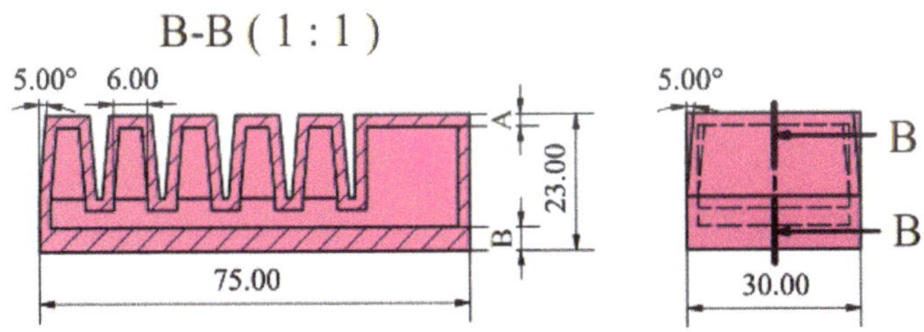

Figure 8. A sectional view of the actuator developed to test the influence of varying wall thickness.

The designed actuator has a total length of 75 mm, a height of 23 mm, and a thickness of 30 mm. It includes five air chambers, each with a width of 6 mm. The rectangular-shaped air chambers have also been incorporated with a draft angle of 5 degrees to facilitate the manufacturing process and reduce the likelihood of errors during demolding.

4.4.1. Varying Active Section Wall Thickness

The predicted bending angles for different active section wall thicknesses, as determined through FEM simulation, are presented in Table A4.

The data presented in Table A4 indicate that increasing the wall thickness of an actuator's active section reduces its bending angle. As a result, higher input pressure is required to achieve the same level of deformation. This trend is consistent with findings from the existing literature [31,32].

Moreover, as wall thickness increases, the equivalent stress decreases significantly, dropping from 0.600 MPa at a thickness of 1 mm to 0.092 MPa at 4 mm. This reduction can be attributed to the increased material resistance provided by thicker walls, which helps to distribute pressure more evenly and minimize stress concentrations.

These results support the notion that greater wall thickness enhances actuators' resistance to bending. Therefore, designers must find a balance between performance—specifically, achieving a greater bending angle—and durability when developing FEAs for various applications.

4.4.2. Varying Passive Section Wall Thickness

Table A5 shows the results obtained from varying the wall thickness of the actuator's passive section while maintaining a constant active section wall thickness of 2 mm.

The results presented in Table A5 show a consistent decrease in the bending angle as the wall thickness of the passive section increases. Similar findings have been reported in the literature [31–33]. Researchers attribute this trend to the increased stiffness of the actuator that occurs with thicker passive sections, which restricts its bending ability.

The stress values in Table A5 illustrate slight fluctuations as the passive section thickness increases. However, there is a general trend of decreasing stress with greater thickness. This stress reduction suggests that thicker passive sections help distribute the applied pressure more evenly, thereby minimizing localized stress concentrations that could lead to material fatigue or failure. These findings imply that thinner passive sections may be more suitable for applications requiring higher bending angles and flexibility. In comparison, thicker passive sections may be more appropriate for applications prioritizing durability.

Table 1 summarizes the effects of key design parameters on actuator performance based on the FEM analysis results.

Table 1. Effect of design parameter variations on actuator performance.

Parameter Varied (Increasing)	Expansion	Bending Angle	Stress
Air Chamber Size	Decrease	† Decrease	Decrease
Number of Air Chambers	Increase	† Increase	Minimal Effect
Active Section Wall Thickness	—	Decrease	Decrease
Passive Section Wall Thickness	—	Decrease	Decrease

† Inferred from expansion measurement.

5. Application of Design Guidelines

Using the design insights from Section 4, a flexible and adaptable soft robotic end-effector was developed specifically for automotive assembly applications where components with irregular shapes and burrs are common. The design prioritized both durability and sufficient bending capability for industrial environments.

For the actuator design, rectangular air chambers were selected based on their superior expansion performance (Table A1). Five chambers were incorporated to enhance bending capability while maintaining structural integrity. The fifth chamber features a nearly triangular shape to improve handling of varied component geometries. A 1 mm spacing between chambers was implemented to minimize stress and prevent self-contact during operation.

To ensure robustness for components with burrs, a passive section thickness of 6 mm was selected, accepting the associated reduction in bending angle. This was counterbalanced by using a thinner 3 mm active section wall to enhance bending performance while still providing adequate durability for continuous industrial operation. The air chambers were designed with a 6 mm width and a 5-degree draft angle to facilitate manufacturing and demolding.

The resulting actuator (Figure 9) measures 100 mm in length, 39.70 mm in width, and 26 mm in height, with a rectangular cross-section at the base to facilitate pneumatic system connection. FEM simulations (Figure 10) confirmed that the design could achieve the required bending performance under operating pressures while maintaining stress levels well below material limits.

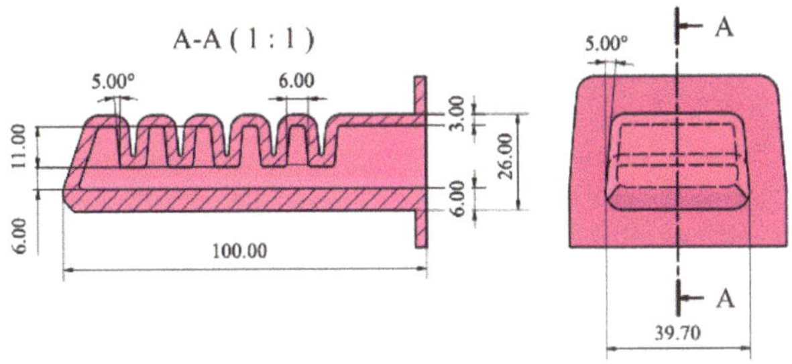

Figure 9. Sectional view of the designed fluid elastomeric actuator.

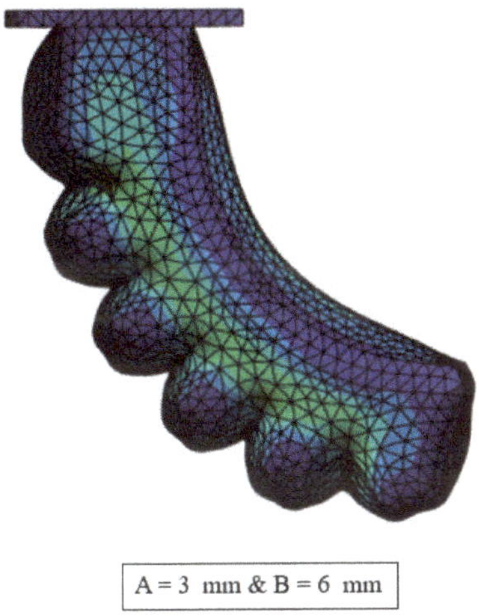

Figure 10. Contour map of the predicted equivalent stress experienced by the designed actuator at 40 kPa.

The final soft robotic end-effector assembly incorporating these optimized FEAs is shown in Figure 11, demonstrating the practical application of our design principles in a functional gripper configuration for industrial use.

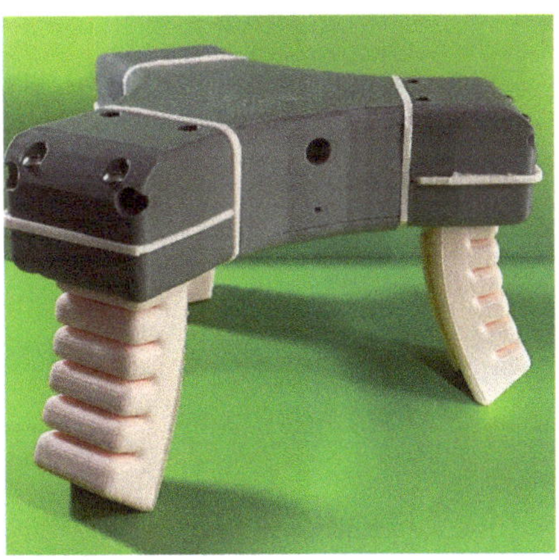

Figure 11. Soft robotic end-effector using FEAs.

6. Conclusions

This study investigated fluid elastomeric actuators (FEAs) for soft robotic applications through a systematic analysis of key design parameters. FEM simulations revealed critical design trade-offs that inform actuator development: rectangular air chambers maximize expansion (50.36%) compared to triangular chambers (21.21%), while smaller chamber widths significantly increase expansion performance (61.41% at 6 mm vs. 50.36% at 8 mm). Increasing the number of air chambers offers marginal improvements in expansion capability (approximately 0.26% per additional chamber) without substantially affecting structural integrity.

Wall thickness analysis demonstrated a consistent inverse relationship between thickness and bending angle in both active and passive sections, with active section thickness having a more pronounced effect. The proposed actuator with 3 mm active and 6 mm passive section thicknesses balances flexibility and durability requirements for industrial applications, achieving sufficient bending performance while withstanding operational stresses.

While FEM simulations provide valuable design insights, real-world performance may diverge from simulations due to material variations and environmental factors.

This work contributes the following: (1) a combined deformation- and stress-based design approach for improved durability assessment, (2) systematic FEM methodology that can reduce the need for extensive physical prototyping, and (3) practical design guidelines derived from both performance and stress considerations for industrial applications such as automotive assembly.

Future work should focus on experimental validation of simulated designs and development of standard testing protocols for soft actuator performance. Additionally, integrating machine learning techniques with FEM simulations could enable automated optimization of actuator geometry for specific application requirements.

Author Contributions: Conceptualization, D.E. and T.v.N.; methodology, D.E.; software, D.E.; validation, D.E.; formal analysis, D.E. and T.v.N.; investigation, D.E.; writing—original draft preparation, D.E. and J.W.; writing—review and editing, J.W. and P.M.; visualization, D.E. and J.W.; supervision, T.v.N.; project administration, T.v.N.; funding acquisition, T.v.N. and P.M. All authors have read and agreed to the published version of the manuscript.

Funding: This research was funded by the Automotive Industry Development Centre Eastern Cape, (South Africa), and the Nelson Mandela University Postgraduate Research Scholarship.

Data Availability Statement: The original contributions presented in this study are included in the article. Further inquiries can be directed to the corresponding author.

Acknowledgments: During the preparation of this manuscript/study, the authors used Claude, Sonnet 4, for the purposes of language editing. The authors have reviewed and edited the output and take full responsibility for the content of this publication.

Conflicts of Interest: The authors declare no conflicts of interest.

Abbreviations

The following abbreviations are used in this manuscript:

FEA	Fluid Elastomeric Actuators
FEM	Finite Element Method
CAD	Computer-Aided Design

Appendix A. Detailed FEM Simulation Results

This appendix provides the detailed numerical results from the FEM simulations discussed in Section 4.

Table A1. Influence of changing the shape of an actuator's air chambers FEM results (12 kPa).

Air Chamber Shape	Expansion (%)	Simulation Results Strain (mm/mm)	Stress (Pa)
Rectangular	50.36	0.17632	2.043×10^5
U-Shape	37.14	0.20426	2.379×10^5
Trapezoidal	32.71	0.21309	2.482×10^5
Triangular	21.21	0.17946	2.037×10^5

Table A2. Influence of changing the size of an actuator's air chambers FEM results (12 kPa).

Air Chamber Size	Expansion (%)	Simulation Results Strain (mm/mm)	Stress (Pa)
$Size_1$ (8 mm)	50.36	0.17632	2.043×10^5
$Size_2$ (7 mm)	55.26	0.1805	2.093×10^5
$Size_3$ (6 mm)	61.41	0.19127	2.238×10^5

Table A3. Influence of changing the number of air chambers in an actuator FEM results (12 kPa).

Number of Air Chambers	Expansion (%)	Simulation Results Strain (mm/mm)	Stress (Pa)
5 Air Chambers	50.36	0.17632	2.016×10^5
4 Air Chambers	50.20	0.17664	2.007×10^5
3 Air Chambers	49.85	0.17638	2.021×10^5

Table A4. Predicted bending angle achieved for varying active section wall thicknesses while keeping passive section constant at 3 mm (12 kPa).

Actuator Models	Simulation Results		
	Active Wall Thickness (mm)	Bending Angle	Stress (Pa)
Model 1.1	1	49.06°	5.997×10^5
Model 1.2	1.5	34.46°	4.872×10^5
Model 1.3	2	24.64°	2.964×10^5
Model 1.4	2.5	18.21°	1.974×10^5
Model 1.5	3	13.81°	1.403×10^5
Model 1.6	3.5	10.15°	0.994×10^5
Model 1.7	4	7.87°	0.926×10^5

Table A5. Predicted bending angle achieved for varying passive section wall thicknesses while keeping active section constant at 2 mm (12 kPa).

Actuator Models	Simulation Results		
	Passive Wall Thickness (mm)	Bending Angle	Stress (Pa)
Model 1.1	1	29.68°	2.925×10^5
Model 1.2	2	27.34°	3.404×10^5
Model 1.3	3	25.32°	2.953×10^5
Model 1.4	4	24.16°	2.964×10^5
Model 1.5	5	23.30°	2.270×10^5
Model 1.6	6	21.42°	2.480×10^5
Model 1.7	7	19.16°	2.031×10^5
Model 1.8	8	16.81°	1.812×10^5

References

1. Pillay, K. *Industry 4.0: Is Africa Ready for Digital Transformation?* Tech. Rep.; Deloitte: Johannesburg, South Africa, 2016.
2. Suleiman, Z.; Shaiholla, S.; Dikhanbayeva, D.; Shehab, E.; Turkyilmaz, A. Industry 4.0: Clustering of concepts and characteristics. *Cogent Eng.* **2022**, *9*, 2034264. [CrossRef]
3. Zhang, H.; Liu, W.; Yu, M.-F.; Hou, Y. Design, fabrication, and performance test of a new type of soft-robotic gripper for grasping. *Sensors* **2022**, *22*, 5221. [CrossRef]
4. Dang, H.M.; Vo, C.T.; Trong, N.T.; Nguyen, V.D.; Phung, V.B. Design and development of the soft robotic gripper used for the food packaging system. *J. Mech. Eng. Res. Dev.* **2021**, *44*, 334–345.
5. Yap, H.K.; Ng, H.; Yeow, R.C.-H. High-force soft printable pneumatics for soft robotic applications. *Soft Robot.* **2016**, *3*, 144–158. [CrossRef]
6. Els, D.; McLaren, J.; van Niekerk, T.; Phillips, R. Design, fabrication and analysis of a soft robotic gripper using fluid elastomeric actuators. In Proceedings of the 2023 14th International Conference on Mechanical and Intelligent Manufacturing Technologies (ICMIMT), Cape Town, South Africa, 26–28 May 2023; pp. 1–7.
7. Walker, J.; Zidek, T.; Harbel, C.; Yoon, S.; Strickland, F.S.; Kumar, S.; Shin, M. Soft Robotics: A Review of Recent Developments of Pneumatic Soft Actuators. *Actuators* **2020**, *9*, 3. [CrossRef]
8. Crowley, G.B.; Zeng, X.; Su, H. A 3D printed soft robotic gripper with a variable stiffness enabled by a novel positive pressure layer jamming technology. *IEEE Robot. Autom. Lett.* **2022**, *7*, 5477–5482. [CrossRef]
9. Connolly, F.; Walsh, C.J.; Bertoldi, K. Automatic design of fiber-reinforced soft actuators for trajectory matching. *Proc. Natl. Acad. Sci. USA* **2017**, *114*, 51–56. [CrossRef]
10. Polygerinos, P.; Correll, N.; Morin, S.A.; Mosadegh, B.; Onal, C.D.; Petersen, K.; Cianchetti, M.; Tolley, M.T.; Shepherd, R.F. Soft robotics: Review of fluid-driven intrinsically soft devices; manufacturing, sensing, control, and applications in human-robot interaction. *Adv. Eng. Mater.* **2017**, *19*, 1700016. [CrossRef]
11. Hoque, M.A.; Petersen, E.; Fang, X. Effect of Material Properties on Fiber-Shaped Pneumatic Actuators Performance. *Actuators* **2023**, *12*, 129. [CrossRef]
12. Li, S.; Vogt, D.M.; Rus, D.; Wood, R.J. Fluid-driven origami-inspired artificial muscles. *Proc. Natl. Acad. Sci. USA* **2017**, *114*, 13132–13137. [CrossRef] [PubMed]
13. Paez, L.; Agarwal, G.; Paik, J. Design and analysis of a soft pneumatic actuator with origami shell reinforcement. *Soft Robot.* **2016**, *3*, 109–119. [CrossRef]
14. Zaghloul, A.; Bone, G.M. Origami-Inspired Soft Pneumatic Actuators: Generalization and Design Optimization. *Actuators* **2023**, *12*, 72. [CrossRef]
15. Brown, E.; Rodenberg, N.; Amend, J.; Mozeika, A.; Steltz, E.; Zakin, M.R.; Lipson, H.; Jaeger, H.M. Universal robotic gripper based on the jamming of granular material. *Proc. Natl. Acad. Sci. USA* **2010**, *107*, 18809–18814. [CrossRef]
16. Amend, J.R.; Brown, E.; Rodenberg, N.; Jaeger, H.M.; Lipson, H. A positive pressure universal gripper based on the jamming of granular material. *IEEE Trans. Robot.* **2012**, *28*, 341–350. [CrossRef]
17. Caro, F.; Carmichael, M.G. A Review of Mechanisms to Vary the Stiffness of Laminar Jamming Structures and Their Applications in Robotics. *Actuators* **2024**, *13*, 64. [CrossRef]
18. Udupa, G.; Sreedharan, P.; Parvathaneni, P.S.D.; Kim, D. Asymmetric bellow flexible pneumatic actuator for miniature robotic soft gripper. *J. Robot.* **2014**, *2014*, 902625. [CrossRef]
19. Pozzi, M.; Miguel, E.; Deimel, R.; Malvezzi, M.; Bickel, B.; Brock, O.; Prattichizzo, D. Efficient FEM-based simulation of soft robots modeled as kinematic chains. In Proceedings of the 2018 IEEE International Conference on Robotics and Automation (ICRA), Brisbane, Australia, 21–25 May 2018; pp. 4206–4213.
20. Ding, L.; Niu, L.; Su, Y.; Yang, H.; Liu, G.; Gao, H.; Deng, Z. Dynamic finite element modeling and simulation of soft robots. *Chin. J. Mech. Eng.* **2022**, *35*, 24. [CrossRef]
21. Suzumori, K.; Endo, S.; Kanda, T.; Kato, N.; Suzuki, H. A bending pneumatic rubber actuator realizing soft-bodied manta swimming robot. In Proceedings of the 2007 IEEE International Conference on Robotics and Automation (ICRA), Roma, Italy, 10–14 April 2007; pp. 4975–4980.
22. Yasa, O.; Toshimitsu, Y.; Michelis, M.Y.; Jones, L.S.; Filippi, M.; Buchner, T.; Katzschmann, R.K. An overview of soft robotics. *Annu. Rev. Control Robot. Auton. Syst.* **2023**, *6*, 1–29. [CrossRef]
23. García-Samartín, J.F.; Rieker, A.; Barrientos, A. Design, Manufacturing, and Open-Loop Control of a Soft Pneumatic Arm. *Actuators* **2024**, 13, 36. [CrossRef]
24. Elgeneidy, K.; Neumann, G.; Jackson, M.; Lohse, N. Directly printable flexible strain sensors for bending and contact feedback of soft actuators. *Front. Robot. AI* **2018**, *5*, 2. [CrossRef]
25. Marchese, A.D.; Katzschmann, R.K.; Rus, D. A recipe for soft fluidic elastomer robots. *Soft Robot.* **2015**, *2*, 7–25. [CrossRef] [PubMed]

26. *ASTM D412-98a*; Standard Test Methods for Vulcanised Rubber and Thermoplastic Elastomers: Tension. ASTM: West Conshohocken, PA, USA, 2002.
27. Marechal, L.; Balland, P.; Lindenroth, L.; Petrou, F.; Kontovounisios, C.; Bello, F. Toward a common framework and database of materials for soft robotics. *Soft Robot.* **2021**, *8*, 284–297. [CrossRef]
28. Smooth-On. *Mold Max Series Condensation Cure Silicone Rubber Compounds*; Smooth-On, Inc.: Macungie, PA, USA, 2024.
29. Ali, A.; Hosseini, M.; Sahari, B. A review of constitutive models for rubber-like materials. *Am. J. Eng. Appl. Sci.* **2010**, *3*, 232–239. [CrossRef]
30. Moseley, P.; Florez, J.M.; Sonar, H.A.; Agarwal, G.; Curtin, W.; Paik, J. Modeling, design, and development of soft pneumatic actuators with finite element method. *Adv. Eng. Mater.* **2015**, *18*, 978–988. [CrossRef]
31. Xavier, M.S.; Fleming, A.J.; Yong, Y.K. Finite element modeling of soft fluidic actuators: Overview and recent developments. *Adv. Intell. Syst.* **2021**, *3*, 2000187. [CrossRef]
32. Hu, W.; Mutlu, R.; Li, W.; Alici, G. A structural optimisation method for a soft pneumatic actuator. *Robotics* **2018**, *7*, 24. [CrossRef]
33. Zhu, Y.; Feng, K.; Hua, C.; Wang, X.; Hu, Z.; Wang, H.; Su, H. Model analysis and experimental investigation of soft pneumatic manipulator for fruit grasping. *Sensors* **2022**, *22*, 4532. [CrossRef]

actuators

MDPI

Article

Torque Oscillation Attenuation in PMSM Using Equivalent-Input-Disturbance-Based Sliding-Mode Control

Ruoyu Jiang [1,2,3], Xiang Yin [4], Jinhua She [1,2,3,5,*], Feng Wang [1,2,3] and Seiichi Kawata [1,2,3]

1 School of Automation, China University of Geosciences, No. 388 Lumo Road, Hongshan District, Wuhan 430074, China; ruoyujiang@cug.edu.cn (R.J.); wangfeng@cug.edu.cn (F.W.); skawata@cug.edu.cn (S.K.)
2 Hubei Key Laboratory of Advanced Control and Intelligent Automation for Complex Systems, No. 388 Lumo Road, Hongshan District, Wuhan 430074, China
3 Engineering Research Center of Intelligent Technology for Geo-Exploration, Ministry of Education, No. 388 Lumo Road, Hongshan District, Wuhan 430074, China
4 School of Electrical and Control Engineering, North China University of Technology, No. 5 Jinyuanzhuang Road, Shijingshan District, Beijing 100144, China; yinxiang@ncut.edu.cn
5 School of Engineering, Tokyo University of Technology, 1404-1 Katakuramachi, Hachioji City 192-0982, Tokyo, Japan
* Correspondence: she@stf.teu.ac.jp

Abstract

This paper presents a torque oscillation attenuation method for permanent magnet synchronous motors (PMSMs) based on the combination of sliding-mode control (SMC) and the equivalent input disturbance (EID) approach. To deal with the changes in PMSM parameters, we explored a continuous-domain ant colony optimization (CDACO) method to design a control system for such a plant. This is the first application of SMC-EID to uncertain PMSM plants, with CDACO enabling robust parameter tuning in continuous spaces. First, we designed an EID estimator to estimate the disturbance caused by torque oscillation. Next, we added the estimated disturbance to the sliding-mode controller to improve disturbance attenuation performance. Then, we extended an ant colony optimization (ACO) algorithm to the continuous domain to optimize controller parameters for an uncertain plant. Finally, a speed control experiment was carried out on a two-mass experimental system for PMSMs to demonstrate the validity of the method. The experimental results show that our method yields better control performance than the SMC.

Keywords: permanent magnet synchronous motor (PMSM); ant colony optimization (ACO); equivalent input disturbance (EID); sliding-mode control (SMC); torque oscillation attenuation

1. Introduction

A permanent magnet synchronous motor (PMSM) has many advantages, including a simple structure, a large torque, and high efficiency and reliability [1]. It has been widely used in industrial robots [2], conveyors [3], drilling machines [4], and many other industrial applications. However, it is difficult to achieve high-precision control of a PMSM servo system because it is affected by disturbances [5].

Torque oscillations are critical for PMSMs because they degrade the control performance of PMSM servo systems [6]. Thus, suppressing torque oscillations is an important figure of merit for PMSMs in the application of electrical vehicle traction, especially in low-speed operations [7]. It has been extensively investigated in the literature. Approaches are mainly divided into two categories: design-based and control-based [8]. The former focuses on optimizing the geometric structures of stators and rotors to minimize torque oscillations [9]. The latter focuses on designing control strategies to smooth the PMSM torque [10].

Control-based approaches can be used to further mitigate torque oscillations for a suitably designed physical structure. Sliding-mode control (SMC) [11], active disturbance rejection methods [12], and many other advanced control strategies have been applied to PMSM control systems to achieve high control precision. The SMC system has been widely used in control practice due to its simple structure, fast dynamic response, and strong disturbance attenuation [13]. However, such a system requires disturbance information in the design of a sliding-mode controller. Moreover, there is a trade-off between disturbance attenuation performance and chattering for SMC systems. More specifically, the high gain of a sliding-mode controller guarantees disturbance attenuation performance, but it leads to a chattering phenomenon. Thus, SMC has been modified to reduce chattering from various angles, such as high-order SMC [14], integral SMC [15], and a reaching-law method [16]. However, these methods have complex structures and relatively difficult design methods. Recent advanced SMC methods, including disturbance-observer-based SMC for synchronous reluctance motors [17] and linear-extended-state-observer-based terminal SMC for PMSMs [18], manage to actively estimate and attenuate disturbances, but these lack handling of uncertainties.

Active disturbance rejection methods use observers to estimate and compensate for disturbances actively. The equivalent input disturbance (EID) approach is one of these methods. An EID is a signal on the control input channel of a plant that produces the same effect on the output as actual disturbances do [19]. Compared to active disturbance rejection control (ADRC), disturbance-observer-based control (DOBC), and other active disturbance rejection methods, the EID approach has its unique advantages. Unlike ADRC, the configuration of an EID system is relatively simple; unlike DOBC, it does not require an inverse model of a plant. It yields satisfactory disturbance attenuation performance in many applications, such as vibration depression of buildings [20], quadrotor control [21], and servo systems [22]. It is worth mentioning that the system stability and control performance of an SMC system are ensured by a low gain of the sliding-mode controller if the system uses the information of a disturbance.

Several studies have demonstrated the potential to improve the oscillation attenuation performance of PMSMs by combining SMC and EID approaches. Refs. [23,24] incorporated EID into a conventional SMC system. These two studies allowed SMC to reject unmatched disturbances and calculate the stability domain of SMC switching gains. However, they did not explain how to choose specific SMC switching gains. Refs. [25,26] used SMC as the controller in an EID system. These two methods improved the dynamic performance of the control system. Nevertheless, they did not give the method for designing SMC parameters. Ref. [27] used an SMC controller to suppress the estimation error that the conventional EID approach neglected. This study managed to ensure system stability with a low switching gain, which produced little chattering compared to conventional SMC. However, it used the method of enlarging and reducing to design the switching gain. There was conservativeness in the process. These studies proved that the combination of SMC and EID improved the control performance and produced little chattering with a low switching gain. In PMSM servo systems, the moment of inertia and the viscous friction coefficient may change with time and operating environments. However, up to now, the SMC-EID method has only been used to handle plants without uncertainties. It is challenging to extend SMC-EID to plants with uncertainties.

Intelligent optimization algorithms provide a way to address the design problem of uncertain systems. Among these algorithms, the ant colony optimization (ACO) algorithm is a reliable population-based algorithm mimicking ant foraging behavior for solving

https://doi.org/10.3390/act15020085

discrete problems [28]. It is a multi-agent meta-heuristic optimization method. ACO algorithms involve simple agents (ants) that cooperate with each other to achieve an emergent unified behavior for a system. They are capable of finding high-quality solutions for problems with a large search space [29]. Note that the solution of the ACO is obtained from a finite set of discrete combinations [30]. The optimization of SMC parameters requires finding a solution in a continuous solution space. How to extend the ACO method to a continuous solution space is a critical problem.

In this study, we combine the EID and SMC methods to attenuate torque oscillation in an uncertain plant. We extend ACO to a continuous solution space and name it continuous-domain ant colony optimization (CDACO) for designing control systems for such plants. Unlike these prior works [23–27], which only handled plants without uncertainties and focused on the switch gain design, we address uncertainties via CDACO, providing a systematic multi-parameter design for real-world variations in PMSM servo systems, where the moment of inertia and the viscous friction coefficient may change with time and operating environments. The basic idea was presented in [31]. This paper is a significant improvement over that paper: the plant is uncertain, and we developed a CDACO algorithm to design the control system for such a plant. In the rest of this paper, first, we analyze the disturbances in PMSM servo systems. Next, we present a system configuration that combines the EID and SMC methods for a PMSM. Then, we develop a CDACO algorithm to design the SMC-EID-based control system for an uncertain plant. Finally, experiments verify its validity. The main contributions of this article are as follows:

(1) We applied the SMC-EID method to an uncertain system. SMC-EID attenuates a disturbance effectively and achieves a fast response while having little chattering.
(2) We extended the discrete ACO to a continuous solution space and developed a CDACO algorithm.
(3) We used the CDACO algorithm to design the control system for an uncertain plant. CDACO optimizes the integral of timed square error (ITSE) of an uncertain system to design the SMC parameters.

In this paper, $||D||_\infty := \sup_{0\leq\omega\leq\infty}\epsilon_{\max}[D(j\omega)]$, and $\epsilon_{\max}(D)$ means the maximum singular value of D. $D(s)$ is the Laplace transform of $d(t)$ and $\mathcal{L}^{-1}$ is the inverse Laplace transform.

2. Torque Oscillations in PMSM Servo Systems

A basic control strategy for a PMSM drive is vector control. The following model of a PMSM (Table 1) is used for this purpose:

$$\begin{cases} \dfrac{di_d}{dt} = \dfrac{1}{L_d}(u_d - Ri_d + n_p\omega L_q i_q) \\ \dfrac{di_q}{dt} = \dfrac{1}{L_q}\left(u_q - Ri_q - n_p\omega L_d i_d - n_p\omega\psi_f\right) \\ \dfrac{d\omega}{dt} = \dfrac{1}{J}(\tau_e - B_m\omega + \tau_o) \\ \tau_e = \dfrac{3}{2}n_p\left[\psi_f i_q + (L_d - L_q)i_d i_q\right], \end{cases} \tag{1}$$

where the subscripts d and q mean the direct and quadrature axes, respectively.

The oscillation torque, τ_o, is mainly caused by a cogging torque and a torque ripple. A cogging torque results from the structure of the motor, also known as a position-dependent torque. A torque ripple is caused by the flux linkage between the permanent magnet and the stator currents. Based on [9], we analyze and model the torque oscillation in PMSMs.

https://doi.org/10.3390/act15020085

Table 1. PMSM parameters ($j = d, q$).

Parameter	Meaning
i_j	Stator current of the j axis
u_j	Stator voltage of the j axis
L_j	Inductor of the j axis stator
R	Reluctance of stator
B_m	Viscous damping coefficient
J	Moment of the inertia
ω	Angular speed of rotor
ψ_f	Permanent magnet flux
n_p	Number of pole pairs
τ_e	Electrical torque
τ_o	Oscillation torque

2.1. *Cogging Torque*

The interaction between the permanent magnet flux and the magnetic conductance generated by the stator slot is the source of cogging torque, which is determined by its fundamental and harmonic waves. The period of the torque is

$$T_1 = \frac{2n_p}{N_1}, \tag{2}$$

where N_1 is the least common multiple between the number of stator slots and pole pairs n_p.

The interaction between the direct current component of permanent magnet flux and the harmonic interaction of magnetic conductance is another source of the cogging torque. Similarly, the period of the torque is

$$T_2 = \frac{2n_p}{N_2}, \tag{3}$$

where N_2 is the least common multiple between the number of stator slots and rotor slots.

As a result, the cogging torque is

$$d_{cog} = d_1^{cog} \sin \frac{\pi N_1}{n_p} \theta_e + d_2^{cog} \sin \frac{\pi N_2}{n_p} \theta_e, \tag{4}$$

where d_1 and d_2 are the amplitudes of the cogging torque and θ_e is the electrical angle.

2.2. *Torque Ripple*

Due to a non-sinusoidal flux density distribution in air gaps, the resultant flux linkage between the permanent magnet and the stator currents contains the harmonics in the *a-b-c* frame.

The corresponding flux linkage appears as the 6th and its multiple harmonics, which are expressed as

$$\psi_{dm} = \psi_{d0} + \psi_{d6} \cos 6\theta_e + \psi_{d12} \cos 12\theta_e + \cdots, \tag{5}$$

where ψ_{d0}, ψ_{d6}, and ψ_{d12} are the DC, the 6th, and the 12th harmonic terms of the d-axis flux linkage and θ_e is the electrical angle.

The effect of a torque ripple is

$$d_{tr} = \frac{3}{2} n_p i_q \sum_{j=0}^{\infty} \psi_{d6j} \cos 6j\theta_e. \tag{6}$$

https://doi.org/10.3390/act15020085

The cogging torque and the torque ripple are both functions of the angle θ_e. Thus, the oscillation torque contains both of them:

$$d_{to} = d_1^{cog} \sin \frac{\pi N_1}{n_p} \theta_e + d_2^{cog} \sin \frac{\pi N_2}{n_p} \theta_e + \frac{3}{2} n_p i_q \sum_{j=0}^{\infty} \psi_{d6j} \cos 6j\theta_e. \tag{7}$$

The disturbance model in (7) is derived from physical principles underlying PMSM torque oscillations. Cogging torque arises from the interaction between the permanent magnet flux and stator slot reluctance, with periods T_1 and T_2 determined by the least common multiple of stator slots and rotor poles, leading to periodic harmonics. Torque ripple originates from non-sinusoidal flux and current harmonics, typically dominated by 6th-order multiples in three-phase systems. It is a disturbance that relates to the angles of the motor. Its frequencies vary with the speed of the PMSM. As the EID approach attenuates disturbances in a wide frequency range, we combine the SMC and EID methods to achieve a fast response and satisfy torque oscillation attenuation.

3. SMC-EID System for PMSM

The configuration of an SMC-EID control system (Figure 1) has four parts: a plant, an SMC, a state observer, and a disturbance estimator. This section explains the basic structure and configuration of the EID approach.

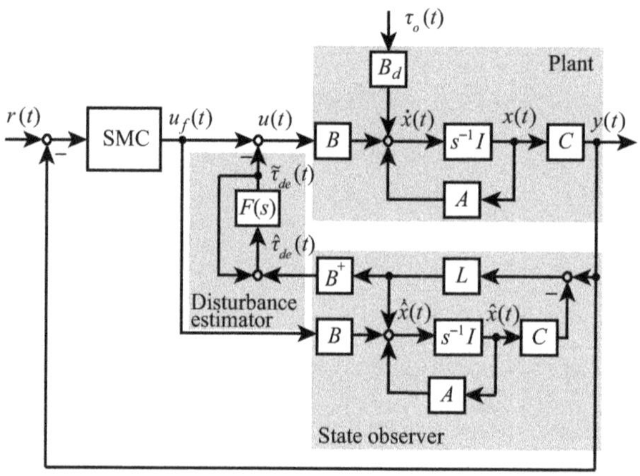

Figure 1. Configuration of SMC-EID.

This study considers a surface-mounted PMSM, in which $L_d = L_q$. Let $i_d = 0$, which is widely used to control such a PMSM. Thus, (1) becomes

$$\begin{cases} \dfrac{\mathrm{d}i_q(t)}{\mathrm{d}t} = \dfrac{1}{L_q}\left[u_q(t) - Ri_q(t) - n_p\omega(t)\psi_f\right] \\ \dfrac{\mathrm{d}\omega(t)}{\mathrm{d}t} = \dfrac{1}{J}\left[\dfrac{3}{2}n_p\psi_f i_q(t) - B_m\omega(t) + \tau_o(t)\right] \end{cases}. \tag{8}$$

Define

state variables and the control input as

$$\begin{cases} x_1(t) = \omega(t) \\ x_2(t) = i_q(t) \\ u(t) = u_q(t) \end{cases} \tag{9}$$

https://doi.org/10.3390/act15020085

respectively. Then,

$$\begin{aligned}\frac{\mathrm{d}x_1(t)}{\mathrm{d}t} &= \frac{1}{J}\left[\frac{3}{2}n_p\psi_f x_2(t) - B_m x_1(t) + \tau_o(t)\right],\\ \frac{\mathrm{d}x_2(t)}{\mathrm{d}t} &= \frac{1}{L_q}\left[u - Rx_2(t) - n_p\psi_f x_1(t)\right].\end{aligned} \tag{10}$$

The inductor of the q-axis is usually small for a small-size PMSM (less than 10 mH). Thus, we assume that

$$u - Rx_2(t) - n_p\psi_f x_1(t) = L_q\frac{\mathrm{d}x_2(t)}{\mathrm{d}t} = 0. \tag{11}$$

This allows us to write (10) as

$$x_2(t) = \frac{1}{R}[u(t) - Kx_1(t)]. \tag{12}$$

Since a voltage compensator is usually used to compensate for $Kx_1(t)$, the model of the PMSM, (10), with a voltage compensator becomes

$$\frac{\mathrm{d}x_1(t)}{\mathrm{d}t} = -\frac{B_m}{J}x_1(t) + \frac{3K}{2JR}u(t) + \frac{1}{J}\tau_o(t). \tag{13}$$

Define the sliding surface as

$$\sigma(t) = c[r(t) - x_1(t)], \tag{14}$$

An SMC controller is given as

$$u_f(t) = \varepsilon_1 \cdot \mathrm{sgn}[\sigma_1(t)] + \lambda\sigma(t). \tag{15}$$

Combining with (21) yields

$$u(t) = \varepsilon_1 \cdot \mathrm{sgn}[\sigma(t)] + \lambda_1\sigma(t) - \tilde{\tau}_{oe}(t). \tag{16}$$

The control input of a conventional SMC system without EID is

$$u(t) = \varepsilon_2 \cdot \mathrm{sgn}[\sigma_2(t)] + \lambda\sigma(t). \tag{17}$$

3.1. Basic Structure of EID Approach

A linear plant with a disturbance

$$\begin{cases}\dot{x}(t) = Ax(t) + Bu(t) + B_d\tau_o(t)\\ y(t) = Cx(t)\end{cases} \tag{18}$$

can be interpreted as a plant with an EID [32], $\tau_{de}(t)$, that is,

$$\begin{cases}\dot{x}(t) = Ax(t) + B[u(t) + \tau_{de}(t)]\\ y(t) = Cx(t).\end{cases} \tag{19}$$

She et al. devised a method to produce an estimate of $d_e(t)$ in a simple way (Figure 2) [19]. A full-order observer is used to estimate it.

https://doi.org/10.3390/act15020085

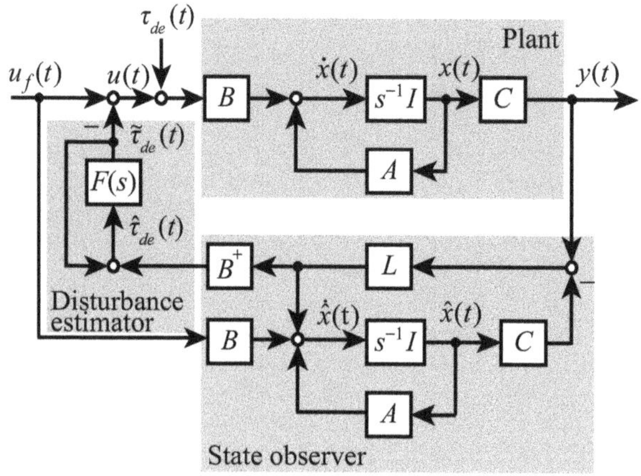

Figure 2. Basic configuration of EID approach.

The observer is

$$\begin{cases} \dot{\hat{x}}(t) = A\hat{x}(t) + Bu_f(t) + L[y(t) - C\hat{x}(t)] \\ \hat{y}(t) = C\hat{x}(t), \end{cases} \tag{20}$$

where $\hat{x}(t)$ is the reconstructed state of $x(t)$, $\hat{y}(t)$ is the output of the observer, $u_f(t)$ is the control input of the observer, and L is the observer gain. Then

$$u(t) = u_f(t) - \tilde{\tau}_{de}(t). \tag{21}$$

An estimate of the EID [19] is

$$\hat{\tau}_{de}(t) = -B^{+}LC\Delta x(t) + u_f(t) - u(t), \tag{22}$$

where

$$\Delta x(t) = \hat{x}(t) - x(t) \tag{23}$$

and

$$B^{+} = \left(B^{T}B\right)^{-1}B^{T}. \tag{24}$$

A low-pass filter $F(s)$ is used to select the estimation frequency band. It satisfies

$$|F(j\omega)| \approx 1,\ 0 \leq \omega \leq \omega_r, \tag{25}$$

where ω_r is the highest angular frequency for disturbance attenuation. $\tilde{\tau}_{de}(t)$ is the filtered signal of $\hat{\tau}_{de}(t)$:

$$\tilde{\tau}_{de}(t) = \mathcal{L}^{-1}\left[F(s)\hat{T}_{de}(s)\right], \tag{26}$$

Substituting (21) into (20) yields

$$\dot{\hat{x}}(t) = A\hat{x}(t) + B[u(t) + \tilde{\tau}_{de}(t)] + L[y - Cx(\hat{t})]. \tag{27}$$

Note that the difference between $x(t)$ and $\hat{x}(t)$ is equivalent to the difference between $\tau_{de}(t)$ and $\tilde{\tau}_{de}(t)$, according to (19) and (27). Thus, a properly designed observer gain, L, can ensure $|\Delta x(t)| \approx 0$, which makes $|\tau_{de}(t) - \tilde{\tau}_{de}(t)|$ small enough.

https://doi.org/10.3390/act15020085

3.2. Stability Analysis

We should design the low-pass filter and the observer gain so that they do not destroy the stability of the whole system. Let

$$\tau_o(t) = 0. \quad (28)$$

The plant (18) is

$$\dot{x}(t) = Ax(t) + Bu(t). \quad (29)$$

Substituting (21) into (22) yields

$$\hat{\tau}_{de}(t) = -B^{+}LC\Delta x(t) + \tilde{\tau}_{de}(t). \quad (30)$$

Combining (27), (29), and (30), and redrawing Figure 1, yields Figure 3.

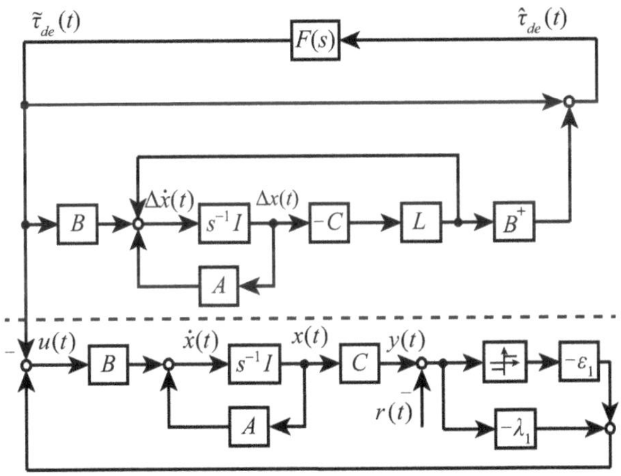

Figure 3. Block diagram of SMC-EID.

The stability conditions for the improved servo system are broken down into two parts. First, the control law (21) guarantees the stability of the EID system. Second, the feedback control system is stable.

The transfer function from $\tilde{d}(t)$ to $\hat{d}(t)$ is

$$G(s) = 1 - B^{+}LC[sI - (A - LC)]^{-1}B = B^{+}(sI - A)[sI - (A - LC)]^{-1}B. \quad (31)$$

Theorem 1. *When the system is under the following conditions,* (21) *guarantees the stability of the control system.*

(1) $A - LC$ is stable.
(2) $||GF||_{\infty} < 1$. Here, $||GF||_{\infty} := \sup_{0\leq\omega\leq\infty}\epsilon_{\max}[G(j\omega)F(j\omega)]$, and $\epsilon_{\max}(G)$ means the maximum sigular value of G.

The feedback system is an SMC system. The sliding surface is defined in (14). The parameters of the state-space Formula (18) for the PMSM (13) are

$$A = -\frac{B_m}{J},\ B = \frac{3K}{2JR},\ B_d = \frac{1}{J},\ C = 1. \quad (32)$$

The derivative of $\sigma(t)$ with time is

$$\begin{aligned}\frac{\mathrm{d}\sigma(t)}{\mathrm{d}t} &= \frac{cB_m}{J}x_1(t) - \frac{3cK}{2JR}u(t) \\ &= \frac{B_m}{J}[cr(t) - \sigma(t)] - \frac{3cK}{2JR}u(t) \\ &= -\frac{B_m}{J}\sigma(t) - \frac{3cK}{2JR}u(t) + \frac{cB_m}{J}r(t)\end{aligned} \quad . \tag{33}$$

The controller is designed as (15). Then,

$$\begin{aligned}\frac{\mathrm{d}\sigma(t)}{\mathrm{d}t} &= -\frac{B_m}{J}\sigma(t) - \frac{3cK}{2JR}[\varepsilon_1 \cdot \mathrm{sgn}(\sigma(t)) + \lambda_1\sigma(t)] + \frac{cB_m}{J}r(t) \\ &= -\left(\frac{B_m}{J} + \frac{3cK}{2JR}\lambda_1\right)\sigma(t) - \frac{3cK}{2JR}\varepsilon_1 \cdot \mathrm{sgn}[\sigma(t)] + \frac{cB_m}{J}r(t)\end{aligned} \tag{34}$$

Define a Lyapunov function

$$V = \frac{1}{2}\sigma^2. \tag{35}$$

The system is stable when

$$\dot{V} = \sigma \cdot \dot{\sigma} < 0,\ \sigma \neq 0. \tag{36}$$

Then

$$\dot{V} = -\left(\frac{B_m}{J} + \frac{3cK}{2JR}\lambda_1\right)\sigma^2 - \left[\frac{3cK}{2JR}\varepsilon_1 \cdot \mathrm{sgn}(\sigma) - \frac{cB_m}{J}r(t)\right]\sigma. \tag{37}$$

In (37), when $\sigma \neq 0$ and $\lambda_1 > 0$, $-\left(\frac{B_m}{J} + \frac{3cK}{2JR}\lambda_1\right)\sigma^2 < 0$. We need to design ε_1 to ensure that

$$\left[\frac{3cK}{2JR}\varepsilon_1 \cdot \mathrm{sgn}(\sigma) - \frac{cB_m}{J}r(t)\right]\sigma > 0. \tag{38}$$

It is obvious that (38) holds when the control gain ε_1 satisfies

$$\varepsilon_1 > \frac{2RB_m}{3K}\|r(t)\|_\infty. \tag{39}$$

Thus, the system is stable when Theorem 1 and (39) hold.

4. Design of EID Estimator and SMC

This section explains the design of the EID estimator and the parameters of the SMC-EID system (c, λ, k, and ε).

4.1. EID Estimator Design

Two elements need to be selected for the EID estimation and compensation: the low-pass filter $F(s)$ and the observer gain L.

An appropriate low-pass filter $F(s)$ ensures disturbance attenuation performance. A first-order low-pass filter is designed:

$$F(s) = \frac{1}{T_F s + 1}, \tag{40}$$

where T_F is used to regulate the angular-frequency band for disturbance attenuation. The cutoff angular frequency $\omega_c = \frac{1}{T_F}$.

To investigate the relationship between the cutoff angular frequency and the disturbance attenuation performance, we chose a set of cutoff angular frequencies of 100, 63, 32,

https://doi.org/10.3390/act15020085

16, 8, and 2 rad/s and drew Bode diagrams (Figure 4). When the cutoff angular frequency of the low-pass filter is five times higher than a disturbance component, the component of the disturbance estimate passes the filter with virtually no gain attenuation or phase lag.

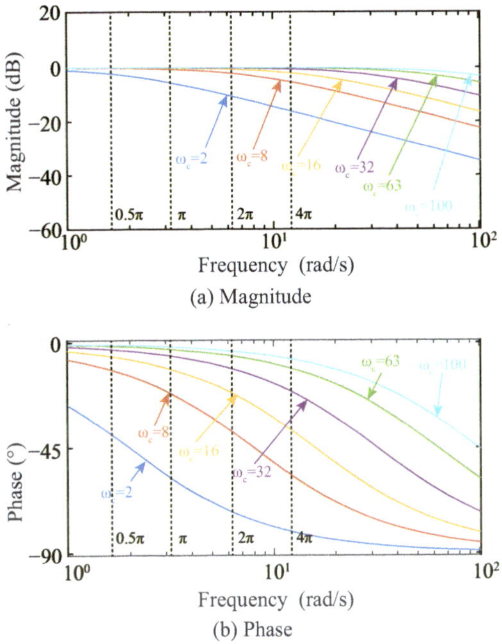

Figure 4. Bode plot of $F(s)$: (**a**) magnitude and (**b**) phase.

But if the angular-frequency band is too wide, it cannot attenuate high-frequency noise in the system. Thus, T_F is chosen as [32]

$$T_F = \frac{1}{5 \sim 10} \frac{1}{\omega_r}, \tag{41}$$

where ω_r in (41) is the highest angular frequency of disturbances. Thus,

$$F(j\omega) \approx 1,\ \forall \omega \in \Omega_r,\ \Omega_r = \{\omega |\ 0 \leq \omega \leq \omega_r\}, \tag{42}$$

where Ω_r is the angular-frequency band for disturbance attenuation.

After we choose an appropriate $F(s)$, we need to design an observer gain L that ensures $|\tau_{de}(t) - \tilde{\tau}_{de}(t)|$ is small enough.

Letting $u_f(t) = 0$ and redrawing Figure 2 yields Figure 5.

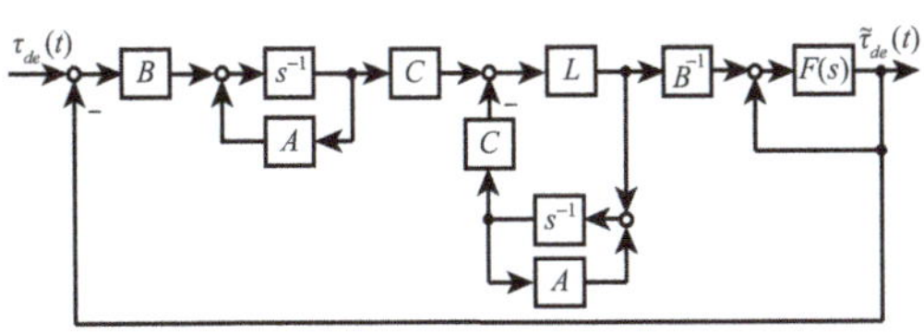

Figure 5. EID configuration for disturbance estimate analysis.

https://doi.org/10.3390/act15020085

The system is a unity feedback system and $F(s) = \frac{1}{T_F s+1}$. The open-loop transfer function is

$$G_o(s) = \frac{CB}{s-A} \cdot \frac{L(s-A)}{s+(LC-A)} \cdot B^{-1} \cdot \frac{1}{T_F s} = \frac{LC}{T_F s^2 + T_F(LC-A)s}. \tag{43}$$

Thus, the closed-loop transfer function from $\tau_{oe}(t)$ to $\tilde{\tau}_{oe}(t)$ is

$$G_c(s) = \frac{LC}{T_F s^2 + T_F(LC-A)s + LC}. \tag{44}$$

Then, the transfer function from $\tau_{oe}(t)$ to $\tau_{oe}(t) - \tilde{\tau}_{oe}(t)$ is

$$\begin{aligned} G_e(s) = 1 - \Phi_\tau(s) =& \frac{T_F s^2 + T_F(LC-A)s}{T_F s^2 + T_F(LC-A)s + LC} \\ =& \frac{s^2 + (LC-A)s}{s^2 + (LC-A)s + \frac{LC}{T_F}}. \end{aligned} \tag{45}$$

According to the Routh Criterion, it is stable when

$$LC > 0. \tag{46}$$

$G_e(s)$ can be interpreted as

$$\begin{cases} G_e(s) = \dfrac{s[s+(LC-A)]}{\left(s + \frac{(LC-A)-\sqrt{\Delta}}{2}\right)\left(s + \frac{(LC-A)+\sqrt{\Delta}}{2}\right)} \\ \Delta = (LC-A)^2 - 4\dfrac{LC}{T_F}. \end{cases} \tag{47}$$

We choose an L that ensures

$$\Delta > 0. \tag{48}$$

Δ can be represented by

$$\begin{aligned} \Delta =& (LC)^2 - 2\left(A + \frac{2}{T_F}\right)LC + \left(A + \frac{2}{T_F}\right)^2 - 4\frac{AT_F+1}{T_F^2} \\ =& \left[LC - \left(A + \frac{2}{T_F}\right)\right]^2 - 4\frac{AT_F+1}{T_F^2}. \end{aligned} \tag{49}$$

When LC is large enough that

$$\left[LC - \left(A + \frac{2}{T_F}\right)\right]^2 \gg 4\frac{AT_F+1}{T_F^2}. \tag{50}$$

Then,

$$\sqrt{\Delta} \approx LC - \left(A + \frac{2}{T_F}\right). \tag{51}$$

Thus,

$$G_e(s) = \frac{s[s+(LC-A)]}{\left(s + \frac{1}{T_F}\right)\left[s + \left(LC - A - \frac{1}{T_F}\right)\right]}. \tag{52}$$

https://doi.org/10.3390/act15020085

As the zero $-(LC-A)$ and the pole $-(LC-A-\frac{1}{T_F})$ are close to each other, and they are both far from other zeros and poles, they result in zero-pole cancellation. Then,

$$G_e(s) \approx \frac{s}{s+\frac{1}{T_F}}. \tag{53}$$

According to (41),

$$\frac{1}{T_F} = m \cdot \omega_r,\ m \in [5,10]. \tag{54}$$

From the Bode diagram of $G_e(s)$ is shown as Figure 6.

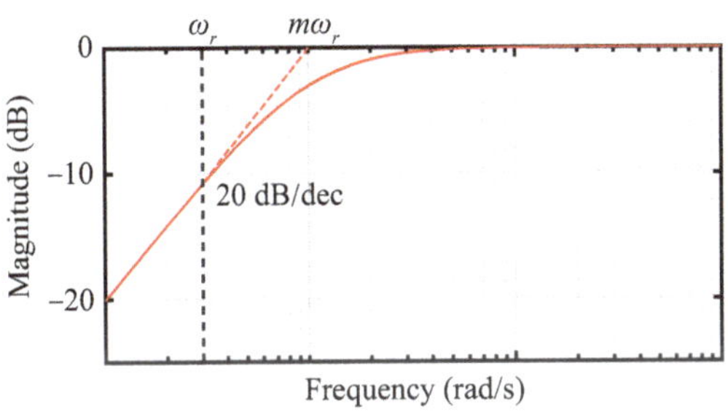

Figure 6. Bode diagram of $G_e(s)$. The solid line represents the exact frequency response, while the dashed line denotes the asymptotic magnitude characteristic indicating the slope and cut-off frequency.

When the disturbance frequency satisfies

$$\omega_\tau \le \omega_r, \tag{55}$$

the inequality below holds.

$$||G_e \cdot \tau_{oe}||_\infty \le \frac{1}{m}||\tau_{oe}||_\infty. \tag{56}$$

We choose the observer gain L that follows (48) and (50).

4.2. *Ant Colony Optimization-Based SMC Design*

We replaced the sign function in SMC with a saturation function to further suppress chattering. As a result, the SMC-EID in this study is

$$u(t) = \varepsilon_1 \cdot \text{sat}\left[\frac{\sigma(t)}{k}\right] + \lambda_1\sigma(t) - \tilde{\tau}_{de}(t), \tag{57}$$

where

$$\text{sat}\left(\frac{\sigma}{k}\right) = \begin{cases} 1,\ \sigma < -k \\ \sigma/k,\ -k \le \sigma \le k \\ 1,\ \sigma > k, \end{cases} \tag{58}$$

and the positive constant k is the linear region of the saturation function. As shown in (14), c should be a positive real number. Thus, $c=1$ is chosen for the design of ε_1, k_1, and λ_1 in (57). As there are uncertainties in the system, we set up a group of systems with different parameters. The number of systems is N. The optimal set of control parameters should show great control performance in all systems.

ε_1 is the switching gain of the SMC and ε_1/k_1 is the gradient of the saturation function. The smaller k_1 is, the closer to a sign function the saturation function is. While the combination of a larger ε_1 and a smaller k_1 yields better disturbance attenuation performance, it

https://doi.org/10.3390/act15020085

produces larger chattering. The chattering is bounded by the saturation function sat$(\frac{\sigma}{k})$, limiting amplitude to $|\varepsilon_1/k|$. λ_1 is a proportional control gain. The larger it is, the faster the dynamic response is and the larger the overshoot is.

This study developed a method of designing the SMC parameters based on CDACO to solve the problem.

A conventional ACO solution space is usually a discrete solution space. However, this study needs to find a solution in a continuous solution space.

Let an ant be $[\varepsilon_1,\ k_1,\ \lambda_1]$. The pheromone of Ant n in the ith iteration is

$$\begin{cases} \mathrm{phe}_m^{(i)} = 0, & \text{for } i = 0 \\ \mathrm{phe}_m^{(i)} = (1-\zeta)\mathrm{phe}_m^{(i-1)} + \zeta e^{-p\overline{E}}, & \text{for } i > 0 \ \&\ \zeta < 1, \end{cases} \tag{59}$$

where p is a weight of the performance index, which is a positive constant indicating the transition performance; ζ is a forgetting factor; and $\overline{E}$ is the weighted average of $E_n\ (n =\ 1,\ 2, \cdots,\ N)$. E_n is the ITSE of the n-th system:

$$E_n = \int_0^2 t(\overline{\omega} - \omega)^2 \mathrm{d}t. \tag{60}$$

E_n is small when

(1) Settling time is short;
(2) Overshoot is small;
(3) Chattering is little;
(4) Static error is eliminated.

E_n evaluates the control performance comprehensively. The smaller E_n is, the better the overall control performance is. The optimal ant has the smallest performance index E_n and the largest pheromone ($\mathrm{phe}_{\mathrm{opt}}$), that is, a small speed fluctuation and small chattering. It searches around itself and looks for a better solution.

Other ants calculate their transition probabilities. For example, the probability of Ant n is

$$P_n^{(i)} = \frac{\mathrm{phe}_{\mathrm{opt}} - \mathrm{phe}_n^{(i)}}{\mathrm{phe}_{\mathrm{opt}}}. \tag{61}$$

Thus, a low pheromone means the ant has a high transition probability of moving toward the optimal solution. If it decides not to do so, then it searches around itself.

We condense the above discussion as follows:

CDACO algorithm:

(Step 1) Set the number of ants M, evaporation rate ρ, initial pheromone τ, and max iteration times I. Generate initial ants uniformly in the search space justified by the lower bound from (39) to ensure stability and the upper bound to avoid excessive gain.
(Step 2) Update the pheromone for each ant using (59).
(Step 3) Calculate the transition probability for each ant using (61) and make a decision by sampling from Gaussian probability.
(Step 4) Go to (Step 2) and repeat until reaching the iteration times.

The flow chart of CDACO (Figure 7) further clarifies the CDACO procedure.

The performance of CDACO depends on the number of ants M, the number of iteration times I, the weight of the performance index p, the forgetting factor ζ, and the performance index E_n. Note that a large ζ means that the pheromone depends strongly on the current state, and a large p means a high transition probability $P_n^{(i)}$.

https://doi.org/10.3390/act15020085

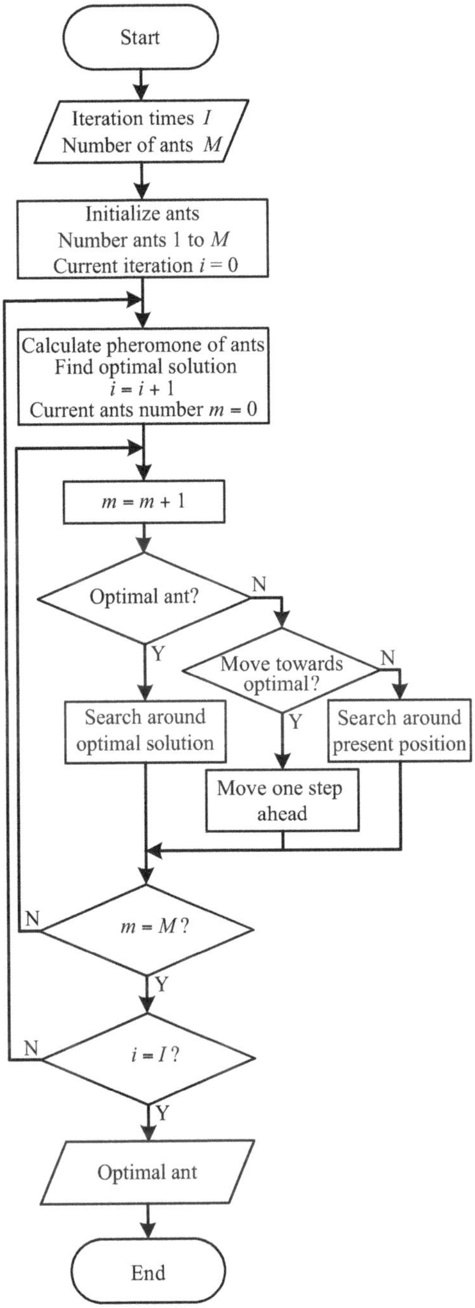

Figure 7. Flow chart of CDACO.

5. Experimental Verification

The values of state-space (32) for Table 2 are

$$A = -25, B = 123, B_d = 250, C = 1. \quad (62)$$

https://doi.org/10.3390/act15020085

Table 2. Nominal PMSM specifications.

Variable	Value
Viscous friction coefficient, B_m	0.001 N/rad·s^{-1}
Moment of inertia, J	0.004 kg·m^2
Flux of permanent magnet, ψ_f	0.072 Wb
Number of pole pairs, n_p	5
Reluctance of stator, R	0.875 Ω
Inductor of q axis stator, L_q	8.5 mH

Let the reference input be

$$r(t) = 500 \times 1(t) \text{ rpm}, \tag{63}$$

and the disturbance model is derived from PMSM torque oscillations (7)

$$d(t) = 0.1 + 0.4 \sin \frac{\theta(t)}{10} + 0.4 \sin \frac{\theta(t)}{5} + 0.4 \cos 6\theta(t) \text{ N} \cdot \text{m}, \tag{64}$$

where $\theta(t)$ is the mechanical angle of the PMSM. Since the target angular velocity of the PMSM is 500 rpm, which is approximately 50 rad/s, the angular frequencies in the disturbance are about 5 rad/s, 10 rad/s, and 300 rad/s. Note that torque oscillations at 300 rad/s or higher have little influence on the PMSM system. ω_r was chosen to be 10 rad/s and the low-pass filter to be

$$F(s) = \frac{1}{0.01s + 1}. \tag{65}$$

According to (48) and (50) and ensuring observer performance, the time constant of the observer should be less than one-tenth of that of the plant. Thus, the observer gain L is chosen to be

$$L = 2000. \tag{66}$$

The CDACO algorithm was established in MATLAB 2020b/Simulink and the parameters optimized offline. We set up different systems with different moments of inertia and viscous friction coefficients, which simulate the uncertainties in the system. Among them, the moment of inertia $J \in \{0.002,\ 0.004,\ 0.006\}$ kg·m^2 and the viscous friction coefficients $B_m \in \{0.0005,\ 0.0010,\ 0.0015\}$ N/rad·s^{-1}. Thus, there are nine different systems with different parameters. The system whose $J = 0.004$ kg·m^2 and $B_m = 0.0010$ N/rad·s^{-1} is the nominal system. Its performance index is E_1. The performance indexes of other systems are E_n, $n \in \{2,\ 3, \ldots, 9\}$. E is calculated as follows:

$$E = 0.2E_1 + 0.1 \sum_{n=2}^{9} E_n. \tag{67}$$

It guarantees the robustness of the system.

The SMC-EID, (57), has three parameters, ε_1, k_1, and λ_1, that need designing. According to (39), we choose $B_m = 0.0015$. Thus, the range of ε_1 is

$$\varepsilon_1 > 1.13. \tag{68}$$

Then, we set the search space to be

$$\varepsilon_1 \in [1.13, 8],\ k_1 \in (0, 3],\ \lambda_1 \in (0, 15]. \tag{69}$$

https://doi.org/10.3390/act15020085

Setting the parameters of CDACO (Table 3), uniformly generating the first generation of ants in the search space, (69), and optimizing ε_1, k_1, and λ_1 in (57) yields

$$\varepsilon_1 = 1.29,\ k_1 = 0.60,\ \lambda_1 = 3.73. \tag{70}$$

Table 3. Parameters of CDACO.

Parameter	Value
Number of ants M	50
Maximum iteration I	100
Forgetting constant ζ	0.9
Weight of performance index p	2

For comparison, we used CDACO to optimize the parameters in a conventional SMC

$$u(t) = \varepsilon_2 \mathrm{sat}\left[\frac{\sigma(t)}{k_2}\right] + \lambda_2\sigma(t). \tag{71}$$

The conventional SMC is designed with the same CDACO, sliding surface (14), and nine identical uncertain systems as the SMC-EID method. Optimizing $[\varepsilon_1,\ k_1,\ \lambda_1]$ gives

$$\varepsilon_2 = 6.40,\ k_2 = 1.52,\ \lambda_2 = 5.01. \tag{72}$$

A two-mass experimental system (Figure 8) was used to demonstrate the validity of the method. It has two identical PMSMs: one is used as a plant, and the other is a disturbance generator. It has a clutch that controls the engagement and disengagement of the flywheel, thereby changing the system's rotational inertia. The nameplate values of the PMSMs are summarized in Table 4.

Table 4. PMSM nameplate values.

Name	Value
Manufacturer	INOVANCE
Model	ISMH1-75B30CB
Rated power	750 W
Rated voltage	220 V
Rated current	4.8 A
Rated rev.	3000 rpm
Rated torque	2.39 Nm
Max rev.	6000 rpm

MATLAB R2020b was used to design a control system. TwinCAT 3 was used to perform system control with a sampling period of 500 μs.

We conducted four sets of comparative experiments with different moments of inertia J and reference speeds r. The four experimental conditions are shown as follows:

Exp. 1: J_1 = 0.004 kg · m^2 and r_1 = 500 rpm;
Exp. 2: J_2 = 0.004 kg · m^2 and r_2 = 250 rpm;
Exp. 3: J_3 = 0.006 kg · m^2 and r_3 = 500 rpm;
Exp. 4: J_4 = 0.006 kg · m^2 and r_4 = 250 rpm.

https://doi.org/10.3390/act15020085

Figure 8. Two-mass experimental system.

The speed response (Figure 9) and speed error (Figure 10) show that the disturbance was attenuated to a lower level for the SMC-EID than for the SMC. A comparison between performance indexes (Table 5) reveals that in all four experiments, the peak-to-peak (PP) speed error and ITSE of the SMC-EID are all less than those of the conventional SMC. The average PP speed error of the SMC-EID is about 37% of that of the conventional SMC, and the average ITSE of the SMC-EID is about 35% of that of the conventional SMC.

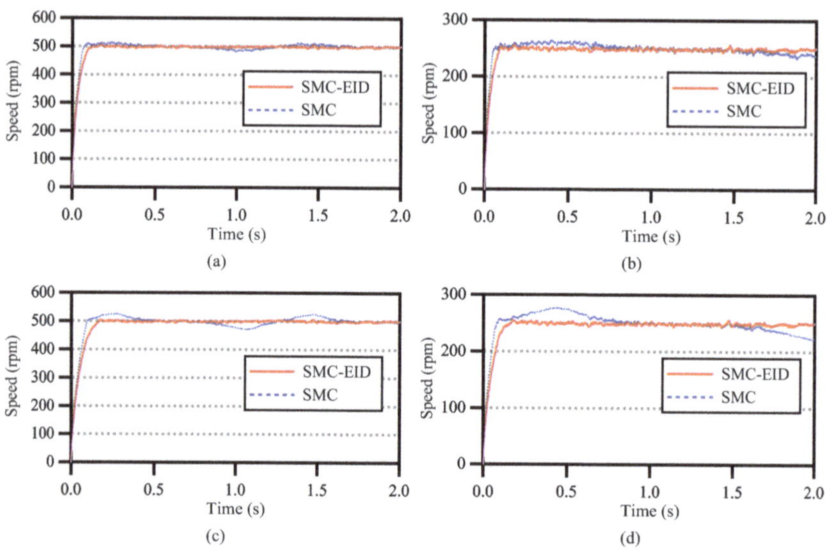

Figure 9. Comparison of speed response between SMC-EID and conventional SMC in different sets of experiments: (**a**) Exp. 1, (**b**) Exp. 2, (**c**) Exp. 3, and (**d**) Exp. 4.

We take Exp. 1 as an example. The torque comparison (Figure 11) between SMC-EID and conventional SMC shows that the chattering is much smaller for the SMC-EID than for the SMC.

https://doi.org/10.3390/act15020085

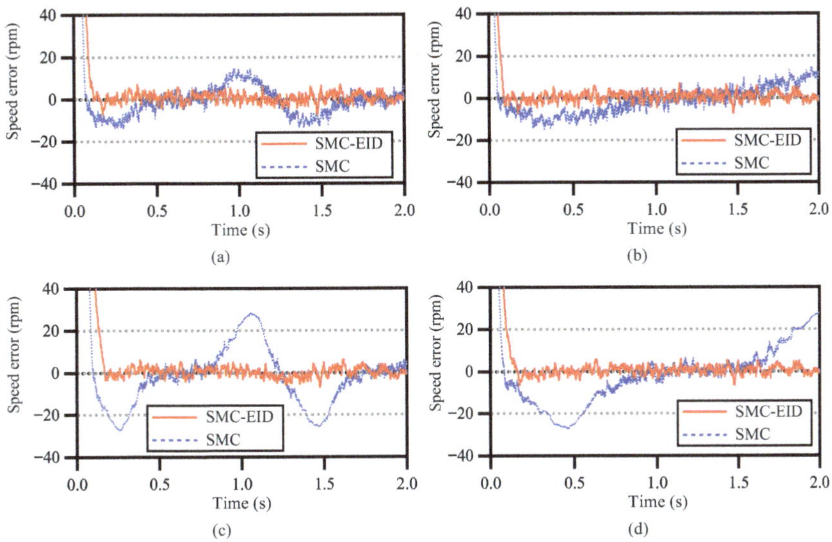

Figure 10. Comparison of speed error between SMC-EID and conventional SMC in different sets of experiments: (**a**) Exp. 1, (**b**) Exp. 2, (**c**) Exp. 3, and (**d**) Exp. 4.

Table 5. Control performance comparison (PP: peak-to-peak).

		PP Speed Error (rpm)	ITSE
Exp (1)	SMC	25.1	1.35
	SMC-EID	11.6	0.96
Exp (2)	SMC	31.5	1.09
	SMC-EID	13.8	0.31
Exp (3)	SMC	56.3	4.60
	SMC-EID	19.5	1.96
Exp (4)	SMC	52.9	3.53
	SMC-EID	15.8	0.50
Average	SMC	41.5	2.64
	SMC-EID	15.2	0.93

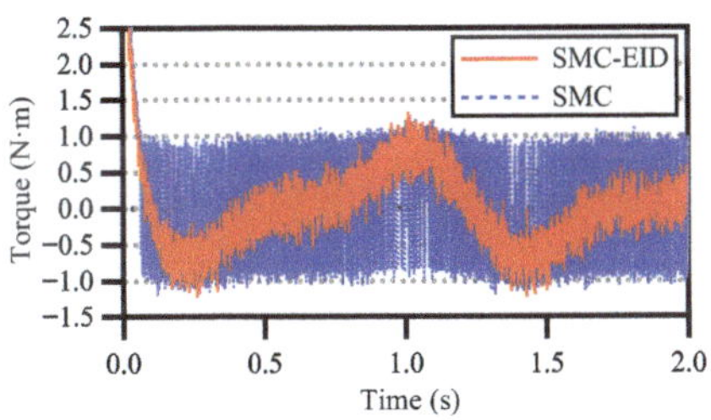

Figure 11. Torque comparison between SMC and SMC-EID in Exp (1).

6. Conclusions

This paper explained a control method that combines the SMC and EID methods for uncertain systems. We added an EID estimate to a sliding-mode controller, not only to improve the disturbance attenuation performance but also to reduce chattering. Then, CDACO solves the design problem of uncertain systems. This method has significant advantages:

https://doi.org/10.3390/act15020085

(1) Our method extends the SMC-EID method to uncertain plants, demonstrated by stability analysis and optimization over nine uncertain systems.
(2) The novel CDACO algorithm handles continuous optimization spaces and solves the multi-parameter design problem for uncertain systems.
(3) Experimental results validate superior control performance compared to conventional SMC. In the four experiments, the average PP speed error of the SMC-EID method was about 37% of that of the conventional SMC, and the ITSE of SMC-EID was about 35% of that of the conventional SMC.

Author Contributions: Conceptualization, R.J. and J.S.; methodology, R.J., X.Y. and J.S.; software, R.J. and F.W.; validation, R.J. and X.Y.; formal analysis, R.J. and J.S.; investigation, R.J. and J.S.; resources, R.J., J.S., F.W. and S.K.; data curation, R.J.; writing—original draft preparation, R.J., X.Y. and J.S.; writing—review and editing, R.J., X.Y., J.S., F.W. and S.K.; visualization, R.J., J.S., F.W. and S.K.; supervision, J.S., F.W. and S.K.; project administration, X.Y, S.K. and J.S.; funding acquisition, X.Y, S.K. and J.S. All authors have read and agreed to the published version of the manuscript.

Funding: This work was supported in part by the Beijing Natural Science Foundation under Grant 4244089; the Fundamental Research Funds for the Central Universities, China University of Geosciences (Wuhan) under Grant CUG240635; the "CUG Scholar" Scientific Research Funds at China University of Geosciences (Wuhan) under Grant 2022029; the Natural Science Foundation of Hubei Province, China under Grant 2020CFA031; and JSPS (Japan Society for the Promotion of Science) KAKENHI under Grants 23K25252 and 24K03325.

Institutional Review Board Statement: Not applicable.

Informed Consent Statement: Not applicable.

Data Availability Statement: Data are available at URL (accessed on 23 Jan 2026): https://pan.baidu.com/s/17rgN6X47Ig7ePO9mR8q28A?pwd=jjii Password: jjii.

Conflicts of Interest: The authors declare no conflicts of interest.

Abbreviations

The following abbreviations are used in this manuscript:

PMSM	Permanent magnet synchronous motor
SMC	Sliding-mode control
EID	Equivalent input disturbance
ADRC	Active disturbance rejection control
DOBC	Disturbance-observer-based control
ACO	Ant colony optimization
CDACO	Continuous-domain ant colony optimization

References

1. Du, H.; Wen, G.; Cheng, Y.; Lü, J. Design and implementation of bounded finite-time control algorithm for speed regulation of permanent magnet synchronous motor. *IEEE Trans. Ind. Electron.* **2020**, *68*, 2417–2426. [CrossRef]
2. Zhang, B.; Zhang, W.; Luo, H. Two-dimensional control of a fractional-order permanent magnet synchronous motor model for industrial robots. *Electr. Power Syst. Res.* **2025**, *249*, 111963. [CrossRef]
3. Zhang, X.; Ren, M.; Wang, H.; Xu, H.; Shi, B.; Gao, M. A Study on a Speed Regulation Method for Mining Scraper Conveyors and a Control Strategy for Permanent Magnet Drive Systems. *Actuators* **2025**, *14*, 106. [CrossRef]
4. Lin, F.C.; Chen, C.S.; Lin, C.J. Adaptive fuzzy observer and control in servo drivers for two-axis drilling machine. *Int. J. Adv. Manuf. Technol.* **2024**, *132*, 3105–3122. [CrossRef]
5. Wu, F.; Li, P.; Wang, J. FO improved fast terminal sliding mode control method for permanent-magnet synchronous motor with FO disturbance observer. *IET Control Theory Appl.* **2019**, *13*, 1425–1434. [CrossRef]
6. Mao, H.; Tang, X.; Tang, H. Speed control of PMSM based on neural network model predictive control. *Trans. Inst. Meas. Control* **2022**, *44*, 2781–2794. [CrossRef]

https://doi.org/10.3390/act15020085

7. Lara, J.; Xu, J.; Chandra, A. Effects of rotor position error in the performance of field-oriented-controlled PMSM drives for electric vehicle traction applications. *IEEE Trans. Ind. Electron.* **2016**, *63*, 4738–4751. [CrossRef]
8. Krishnamoorthy, S.; Sanjeevikumar, P.; Holm-Nielsen, J.B.; Bo, J.; Singh, R.R.; Rajesh, M. Torque ripple minimization of PMSM using an adaptive Elman neural network-controlled feedback linearization-based direct torque control strategy. *Int. Trans. Electr. Energy Syst.* **2021**, *31*, 1.
9. Nakao, N.; Akatsu, K. Suppressing pulsating torques: Torque ripple control for synchronous motors. *IEEE Ind. Appl. Mag.* **2014**, *20*, 33–44. [CrossRef]
10. Guo, Y.; Jiang, F.; Wang, S.; Cheng, S.; Hu, Z. Variable Control Period Model Predictive Current Control with Current Hysteresis for Permanent Magnet Synchronous Motor Drives. *Actuators* **2025**, *14*, 517. [CrossRef]
11. Suhail, S.A.; Bazaz, M.A.; Hussain, S. Adaptive sliding mode-based active disturbance rejection control for a quadcopter. *Trans. Inst. Meas. Control* **2022**, *44*, 3176–3190. [CrossRef]
12. Xie, F.; Hu, T. Speed fluctuation suppression of PMSM based on composite quasi-resonant active disturbance rejection control. *J. Power Electron.* **2025**, 1–11. [CrossRef]
13. Wang, S.; Tao, L.; Chen, Q.; Na, J.; Ren, X. USDE-based sliding mode control for servo mechanisms with unknown system dynamics. *IEEE/ASME Trans. Mechatronics* **2020**, *25*, 1056–1066. [CrossRef]
14. Zeng, Q.; Jiang, M.; Hu, J. The Second-Order Sliding Mode Control Algorithm for Fixed-Time Stability of Nonlinear Systems. *Circuits, Syst. Signal Process.* **2024**, *43*, 5507–5531. [CrossRef]
15. Abbas, S.J.; Husain, S.S.; Al-Wais, S.; Humaidi, A.J. Adaptive integral sliding mode controller (SMC) design for vehicle steer-by-wire system. *SAE Int. J. Veh. Dyn. Stab. NVH* **2024**, *8*, 383–396. [CrossRef]
16. Ahn, H.; Kim, S.; Park, J.; Chung, Y.; Hu, M.; You, K. Adaptive Quick Sliding Mode Reaching Law and Disturbance Observer for Robust PMSM Control Systems. *Actuators* **2024**, *13*, 136. [CrossRef]
17. Liu, Y.C. Disturbance-Observer-Based Sliding-Mode Speed Control for Synchronous Reluctance Motor Drives via Generalized Super-Twisting Algorithm. *Actuators* **2024**, *13*, 233. [CrossRef]
18. Yao, Y.; Zhuang, Y.; Xie, Y.; Xu, P.; Wu, C. Prescribed Performance Global Non-Singular Fast Terminal Sliding Mode Control of PMSM Based on Linear Extended State Observer. *Actuators* **2025**, *14*, 65. [CrossRef]
19. She, J.H.; Xin, X.; Pan, Y. Equivalent-input-disturbance approach-Analysis and application to disturbance rejection in dual-stage feed drive control system. *IEEE/ASME Trans. Mechatronics* **2010**, *16*, 330–340. [CrossRef]
20. Miyamoto, K.; She, J.; Nakano, S.; Sato, D.; Chen, Y. Active structural control of base-isolated building using equivalent-input-disturbance approach with reduced-order state observer. *J. Dyn. Syst. Meas. Control* **2022**, *144*, 091006. [CrossRef]
21. Cai, W.; She, J.; Wu, M.; Ohyama, Y. Quadrotor waypoint-tracking control under exogenous disturbances based on equivalent-input-disturbance approach. *J. Frankl. Inst.* **2020**, *357*, 4709–4741. [CrossRef]
22. Mei, Q.; She, J.; Liu, Z.; Wu, M. Estimation and compensation of periodic disturbance using internal-model-based equivalent-input-disturbance approach. *Sci. China Inf. Sci.* **2022**, *65*, 182205. [CrossRef]
23. She, J.; Pan, Y.; Hashimoto, H.; Lei, Q.; Wu, M. Equivalent-input-disturbance approach enables sliding-mode control systems to reject unmatched disturbances. In *Proceedings of the 2015 IEEE International Conference on Industrial Technology (ICIT), Seville, Spain, 17–19 March 2015*; IEEE: New York, NY, USA, 2015; pp. 226–231.
24. Jiang, Z.; Zhou, L.; Zhang, Z.; Jiang, F.; Cheng, L. An equivalent-input-disturbance-based sliding mode control approach for DC-DC buck converter system with mismatched disturbances. In *Proceedings of the 2019 Chinese Control Conference (CCC), Guangzhou, China, 27–30 July 2019*; IEEE: New York, NY, USA, 2019; pp. 639–644.
25. Mahapatro, K.A.; Gundecha, A.D.; Chavan, A.D.; Suryawanshi, P.V. Disturbance rejection in motion control based on equivalent input disturbance approach with experimental validation. In *Proceedings of the 2019 4th International Conference on Recent Trends on Electronics, Information, Communication & Technology (RTEICT), Bangalore, India, 17–18 May 2019*; IEEE: New York, NY, USA, 2019; pp. 1247–1251.
26. Fang, M.; Liang, X.; Chen, Y.; Cheng, J.; Qian, H.; Huang, S.; Du, Y.; She, J. Speed control of permanent magnet synchronous motor using sliding mode control combined with equivalent input disturbance control. In *Proceedings of the 2022 41st Chinese Control Conference (CCC), Hefei, China, 25–27 July 2022*; IEEE: New York, NY, USA, 2022; pp. 2860–2863.
27. Yin, X.; She, J.; Wu, M.; Sato, D.; Ohnishi, K. Disturbance rejection using SMC-based-equivalent-input-disturbance approach. *Appl. Math. Comput.* **2022**, *418*, 126839. [CrossRef]
28. Zhao, D.; Liu, L.; Yu, F.; Heidari, A.A.; Wang, M.; Oliva, D.; Muhammad, K.; Chen, H. Ant colony optimization with horizontal and vertical crossover search: Fundamental visions for multi-threshold image segmentation. *Expert Syst. Appl.* **2021**, *167*, 114122. [CrossRef]
29. Paniri, M.; Dowlatshahi, M.B.; Nezamabadi-Pour, H. MLACO: A multi-label feature selection algorithm based on ant colony optimization. *Knowl.-Based Syst.* **2020**, *192*, 105285. [CrossRef]
30. Stützle, T.; Dorigo, M. ACO algorithms for the traveling salesman problem. *Evol. Algorithms Eng. Comput. Sci.* **1999**, *4*, 163–183.

https://doi.org/10.3390/act15020085

31. Jiang, R.; She, J.; Liu, Z.; Li, D.; He, W. Attenuation of PMSM Torque-Ripple Based on Combination of Equivalent-Input-Disturbance Approach and Sliding-Mode Control. In *Proceedings of the 45th Annual Conference of the IEEE Industrial Electronics Society (IECON), Lisbon, Portugal, 14–17 October 2019*; IEEE: New York, NY, USA, 2019; Volume 1, pp. 6838–6841.
32. She, J.H.; Fang, M.; Ohyama, Y.; Hashimoto, H.; Wu, M. Improving disturbance-rejection performance based on an equivalent-input-disturbance approach. *IEEE Trans. Ind. Electron.* **2008**, *55*, 380–389. [CrossRef]

https://doi.org/10.3390/act15020085

MDPI AG
Grosspeteranlage 5
4052 Basel
Switzerland
Tel.: +41 61 683 77 34

Actuators Editorial Office
E-mail: actuators@mdpi.com
www.mdpi.com/journal/actuators

www.ingramcontent.com/pod-product-compliance
Lightning Source LLC
Chambersburg PA
CBHW061446210926
51850CB00008B/1634